Final Cut Pro 2 for FireWire DV Editing

Final Cut Pro 2 for FireWire DV Editing

Charles Roberts

Focal Press
An imprint of Butterworth-Heinemann

OXFORD · AUCKLAND · BOSTON · JOHANNESBURG · MELBOURNE · NEW DELHI

Focal Press is an imprint of Butterworth–Heinemann.

  A member of the Reed Elsevier group

∞   Recognizing the importance of preserving what has been written, Butterworth–Heinemann prints its books on acid-free paper whenever possible.

Library of Congress Cataloging-in-Publication Data

Roberts, Charles, 1967–
  Final Cut Pro 2 for FireWire DV editing / Charles Roberts.
    p.cm.
  ISBN 0-240-80499-6
    1. Video tapes—Editing—Data processing. 2. Digital video. 3. Final cut pro. I. Title.

TR899 .R56 2001
778.5'235--dc21   2001054497

British Library Cataloguing-in-Publication Data

A catalogue record for this book is available from the British Library.

The publisher offers special discounts on bulk orders of this book.

For information, please contact:

Manager of Special Sales
Butterworth–Heinemann
225 Wildwood Avenue
Woburn, MA 01801-2041
Tel: 781-904-2500
Fax: 781-904-2620

For information on all Focal Press publications available, contact our home page at:
http://www.focalpress.com

10 9 8 7 6 5 4 3 2 1

Printed in the United States of America

This book is dedicated to anyone that has made or will make video or film for love of the game, regardless of the paycheck. Also to all the telemarketing bozos that did not call while I was busy trying to write.

Contents

Foreword

I can still vividly recall that bluesy, billowy, caffeinated Cupertino night in August of 1999 when a mystery poster named "chawla" dropped in after hours at the still rocking Final Cut Pro online discussion forum at 2-pop.com. Here it is a bleary-eyed 3 a.m., the History Channel is into infomercials by now and I'm not even noticing, and somebody I have never heard of, with a name straight out of the bar scene in Stars Wars, jumps in ahead of me in the discussion forum and responds to this post with the burning question of the day in the subject line: "How professional is this new thing called DV, and how professional is new Final Cut Pro, anyway?"

Now, in those very early days of 2-pop, I thought I knew just about anybody who knew anything about DV and FCP, and who was also of a generous enough spirit to come by 2-pop and contribute to the online discussion. So I checked out this chawla person real fast, to head off any possible infusion of misinformation. Good answers were hard to come by back then, and so the black market trade in misinformation in DV books and the online forums was measured by the gross ton, and business was brisk. But lo and behold, something struck me about the way this "chawla" person was able jump right into the discussion and add valuable information to the dialog with both feet jogging away in place, and still challenge us to stay honest and on track at the same time. I said to myself, "Aha, now here's a new voice of reason and energy in this whole desktop video conundrum. I sure hope s/he sticks around 2-pop and the DV scene for awhile."

Fortunately for me, and for thousands and thousands of other early DV pioneers, chawla did stick around 2-pop for quite some time, and became a strong and steady contributor to 2-pop's vibrant discussion forums and online library of DV How-To articles. He went on to lead a valiant FCP bug-chase effort and worked closely with the Apple/FCP development team to track down some of the toughest and most pernicious of the bugs in the early versions of FCP. And fortunately for you the reader, the mysterious "chawla" has now been unmasked in real life as none other than Charles Roberts, and he has written this most valuable field manual for the aspiring DV revolutionary. You will find it stuffed from cover to cover with in-depth answers to all those hundreds and hundreds of questions we all struggled with in the early days of Final Cut Pro, 2-pop, and Desktop Video Revolution.

Before we proceed with the rest of this story, however, for those of you who may not be familiar with 2-pop and with the early days of Final Cut Pro's development, I will digress a bit. It's spring 1999, I'm working at Apple Computer on the Final Cut Pro development team, and FCP 1.0 has finally been released at NAB. My boss at Apple/Final Cut Pro, Will Stein, asked me if I wanted to "volunteer" to be the one on the team to go out and monitor the online discussion forums and e-mail lists where people might be talking about and beginning to use Final Cut Pro 1.0.

Now, at this very same time, fortunately for FCP and DV revolutionaries everywhere, a guy named Larry Jordan woke up crazy bonkers one morning and started a new Web site with an online discussion forum format dedicated to FCP, in his bedroom on a Mac laptop. Larry was an established film editor and had been an early workflow consultant to the FCP application designers at Macromedia. Larry's Dad was a film editor from way back, when they made their cuts with a real razor blade on real celluloid, and put the familiar numerical countdown we all know with an audible "pop" at the number 2 at the head of their carefully crafted final edits. Hence "2-pop.com," in honor of Larry Jordan's father.

And that was the start of 2-pop: Larry Jordan on a Mac laptop in his bedroom in LA basically 24/7, figuring out Adobe GoLive and WebBBS and all that gooey Perl/CGI gunk, putting up juicy DV news flashes on the front page, and constantly haranguing his Web host to get the site back up online, please! The hardest part of Larry's job in those early 2-pop days was just to keep up with the sheer volume of posts that relentlessly overflowed the single discussion forum. The news was out that FCP was cheap, solid, empowering, and easy to pick up and start using, and the HELP!!! questions poured in from the brave souls who were starting to use it and from the curious folks who hadn't taken the plunge yet, but wanted to know if it was really for real.

My job for Apple/FCP was to answer all those HELP!!! questions online—HELP!!!! Can't Print to Video!!! HELP!!! Error 22!!! HELP!!! My audio is out of sync! HELP!!! My video gets dark when I render! I had my hands full just keeping up, and the more questions I answered at 2-pop, the more people posted even more challenging and interesting questions! Pretty soon, I was in the same shape Larry was in: me in my bedroom in Cupertino, with Larry's burgeoning 2-pop Web site open on my home system pretty much 24/7, thinking HELP!!! I need a few folks who know about DV and FCP to help me out here! And then secretly (don't ever tell anybody this), I was thinking HELP!!! LARRY!!! Can't you get the 2-pop Web site to crash for a few weeks again, so I can get some sleep!!!

Which is where chawla comes into this story, who I soon learned was actually none other than Charles Roberts in disguise, a mild mannered media arts teacher from Fitchburg State College in Fitchburg, MA. To my great relief, Charles began to help out in the forum and answer many of the FCP and DV questions that came in to 2-pop. All the stuff about timecode and sample rates and codecs and firewire drives and nesting that people wanted to know about, day after day. And in a magical way that I still have no explanation for, just as chawla started to pitch in and help out at 2-pop, so did a whole big bunch of other folks: Ken Stone, Michael Horton, Kevin "telly" Monahan, Randall Dearborn, Adam Wilt, Andrew Balis, Dan Brockett, Tom Wolsky, Lisa Brenneis, Bill Smith, and Ross and Miles and Phil and Jose and Ned and Joel and Marc and Charles and Gretta and I hope you get the picture here—a whole gang of folks joined with Charles and started answering more questions than they asked. Whew . . . So I finally got a little sleep.

Then something funny and wonderful happened. This whole ad hoc, ragtag bunch of video folks at 2-pop became greater than the sum of its parts, and we became "2-poppers." More importantly, we discovered that we had one thing in common—as we gained a certain mastery of DV and FCP, we all felt like revolutionaries on the very front lines of a major media revolution world-wide. With DV equipment and FCP so cheap and easy to pick up and use, we finally had a way to tell our own stories, without having to pander or sell out to the gate-keepers of the major media outlets.

We finally had tools we could afford to buy on our own, without permission from the banks, to tell the stories WE wanted to tell, distributed directly the way we wanted to distribute them, without permission from some lowest-common-denominator-type programming exec or some nervous agency person at Pepsi or General Motors or General Electric or the General Secretariat, for that matter. And we felt pretty heady at 2-pop in those early days, let me tell you. As a few 2-poppers started sharing their early success stories in the forums, on more than one occasion you could log on to the general discussion forum late at night, and find a post here and there with the exuberant subject line: "Viva la Revolution!" I know, because I posted a few of those rallying cries myself!

One of the most rewarding parts of my job at Apple in those early FCP days was to gather together a summary of the worst problem reports and the best success stories about FCP from around the Web, for review by the entire FCP team. The FCP team relished hearing those early success stories, because the people who told them almost always talked about how amazing it was that FCP didn't crash (very much!) and almost always did what you asked it to do, for very little money. And in particular the FCP development team was proud and encouraged to hear about the revolutionary spirit that was growing up and being voiced by 2-poppers all over the world around FCP and DV.

Because that's exactly what they set out to do, some four to six years earlier: Randy Ubillos, the visionary code slinger behind FCP; Will Stein, the team-honcho who made the tough decisions; Brian Meaney and Michael Wohl, the workflow designers who raised Drag and Drop and Tabs to a whole new level of elegance; and the entire FCP development team set out from the start to do one simple thing—put the tools of visual storytelling into the hands of the folks with the actual stories to tell. They worked long and hard and uncompromisingly to make the application easy to use for the beginner, deep and robust for the seasoned pro, affordable, and based on an open architecture so it could grow and adapt quickly when something extraordinary like the DV Revolution comes along, for instance.

Two years after its first release, FCP is proving to be one of the major creative enabling tools of this incredible Desktop Video Revolution we are participating in. I used to call FCP "The Killer App," and "The Kalashnikov assault rifle" of the DV Revolution. But I have thought better of using those military metaphors from a previous era, because it's clear there's a real velvet revolution going on around the world, in terms of real folks being able to tell their stories directly to other real folks, with video, print, e-mail, Web sites, and whatever else we can come up with from here on out. So I think the military metaphors for this emerging revolution/evolution and its tools and participants are out of place. If there's any worthy and enduring purpose at all to this thing we call the DV Revolution, it's to help make a world free from the need to use Kalashnikov assault rifles and dumb things like smart bombs in order to share our stories and get our points across to a wider audience in our human family.

Which brings us back to Charles Roberts, 2-pop, this field manual you have in your hand, and the empowering editing tool called FCP you are about to explore. The Desktop Video Revolution is here and well under way, and Final Cut Pro and DV no longer need to prove that they are "professional." Those very early HELP!!! questions at 2-pop have all been solved by the FCP team, and FCP continues to evolve with each new release as a world-class editing tool. Tens of thousands of people currently use FCP to create video pieces for broadcast and cable, direct to video casette, the Web, DVDs, and CDs, theatrical motion pictures, and for heartfelt pieces of personal and artistic expression. People with stories to tell just like you have proven beyond a doubt that they can take FCP and their source footage, roll up their sleeves and clean out their mouse, get right on down and seriously involved about the whole thing, and produce high quality, polished programs that tell their stories just exactly the way they want to tell their stories.

So then, that only leaves one final question to resolve for you the reader: Now that I have FCP, how do I actually use this great new tool?

To get you started on that tale my friends, I leave you in the very capable hands of Charles Roberts, aka chawla, who will take it from here. And when you're done with a chapter or two, stop on by 2-pop.com, pull up a log in the clearing by the fire and chat for a while. Share your answers and questions and ideas and opinions and successes, and your occasional frustrations too, about telling your very own visual stories with video and sound and Final Cut Pro.

Ralph Fairweather
Strawberry Fields Hot Springs, CO
July 2001

Preface

The old cliche "What an age we live in . . ." is one I never thought I'd find myself muttering, and yet here I stand today looking around at the video editing possibilities on the market. Never before in the history of video have the tools been within reach for so many individuals. Cheap, high-resolution cameras and desktop computers have removed the dollar sign from the list of practical reasons for not "makin' movies." The tools aren't free, but they aren't ruinously expensive anymore either.

It used to be that you had to be in school or working in the industry to get your hands on video cameras and editing stations. To a large degree, that meant you got the perspective of two select, though admittedly broad, groups of people: the artists, the artisans, or some strange, delightful mixture of the two. That is not to say that video and film production in the past century have been lacking in quality—far from it. There's been lots of great stuff to watch, from the Lumiere Brothers up to the present day. What there really hasn't ever been is a truly wide range of perspective and subject matter in video and film. The reasons are pretty simple: if only a small section of society can afford the time and money to acquire equipment and technical skills, then we are doomed to watch that small section of society's vision of the way things are.

We've had plenty of genre and documentary work about the Working Class and the Common Man, but not nearly enough Working Class– and Common Man–produced work. What a shame. And now it's more possible than ever to change that scenario. I believe the man said, "the means of production in the hands of the people . . ." We can stop letting the establishment make our entertainment for us and start making it for each other. We can stop waiting for people to address subject matter that is important to us; we can make it ourselves. That is the style and substance of a revolution; it's self-empowerment on a grand scale.

But with revolution and self-empowerment come responsibilities. If we intend to make our own video and film work, we owe it to ourselves, our viewers and the medium to become skilled with the tools of the trade. Digital video editing isn't really all that difficult a task, but there are some skills and habits that make it a lot easier to concentrate on the editing rather than fighting with a computer to make it do what you want. Most people in the past picked up these skills in school or in the field while working in the broadcast or film industries. For those who did not (and for a few who did, but are new to Final Cut Pro), this book is for you.

The purpose of this book is to introduce new editors to digital video, the Macintosh platform and the Final Cut Pro application. Since it is assumed that the user is new and very green, capturing, editing, special effects and output-to-tape techniques are covered.

The aim is to get the user from square one, which may be a blank slate, to square whatever, which means having set up a project correctly and gotten some editing and compositing techniques down, and then putting it all back out safely to tape.

How should you use this book? Read the first two chapters and get some perspective on the system you are working with. The more you understand about the inner workings of DV and the Macintosh, the easier it will be to understand what your system is doing as you edit, and the better prepared you will be if the time comes to upgrade, repair, or shop for add-ons to your system. Then work through the rest of the book side by side with your Final Cut Pro editing station. If you don't understand an exercise, do it again until it becomes clear. That's how you learn.

No CD-ROM accompanies this book, because I want users to work with their own video materials from the get-go. If you do the exercises using your own video footage, you will be very confident after having completed all seven chapters. Take the offered and suggested techniques as the hard-won wisdom of someone who has made most of the mistakes, some of them a few times. It's always better to learn from the painful mistakes of others and DV editing is no exception.

This book, while comprehensive in some respects, is not the last word in DV editing. There are many different ways of approaching editing tasks, and it would be impossible to include them all in one volume. It is also not a "Quick Reference" guide that covers every single drop down menu item in the application. It is not a silver bullet, promising that you'll be editing in Hollywood next week after you finish Chapter 2.

It is, however, a couple of things. It is honest, and intends to inform the reader as thoroughly as possible about the basic techniques of Final Cut Pro and Firewire DV Editing. It will establish a context for the reader to learn in, but it will not promise to learn for you. Like violin lessons, this book can't make you better without practice. It will assume that the user knows nothing about video beyond the On switch of their television. My mother is the litmus test here, and to her a "cold boot" is a heavy shoe in the freezer. By the end of the book, the user should not only be able to navigate around the Final Cut Pro editor, they should have become familiar with video itself. It will enable the new user to move on to the next step, which is the act of editing unencumbered by confusion about what is happening under the hood of their editor. Once you master the tools, you can forget about them and REALLY get to work.

Acknowledgments

First and foremost, I have to thank the 2 Pop Final Cut Pro Web site, home of the greatest and friendliest community I have ever encountered on the Web. It's one of those places where you return to a discussion forum not just because you have a burning question about DV, but because it's full of friendly and easy-going users. If the rest of the Web followed the 2 Pop model, you wouldn't have to worry about your kids surfing . . . To all the denizens there, thanks for making it what it is.

A special thank you goes to Larry Jordan for starting it, Ralph Fairweather for inspiring and leading it, Ken Stone for not sleeping until every newbie has an answer (he also tech edited this volume; no small misery, thanks, bud), and all the 2 Pop Guides that keep the boards rolling, unsung heroes one and all, the reason that dot com DIDN'T fail.

Thanks to Michael "headcutter" Horton for giving direction to the LA Final Cut Pro User Group, whose meetings make us East Coast folks green with envy, but whose Web site also provides a great source of information and news. Thanks for letting me be an honorary member without an LA address—will a mail drop do?

Thanks to my colleagues in the Fitchburg State College Communications Media Department in Fitchburg, Massachusetts for making our program so great to teach in. It's the only program I've seen that has such a strange mix of heavy aesthetic and theoretical training right alongside the rigorous production requirements. Enough to warm the heart of a cranky, old MFA.

Special thank you there to Gunther Hoos, who warned me about how difficult it would be, then commiserated just enough to keep me from losing my mind and not finishing. Also to Jeff Warmouth, who returned from a honeymoon to get guilt-tripped into doing heavy editorial work. I owe you lots and lots. And to the students of our program—couldn't be prouder if I could claim you all as my own.

Thanks to Terri Jadick, who calmed me down early on about this, and to Amy Jollymore of Focal Press who has helped me negotiate this thing. To be a better book, that is the purpose.

1 What Is Digital Video Anyway?

How video works

Analog video

In order to understand digital video, we have to get a clear grasp of what video itself is, and how digital video differs from analog video, the common format used in most televisions and consumer video tape recorders. Like film, video is simply the transmission of time-based visual information accomplished by the rapid sequence of many still images. We see the illusion of real motion because the still images replace each other very rapidly on the screen. In film, the rate of these images is 24 frames, or individual images, per second. In NTSC (National Television Standards Commission) video, the system used in the United States and many other countries, the rate is 29.97 frames per second. In PAL (phase alternating line) video, the system used in most of Europe and elsewhere, the rate is 25 frames per second.

This concept of illusory motion is where the similarity between film and video ends. Film manages to create the illusion of movement by directly shining light through celluloid material to create a projected image. Each individual frame is created on a screen as a result of light being blocked by the material coating the film. Each frame is projected through the lens of a projector completely at once, top to bottom, side to side.

Video, on the other hand, is an indirect system of display. A video signal is actually an electrical wave that travels through a metal cable. This electrical wave carries the information for each frame to your television. On receiving the wave, the television—or video monitor—interprets it and uses an electron gun to shoot a beam of electrons at the inside of the television screen. The electron beam charges up phosphors on the back of the television screen, producing the glowing images you see on the front. It does this at the frame rate established by the video standard in use, 29.97 or 25 times per second for NTSC and PAL, respectively. The origin of the term *analog* comes from the fact that video's electrical waveform is an analogy, or corresponding likeness, of the original image. A video monitor can interpret this waveform and produce a new image to which the electrical signal corresponds.

The process of directing the electron beam all the way through the television screen is called scanning. Instead of illuminating the entire screen at once as film does, the electron gun must draw thin horizontal lines, one after another, all the way down the screen. When the beam has crossed the video screen once, we say it has scanned one line. After it scans a single line, it moves to where the next line is to be drawn a little further down the screen.

A single frame on an NTSC video screen is composed of 525 separately scanned horizontal lines. When the scanning of 525 lines on the screen is complete, we say one frame of NTSC video has been scanned. Then the process begins again, repeating this action at the rate of 29.97 times per second for NTSC and 25 times per second for PAL, which is composed of 625 lines.

To make matters a little more confusing, a video frame is actually scanned in two sweeps, or two separate scans of the video frame. The two sweeps are referred to as fields. The electron gun in the television monitor does not go directly from the first scanned line to where the second one should be. Instead it skips a line before it performs the horizontal scanning process again. While scanning a single frame, the electron gun actually scans all the odd lines first, and then on reaching the bottom of the screen, it returns to the top to scan all the remaining even lines.

This process is referred to as field interlacing, because the two separately scanned fields when viewed together form one interlaced frame. Although the precedence of one field being drawn before the other, or *field dominance*, can vary between digital video systems, for the moment it is enough to know that electricity is driving the video signal, that the electrical waveform for each field of the frame determines what you see on the screen, and that this scanning action occurs at the rate of the video standard in use in your area. For NTSC, there are 29.97 frames per second, each frame being composed of two separate fields, yielding a field rate of 59.94 fields per second. For PAL, there are 25 frames per second, each frame being composed of two separate fields, yielding a field rate of 50 fields per second.

Knowledgeable readers will recognize that this field rate corresponds to the system of electricity in use in the country in question. Countries like the United States, which have a standard electrical current of 60 Hz, have a field rate of 59.94 times per second, while countries like Great Britain which use an electrical current of 50 Hz, use PAL, which has a field rate of 50 times per second.

Generation loss and digital video

"Great," you say, "it works, and that's all I care about." What is the need for digital video if we have a perfectly good analog system? The answer lies in what is referred to as generation loss. Because the analog video signal is an electrical wave, it is a physical thing; physical electrons move up and down in a metal cable to transmit the necessary information so the video monitor can display images.

The problem is that every time the signal passes through a medium—for example, when it passes through a cable from your consumer VCR to your video monitor—the electrical wave changes a little. The electrical wave that makes it to your television is never quite the same as the electrical wave that was magnetically stored on the videocassette and played back. The slight distortion in the wave from its transmission through the cable results in a change for the worse in the picture quality generated by the wave.

We call the decay of the electrical wave when it passes from medium to medium generation loss. Every time you make a copy of a videotape, or even simply pass it through a cable, you lose a generation. After a few generations of passing between even the highest quality analog equipment, the electrical wave will deteriorate until the quality of the video image is poor indeed.

Here digital video comes to the rescue. At some stage in the game, the analog video electrical wave is converted into a sequence of binary numbers that can be stored as digital data, essentially no different from a word processing file. Unlike analog video, which is an electrical wave and is therefore subject to waveform deterioration, the digital file has only a numerical value, which never alters until we ourselves choose to change the numbers. The quality of the digital video data will remain intact regardless of how many times it is copied.

Think of it this way: The number one has the value of one, no matter what it is counting. Copying the number one always results in an exact copy, because it is a value and not a physical thing. An electrical wave (or a zebra or a goldfish), on the other hand, can never be exactly copied, modern cloning processes notwithstanding. Some subtle difference will always be introduced in the copy. In analog video, those subtle differences turn up as signal noise, the evil result of generation loss. The digital copy, however, being numerical and immaterial, is never subject to such loss.

Another bonus of the digital video system is that the technology is remarkably plug-and-play, provided your digital video system is configured correctly. Broadcast quality video editing prior to the advent of the new Firewire-based DV editing technology involved expensive and difficult to configure analog components that required a relatively high degree of technical experience to run properly. You will find that Apple Final Cut Pro software, the Apple Macintosh desktop computer, and a DV deck or camera constitute an inexpensive, elegant, and simple solution that produces quality results possible only with an astronomically expensive solution a few years ago.

Before we lose our heads in revolutionary hyperbole, it is important to note that we will not be completely free of analog devices for quite a while. While all-digital DVD technology is rapidly ascending, most consumer televisions and VCRs are still analog video devices, and that situation is unlikely to change any time soon. So chances are that your new Final Cut Pro editing station will contain a mix of analog and digital video devices, since most of the people you will want to distribute your edited products to will be using analog devices as well.

Your television (or a dedicated NTSC or PAL video monitor if you have one) and your VCR are the analog devices that are still necessary; we will use them to monitor what we have edited and to distribute the final product to a world that does not yet widely own DV decks. The Macintosh desktop computer and the DV device (e.g., a DV deck or camera) will constitute the digital devices in your editing system. Between the Macintosh and the

DV device, you will be able to capture and edit video without incurring generation loss. When you have finished editing, you will be able to output your edited material to a DV tape, which will keep a perfect digital copy always available for further work and for archiving. Finally, you will be able to pass video out from the DV device to analog video-tape so that you can share your handiwork with the rest of the world.

Your DV device is the hardware center of your editing system. It is more than simply a record/playback unit for DV tapes. It can pass digital footage directly to the Macintosh through its Firewire connection. In addition, the deck or camera acts as a conversion device from digital video to analog video. That means that while editing digitally in Final Cut Pro, you can watch the results on your NTSC or PAL video monitor instead of squinting at a postage stamp-sized box in the computer monitor.

Where does digital video originate?

How light becomes numbers

How does DV work? How does the video become digital in the first place? How exactly does light become the series of numbers that are safe from generation loss and deteriora-tion? Let's take a look at a video camera. When light passes through the lens of the camera, it is evaluated for its luma and chroma, or bright/dark values and color values respectively. The luma and chroma information is collected on light sensitive chips called "charged cou-pled devices" (CCDs). The luma and chroma information is gathered from the chips at the same rate and in the same manner as we saw with the electron beam in the back of the tele-vision set. It is essentially the same operation in reverse, the television generating light from electrical information and the CCD generating electrical information from light.

This much of the process is really no different whether a digital or an analog video camera is being used. Both generate luma and chroma information from CCDs. The dif-ference is in what happens after that information has been generated. In the analog camera, generally speaking, nothing more is done with the signal. The electrical wave created by the CCD passes directly to the destination videotape, where it is stored magnetically for later playback.

The digital camera, on the other hand, digitizes, or assigns numerical values to the var-ious qualities of the light, translating the luma and chroma values into data that can be stored on DV tape. As a result, although both analog video and digital video end up being recorded to tape, the digitally recorded video is never recorded as a raw electrical waveform. It passes from CCD to the recording tape as pure data.

Sampling and compression

It sounds very easy. The camera magically changes light into numbers. In reality, the process is quite mechanical and predetermined. First, the luma and chroma values are subjected to a process called sampling. It turns out that, perceptually speaking, bright/dark values are far more important to the naked eye than color information. You need a lot more luma information to make out detail in an image than chroma. Consequently, the chroma values are recorded in camera less frequently than the luma ones.

As the camera scans a video line and gathers luma and chroma values, it does so at a ratio of luma values to chroma values. Since the luma values are far more important to the image than the chroma values, all the possible luma values are recorded, or sampled. But because the chroma values are less important, they are sampled less frequently. The actual ratio between the luma and chroma sampling depends on the camera and the system of processing the digital video data that follows this sampling.

There are three common ratios in use in digital video. Since ratios generally work best in whole numbers, the ratios list the luma value first, designated as Y, and assign it a value of 4. This yields the following three ratios: 4:4:4, 4:2:2, and 4:1:1. In these ratios, the 4 designates that the luma value is always sampling all the possible values for bright/dark. The other two numbers in each ratio refer to two of the primary colors for video, red and blue.

Thus, in a 4:4:4 sampling ratio, all luma and chroma are sampled: four Y, four red, and four blue. In a 4:2:2 ratio, however, the red and blue values are only sampled every other time in comparison to the full four samples of the Y, resulting in four Y, two red, and two blue. And finally, in the 4:1:1 ratio, the red and blue values are sampled only every fourth time. The third primary color, green, never needs to be sampled, because its value can be determined easily by calculating by process of elimination using the two sampled colors red and blue and the luma value, Y.

Few systems are robust enough to sample at the full 4:4:4 ratio, and there is very little need to do so in nearly all circumstances. A 4:2:2 sample ratio is nearly indistinguishable from a 4:4:4 ratio for most purposes. 4:2:2 ratios characterize the bulk of the high-end equipment in use in the broadcast industry, where very high quality is demanded. Our DV prosumer systems (prosumer, an informal term concocted from the words professional and consumer, is used to describe the many devices on the market that are of substantially higher quality than consumer gear, but which are not considered to be of the absolute highest quality) use the 4:1:1 sampling ratio. Although not nearly as detailed an image system as the far more expensive 4:2:2-based cameras and decks, 4:1:1 provides a fairly high-resolution sampling at a fraction of the price. In comparison with the video quality of prosumer video gear from years past, DV's 4:1:1 sampling ratio is vastly more detailed and ensures much higher fidelity to the image being recorded.

The digitizing process is not complete with the sampling process, however. The next key element to the process of digitizing video is the application of the codec, a basic fact of digital life that has important implications in every aspect of digital video. The codec is the key to efficient digital video acquisition and editing. As we shall see, the codec is the link that makes all digital video possible in the camera, the desktop editor, and in every other use you can create for it.

The previously described sampling process generates an astounding amount of information. In order to create a single scanned line of a digital video frame, we must sample 720 individual values, or pixels, for luma. And then, depending on the sampling ratio involved, we must also sample a large number of values for red and blue. Beyond this, we must gather 480 or more such lines per frame and then repeat this 29.97 times per second for NTSC and 25 times per second for PAL. Even with the great speed of today's advanced cameras and computers, this is far more information than can be juggled by prosumer electronics and computers.

The amount of data generated by the sampling process is its *data rate*, which is expressed in terms of megabytes per second. The excessively large data rate coming directly from the sampling process can exceed 24 megabytes per second. To put that number into perspective, the sustained data rate of the stock hard drives in a Macintosh G4 generally clocks in at around 10 to 14 megabytes per second and peaks at around 16 megabytes per second. The hard drive solutions of the standard inexpensive desktop computers we intend to use for DV editing cannot match the data rate of the video coming right out of the sampling process.

There are two ways to deal with this situation. One solution is to speed up the data rate of our computer with faster hard drive systems. In higher-end editing systems in which the absolute optimum quality is necessary, hard drive storage solutions are employed that can match or exceed the requirements of enormous data rates. The downside, of course, is that these super-fast drive solutions are also radically more expensive and probably far out of the price range that most of us can afford.

For our low-cost system, however, there is a better way of dealing with the issue—reduce the video data rate to something more manageable. In the Firewire-based Final Cut Pro DV editing system, this latter method is employed through a process called *compression*. The initial staggeringly large video data rate is brought down to a respectable 3.6 megabytes per second, which is well within range of the inexpensive stock drives of your Macintosh.

This data rate reduction is made possible by the *codec*, a term that combines the words compression and decompression. A codec's main function is to reduce the amount of data in a video frame while at the same time maintaining the integrity of the image. When a video frame is initially compressed, the codec throws away excess digital information, retaining only what it deems will be necessary to decompress, or reproduce the image again later. The result is a much smaller amount of data for each individual frame, a correspondingly lower data rate, and, hopefully, an image that more or less resembles the original uncompressed source footage.

Types of compression

There are many different codecs, and they accomplish the previously described task in as many different ways. A discussion further on in this book will investigate the primary methods by which various codecs do their compression magic. But for the moment, there are two main criteria to judge a codec: (1) how effective the codec is at reducing the video file's data rate and (2) how well it maintains the image's integrity while doing so. To illustrate the relationship between these factors, we can take as examples three popular codecs available through your Macintosh operating software and Apple QuickTime: Animation, Cinepak, and the DV codec.

The Animation codec is a completely lossless codec. When video is compressed and decompressed using it, there is zero loss of image detail. In terms of image integrity, it is as if no codec had been used. Indeed, the size of the compressed file is only slightly smaller, and the corresponding data rate is still very high. Thus we say that Animation is an impractical codec, even while admiring the image integrity. Great image quality with large data rates will not serve our general DV editing purposes.

The Cinepak codec, on the other hand, discards huge amounts of image data, reducing the data rate enormously in the process. In fact, the data rates are so low that it is valuable as a codec for Web distribution of video content. But as a consequence of the heavy compression, the image quality suffers terribly, resulting in horrible visible artifacting and loss of detail. Thus, we say that Cinepak is also impractical, although we might accept the low image quality in exchange for the very low data rates that will make CD-ROM and Web applications easier to implement.

The DV codec is our happy medium for editing. It discards a vast amount of image data on compression, generating a very satisfactory data rate that is well within the limits of what our inexpensive Macintosh's hard drives can handle. And decompression (i.e., playback of the compressed material) reveals that we are left with very acceptable image quality. The DV codec is, in many ways, the perfect codec for the DV editor, because it generates very good image results with a very low data rate and it doesn't break the bank doing it. Although the DV codec is not perfect (and the trained eye can find evidence of the codec's artifacting quite easily), it is in the ballpark with the codecs used by editing systems costing many times more than the amount you will spend on your Final Cut Pro editor.

After comparing lossless, high data rate codecs such as Animation, with lossy, low data rate codecs such as Cinepak and DV, we can further categorize codecs into three categories: software-only, hardware-only, and software/hardware hybrids. This simple division explains much about the way codecs function within computer systems. The proper performance of your Final Cut Pro editing system and the ability to maintain the highest possible quality in your video productions depends on how well you grasp what happens and when.

The first type of codec, software-only, is a compression/decompression process that is accomplished entirely by software calculation. The Animation and Cinepak codecs described previously are good examples of software-only codecs. Video is analyzed by the software codec, and data is discarded, redistributed, and reformulated according to the calculation method used by the codec. The calculation method is generally very flexible for software codecs, allowing the user to determine how lossy the codec will be and how low the data rate will be pushed.

The major trade-off is in processing speed. Software codecs are, without exception, the slowest at compressing and decompressing video. The computer must process each pixel of each frame and decisions must be made as to whether the data must be kept, changed, or discarded. Such processing is slow and relies entirely on the desktop computer's built-in hardware. At lower compression levels, where the quality and data rate are higher, the built-in hardware may not be able to process the video fast enough to achieve the high frame rate, a phenomenon known as choking. At higher compression levels, the image quality is sacrificed beyond what we would regard as acceptable for broadcast video.

Most software-only codecs are used exclusively for Web and CD-ROM applications, because the files they produce must be accessible by computers that do not have specialized software or hardware. Any file compressed by a codec must be decompressed by the same codec to be played back, so the same codec must live on both the computer that compresses the video and the computer the video is distributed to. Whether PC or Macintosh, using a dedicated editor or a very limited word processor, most computers have such software-only codecs preinstalled in their system files and are able to decompress these files. This makes software-only codecs very attractive to the Web and CD-ROM distribution markets.

Hardware-only codecs are not so common. A hardware codec is a specialized piece of electronic circuitry that takes an analog video signal and converts it to or from digital video. There is little or no variability involved. The video is digitized with a set of parameters that are usually optimized for a specific level of quality whereby little or no editing or manipulation of the video will take place before playback. Since little variation is built into the system, the hardware is simply converting electrical values into digital ones and back in a rigid fashion.

One example of a hardware-only codec system is the digital video conferencing system, such as those produced for corporate and industry use. With such devices, the only objective is getting digital video data from point *A* to point *B* intact, usually at fixed, relatively high data rates. Although the image quality of these solutions is quite acceptable, they are constructed expressly for data transfer and cannot be made to work with an editing station.

Another example is the so-called all-in-one-box video editing solution, in which the editor gives up the wider options of codec quality control for the ability to simply plug analog video into the connectors of a box and edit away. Such systems usually offer limited transitions and effects but are not built to function as completely expandable post-production solutions. The lack of flexibility makes them useful in some situations but impractical for many scenarios, such as being integrated into an editing situation in which many different editors are working on the same project.

Hardware solutions function in a manner referred to as *real time digitizing*, meaning that unlike the software codec, there is no appreciable delay in the conversion of the analog video signal to digital data. The circuitry of the hardware codec is designed to compress or decompress the video electronically rather than mathematically, almost instantly producing a digital video image at a reasonable data rate that the hardware in the "box" can handle. Since the hardware is constructed specifically with this purpose in mind, the quality depends on the hardware system, but you can expect to pay heavily for acceptable quality.

The third sort of codec is a hybrid of software and hardware. This type of codec uses both the high-quality and speedy results of hardware codecs with the variability and control of software interfaces. These are usually more expensive solutions, such as Avid and Media 100, although third-party manufacturers produce similar hardware capture cards for use with Final Cut Pro as well. The vendors of the Aurora, Digital Voodoo, and Pinnacle Systems all sell hardware cards that Final Cut Pro can access for capturing video.

Generally speaking, regardless of who manufactures the analog capture card and its corresponding software, the function of the hardware/software combination is ultimately the same. A hardware digitizing card is used to convert analog video from a source such as a video deck into digital data using a flexible compression scheme; it is usually capable of generating very low compression and high data rates. The software interface that works with the hardware card allows the user subtle control over how much compression takes place, giving the option of better image quality or smaller file size, depending on the editor's needs.

Much less lossy compression and higher image quality schemes can be instituted, because the hardware codec is taking care of the processor-intensive actions that would otherwise have bogged down the system. This sort of system has been considered the standard in broadcast digital video editing for a number of years; such systems have offered the best quality and highest reliability available. Even today, where the absolute highest image integrity is necessary, software/hardware hybrids offer the only real solutions.

The problem of course is that such a hybrid of hardware and software is often strictly proprietary and generally very expensive. If you use the hardware codec from an Avid or Media 100 system, you can decompress only using that same software/hardware codec. Since these expensive software/hardware combinations will generally burst the seams of most modest budgets, the likelihood of distributing files created with the expensive systems' codec is slim to none. The few smaller-scale editors who can afford such an editing solution operate as stand-alone editors, digitizing material from analog tape and then outputting to analog tape when the editing is complete.

Some analog capture cards like the modestly priced Aurora Igniter can use a more universal M-JPEG compression format that could be accessed without the Aurora's special hardware codec, but being able to decode and decompress the hardware codec is only part of the expense. Even the more universal M-JPEG compression format carries a very high data rate if the quality is set to an acceptable level. And high data rates equal the fast-drive solutions that, in turn, equal sticker shock as the price of your turnkey DV editing station grows.

And expensive these storage solutions can be. Although prices have dropped drastically during the past few years, the price tag of these software/hardware solutions remains out of range for most users. After you tack on the price of the super-fast storage solutions necessary to take advantage of the low compression and high data rates, you end up with a system that most of us simply can't afford to own. Even at the most cost-efficient end of the spectrum, $1,000 for an appropriate capture card, $300–400 for a fast SCSI card, and another $1,000 or so for the drives themselves really add to the overall cost of the investment.

There is another game in town these days, however. The DV codec used by Firewire-based DV editing applications is a unique hybrid. It exists as both a software and a hardware codec, but independently. Material can be compressed and decompressed using either a software-only version of the codec, such as the Apple DV codec that Final Cut Pro uses, or a hardware version of the codec. Material that has been compressed using the hardware codec can be played back with the software version and vice versa.

The hardware version of the codec is closer at hand than you think. You may already own it. It lives in your DV deck or camera. As you are shooting and recording video onto DV tape, the video is being sampled using the 4:1:1 ratio and is then immediately compressed by the hardware DV codec in the DV deck or camera. When we need to edit, instead of having to purchase an expensive digitizing card to access a special hardware codec, we simply plug the Firewire port of the camera into the Macintosh, accessing the camera's built-in hardware codec for our editing purposes.

The Firewire cable going between the DV deck or camera and the Macintosh is simply a data tube that transmits digital information from one device to the next. The potential data rate for this tube is more than high enough to transmit the DV codec-compressed video from deck or camera to Macintosh and back. It is this Firewire connection between the Macintosh and the DV deck or camera that allows digital video to be translated to analog for viewing on an NTSC or PAL monitor. It also allows the digital video previously recorded on DV tape to be passed through to the Macintosh where it will be accessed by the Macintosh's QuickTime software version of the codec, making it possible for you to play back and edit the material.

What about digital audio?

How sound becomes audio

We've described how video is compressed and sent down the Firewire tube to be accessed by a DV codec on the other side, but obviously there is another piece of the puzzle. Video without audio would make for very uninteresting viewing. Digital audio is also included in the Firewire data stream. In order to understand how it is available to the Macintosh, we need to understand how sound becomes *digital audio*.

Sound becomes *analog audio* when sound waves in the air encounter a transducer (e.g., a microphone), thereby producing an electrical waveform that is an analog of the original sound wave. This process is theoretically similar to that of analog video signals, which are also produced by generating an electrical wave that is an analog of light waves. Analog audio at this stage is usually directly recorded to analog tape. When the tape plays back, the electrical waves recorded on the tape reproduce the original sound.

As with analog video, the audio waveform experiences deterioration through generation loss, and digitizing audio eliminates this problem just as digitizing video does. Digitizing audio also involves a procedure called sampling. In audio sampling, the digitizer analyzes the audio wave and assigns numerical values to describe each variation of level and frequency in the audio waveform.

Of course, as we saw with sampling video, sampling audio is a variable task. You can sample the video or audio more or less intensively. Unlike video, however, audio data rates are far lower than video ones, so audio really never needs to be compressed for acceptable playback performance, with the exception of Web applications, for which extremely low data rates are necessary. Digital audio for use in editing DV footage is never compressed, and its quality is determined by the frequency of the sample rate.

Sampling and digital audio

This sample rate is described in kilohertz or thousands of cycles per second (one hertz equals one sample or cycle per second). The higher the sample rate or KHz, the more finely detailed the information about the wave that is gained per second and thus the higher the integrity of the digitizing process. It follows that the higher the number of samples per second, the larger is the amount of information about the recorded audio, and thus the better the quality and the higher the necessary data rate. There is a corresponding increase in the data rate of digitized audio when the sample rate is increased.

There are standard sample rates that are most commonly used in the world of media production today. Audio compact discs, for example, use a standard sample rate of 44.1 KHz or 44,100 samples per second. This is adequate definition for high quality two-channel stereo recording. Another common sample rate is 22.05 KHz, which is a lower quality sample rate used in multimedia CD-ROMs and for the Web because its file sizes and data rate are small enough for delivery over low-bandwidth Internet connections. The quality is quite horrible, better than the quality of telephone audio, but not much better.

The two sample rates that concern us here are the native sample rates for DV decks and cameras. These two sample rates are 48 KHz and 32 KHz. 48 KHz is a very high quality sample rate. It exceeds the definition of audio CDs and provides very accurate reproduction. It is a fully professional sample rate, and not long ago it was available only through expensive digital audio tape (DAT) recorders. It is a testament to the power of the prosumer market that such quality acquisition is available at such low prices today. Currently, 48 KHz is the native sample rate of nearly all DV cameras and decks, and it is your best bet for recording audio.

Why then, you ask, would we use a lower sample rate of 32 KHz, when 48 KHz is so much better and available on the same camera or deck? Some cameras and decks allow you a really interesting option in recording your audio. It has to do with the different *bit depths* that are allowed with the use of the two sample rates; 48 KHz and 32 KHz are referred to as 16-bit and 12-bit respectively in DV cameras. Each sample of the audio wave records a certain number of bits, which is the raw data of the sample. Bit depth describes how many of these bits are recorded at each sample, and, therefore, how much information is gathered at each sample.

48 KHz gives us 16 bits of audio per sample, which is enough to provide for two tracks of very high quality stereo audio. Most DV cameras can also be set so that the audio is

recorded at 32 KHz, capturing 12 bits of audio per sample. The twist is that this distribution of 12 bits can be organized in the sample so as to allow four separate tracks of 32 KHz 12-bit audio instead of the two allowed at 48 KHz 16-bit. This means that if we shoot video and record the audio at 32 KHz, we can simultaneously record four tracks of 32 KHz audio, or even go in later and overdub the two extra tracks. Multitrack audio recording on digital videotape is a highly underutilized feature, one that many people aren't even aware exists on their camera. But it can be highly useful in event videography, such as weddings and ceremonies, where more than two microphones must be used and where separation must be maintained between sound sources. And despite what you might assume, 32 KHz is of fairly high quality and may be acceptable for such applications.

What are timecode and device control?

We mentioned Firewire earlier in reference to transmission of DV data between the DV deck or camera and your Macintosh desktop computer. But what exactly is Firewire? Firewire is Apple Computer's proprietary name for an industry-agreed-on engineering specification called IEEE 1394. The IEEE 1394 standard was developed so that the many companies developing high-bandwidth digital interfaces would be more likely to construct machines and software that could talk to each other. Sony's proprietary name ILink refers to the same IEEE 1394 specifications. This is why you can use so many different decks and cameras with Final Cut Pro. Although not all DV devices are officially tested and supported by Apple for use with Final Cut Pro, they are all based on the same technology. The engineering specification allows for a certain amount of digital information to pass from one device (say, your deck) to another device (say, your Macintosh) at a specific rate and using certain wiring and data transfer protocols.

What that means is that through one cable you can pass digital video and audio between the deck or camera and your Macintosh. But it also allows for other less obvious data to be transferred between the two. You can also pass *timecode* and *Device Control* data between computers and decks and cameras. Final Cut Pro is capable of manually controlling any of the devices that Apple Computer specifies is supported, as well as a good many more that they simply haven't gotten around to thoroughly testing. How it does this is really not a new process for those experienced with video production, but it is very important information for new users who have never encountered the concept of timecode.

Timecode: analog and DV

We said earlier that there are 29.97 frames per second in NTSC Video and 25 frames per second in PAL. But how can we track each and every frame when they pass that quickly? When we get into the editing process, we will see that sometimes the difference between making a cut plus or minus one frame can make a serious difference in the success of the edit. Clearly, it is important to be able to access and manipulate video at the level of frames rather than seconds. But anyone who has ever tried to edit with a VCR using a time scale based on seconds will tell you that it is an impossible task. How can you expect to do exact editing when there's no way to get a deck to record or play with any degree of accuracy by pushing buttons on the fly?

The answer lies in *timecode*. Timecode uses what is referred to as video frame addressing. When the video footage recorded on a tape is accompanied by timecode, we say that each frame of the video has a specific and unchanging address. When you insert a tape with timecode into a deck that can read timecode, the deck will display the frame number of the tape that is currently cued up on the play head of the deck. Each frame of video on the tape has a frame number that is unique to it.

For simplicity, it is a standard that this numerical frame address follows the universal convention of the digital clock, that is, in the form of hours:minutes:seconds:frames. Although the real-world clock does not contain frames as a unit, video does—at a fixed rate to the other elements of hours, minutes, and seconds. Thus the NTSC timecode clock runs up 29 frames and then racks up a second, and the PAL timecode clock racks up 24 frames before winding up a second in the same way that a real world clock runs up 59 seconds before racking up a minute.

The benefits of this system should be obvious. If we wish to make an edit at such and such a frame, we simply start recording and playing based on that frame's exact timecode address. Fine and dandy, but we are still obstructed by the fact that our fingers can't move with that degree of accuracy. Luckily, the frame address system of timecode confers another bonus on our tired fingers. Because the frames we want have immutable assigned numbers as addresses, it is easy for computers and decks to automate the process of play, record, and capture. As long as there is timecode, your computer and decks can do this precise work for you, getting exactly the results you want without your lifting a finger.

Now, timecode itself is neither a new nor a digital innovation. It has been around for quite some time and has been as beneficial to the analog video editing community as it is now to the digital one. The real difference for us here is where the timecode is and how DV timecode is accessed compared with the venerable systems of analog timecode. There are two types of timecode in the older analog systems that use ingenious, if problematic, methods of including timecode information on videotape.

The first system is called *longitudinal timecode* (LTC) and is simply an audio signal that occupies one of the two audio tracks on your videotape. The high-pitched audio pulse that is the LTC timecode signal can be interpreted by LTC-savvy decks as frame numbers, thus giving a corresponding numerical address to the frame of video that crossed the play head at the same time as the LTC number. It is a pretty efficient system and is still in heavy use, although an audio track must be sacrificed in order to use it.

A second system that was developed is vertical interval timecode (VITC, pronounced "vit-see"). Instead of including the timecode information in an audio track, the VITC system includes the timecode address information in the video portion of the signal on the tape. As described earlier, the 29.97 frames per second of NTSC Video and 25 frames per second of PAL Video are actually not scanned one line immediately following the next down the screen. A frame of video is scanned in an interlaced fashion, meaning that one field consisting of every other line in the frame is scanned, after which the second field containing all of the remaining lines is scanned.

The period in which the electron gun returns from the bottom to the top of the screen between the first and second fields is referred to as the vertical blanking interval, because the electron gun is not firing during this time. In between drawing the first field and second field of each frame, VITC uses the scan gun to add frame address information. Although you can't see it, the result is that the timecode information is present in the video signal, integrated into the electrical waveform without occupying an audio track. As with LTC, there is a con to this form of timecode address. Because the information is integrated into the video signal, there is no way to change the numbers without rerecording and losing a generation. With LTC, one could simply audio dub onto the timecode track to alter the timecode numbers if they were incorrectly laid down. With VITC, you get what you get.

This analog timecode information is transferred from the deck to the edit controller (the device with which the editor selects the frames to be edited) via various electrical interfaces and protocols. In addition to use with linear editing systems based on direct tape-to-tape transfer, this system of timecode Device Control editing is also used extensively with nonlinear editing solutions such as Final Cut Pro (mostly when using an analog capture card), Media 100, and Avid, in which timecode accuracy and Device Control are necessary for precise digitizing from decks.

There is another distinction, this one between two types of timecode, that applies only to NTSC. This is the difference between *drop frame* and *non–drop frame*. What are drop frame and non drop frame timecodes? The difference has to do with the slightly odd frame rate of 29.97 frames per second of NTSC video. Remember that this is the rate of frames being seen per second. But there is no such thing as a fraction of a frame; all video frames are whole intact entities. And the function of timecode is to count their progression over time using a clock of hours:minutes:seconds:frames.

Unfortunately we need a whole number system to do this. We cannot use the fractions that exist in the actual NTSC frame rate, which would result in a time scale using increments of .999 frames instead of 1! This is not a problem for PAL users, whose frame rate is exactly 25 frames per second, but for NTSC users the issue is critical. If we take the most obvious path and just round off the frame rate from 29.97 to 30 frames per second, at the end of an hour you end up with a disparity of 108 frames between the real-world clock and the timecode clock, which at a frame rate of 29.97 frames per second equals more than three seconds!

Here's the math:

60 minutes × 60 seconds × 30 frames = 108,000 frames
(if we use the rounded-up, incorrect 30 fps)

60 minutes × 60 seconds × 29.97 frames = 107892 frames
(if we use the correct 29.97 fps)

108,000 frames − 107892 frames = 108 extra frames

Imagine the consternation at the television broadcast station. The timecode standard of counting 30 whole frames per second slowly and progressively throws off the relationship between when the show should end and when in fact it does end. As the timecode values get further away from the real-world clock values, hours get longer, and if the union watchdogs weren't awake, workers at the station would be putting in an extra 30 seconds of free labor on an eight-hour day! But, more important, broadcast schedules would go completely out of whack after only a few hours.

Think of it this way. There have been many calendars in the history of the world, but they haven't always been as accurate as our 365 1/4 day calendar. We know that a solar year is the amount it takes for the earth to go around the sun once. Unfortunately, that one revolution doesn't fit exactly right with the number of days it takes to complete that revolution. It takes that extra 1/4-day to complete the revolution.

How does the calendar compensate for one quarter of a day? Does the sun stop rising once a year at 6:00AM and start over? Do we have one day of the year marked with only a quarter-of-a-day marker? No, instead we use a leap year. Every four years, we add a day to the calendar to make up for four years of missing one-quarter days. That way, over many years, the calendar doesn't go completely out of order and we don't watch the Superbowl in the middle of the summer.

So how do we get an exact whole number for each individual frame of NTSC video without the numbering system going out of whack? The Society of Motion Picture and Television Engineers (SMPTE), developed an ingenious system that allows all NTSC video frames to fall on a whole number for a frame address without losing sync between the real-world clock and the timecode clock. Instead of adding a leap year, we are periodically subtracting a "leap frame" from the timecode.

The system is called *drop frame timecode*. In drop frame timecode, the numbering system is based on the frame rate of 30 whole frames per second. Then, over the course of one hour, exactly 108 frame numbers are left out. Although the real-true-honest-to-Joe frame rate of the video is 29.97 frames per second, the drop frame timecode numbering system actually labels the frames with the progression of numbers at the whole number rate of 30 frames per second. Then to keep the frame address numbers from exceeding the actual clock time of the hour, it periodically skips numbers.

Which numbers are left out of the count? It isn't random at all. Frame numbers 0 and 1 are skipped in the first second of each minute, unless the minute number is divisible by ten. Thus after ten minutes, 18 frame numbers have been skipped, leaving a progression of numbers like 00:01:59:28, 00:01:59:29, 00:02:00:02, 00:02:00:03.... It is important to remember that no actual video frames are ever left out; there is only a skip in the number chosen for its address. Just as street numbers for houses are rarely in perfect sequence but the houses are all present, timecode address numbers are skipped, though each frame has a numerical address.

Non–drop frame timecode does not take the 108 frames per hour discrepancy into account and always uses 30 frames per second as its numbering system, never skipping any frame address numbers. Bear in mind that as long as the minute number of the timecode is divisible by ten, there is no difference between drop frame and non–drop frame timecode, since no frame numbers are dropped during those minutes. And remember that the actual video frame rate of both drop frame and non–drop frame timecode video is 29.97 frames per second. Of course, as mentioned previously, this issue relates only to NTSC, since PAL does not suffer a frame rate based on fractional values.

Drop frame and non–drop frame timecode are easily distinguished in the timecode number. Drop frame timecode always uses a semicolon instead of a colon to separate seconds and frames—for example, 01:02:03;15. All NTSC DV timecode is drop frame and exhibits this. PAL DV, not having a drop/non–drop frame issue, always uses non–drop frame signatures.

Device control

DV timecode utilizes the revolutionary new protocol that made lossless digital data transfer available on the consumer level. This protocol, the IEEE 1394 Firewire specification, includes video, audio, and timecode streaming data along with Device Control. Thus, when you connect the Firewire cable from your camera or deck to the Macintosh, you not only enable the Macintosh to access the video and audio on the tape, you also gain access to the timecode information of that video, and you get remote control of the deck through the Device Control data. That's a lot of control when you get into the process of production.

Why is that? Because, in addition to other benefits we'll encounter later, it adds the element of safety to your production. Timecode and Device Control allow you the power of automation when you are working. Once you have arrived at the perfect edit of your video, you will want to record it out to tape. But what would happen if your drive failed or any of a number of horrible accidents happened, wiping all your hard work away? Without the timecode numbers from all those edits, you'd have nothing left but the fond memory of your project. But since you are working with a computer, you can use the timecode numbers embedded in all those edits to retrieve the appropriate material from the original tapes and place it in order in your project exactly as you originally had it. This is available only when using the full resources of Firewire: video, audio, timecode, and Device Control.

DV decks and cameras

Using a DV camera for a deck

"Do I need a deck or is using the camera OK?" This is probably the most frequently asked question about Firewire and Final Cut Pro, by new users and experienced editors alike. One of the grandest accomplishments of the DV revolution is that for such a small amount of money, you can get a turnkey, or complete video editing, solution that also includes a high-quality three-chip camera. This is about a third of the price of a system with similar functionality a few years ago, minus the camera! So it's a pretty good deal all around. Still, there is faint grumbling among initiated Final Cut Pro users that using a camera as a deck is not such a great idea. Strictly speaking, they are right. In the long run, using a DV camera for a deck can be a bad idea. Let's take a look at the arguments for and against using a camera as a deck.

The arguments for using a camera as a deck are pretty obvious. The cost savings of using a camera both for footage acquisition and as an editing deck are enormous. A decent quality DV deck runs into the thousands of dollars, so if we can accomplish the exact same deck functions with a camera we already possess, we drive the total price of our system down into the bargain basement. Most would find this a hard argument to beat, especially those for whom video editing is simply an interesting activity and who don't want to break the bank doing it.

There are, of course, other good reasons for using a small DV camera as a deck that have little to do with saving money. There's nothing quite like the luxury of logging and editing freshly shot tapes in the field. The convenient size of a DV camera and the fact that most can run on batteries instead of having to be plugged into a wall socket make editing possible in ways that were unthinkable a few years ago. The staggering number of consumer and prosumer level cameras that have been approved for use with Final Cut Pro makes this an attractive potential use. It's hard to believe that you can fit a fully functioning nonlinear editor and deck on the dinner tray of a coach-class airline seat (assuming you have the Powerbook laptop with you and not the minitower!).

The rough-and-tumble videographers often tell a different story, and for those who intend to use their equipment very intensively, paying attention to their logic and experience would be wise. Any videographer who has experience with intensive use of equipment will tell you that portable camera packages, and especially prosumer and consumer packages, are not designed for the heavy logging and capturing work that is a major part of editing. Camera tape transports, the little motors and parts that the tape moves around on, are designed for long life, assuming that the majority of that life is spent rolling at a consistent playing speed or performing high-speed rewind and fast-forward.

The "stop, start, rewind, frame-advance, play, stop, fast-forward, pause" actions that typify the log and capture editing process really stress out the transport mechanism of the camera. Think of it like this; your car can be put into reverse and then back into a forward gear in order for you to operate it on public roads in normal use. Normal use means shifting it to reverse perhaps a few times a day, or at most a small percentage of your actual mileage on the car. You can drive for a hundred thousand miles in normal use without running into transmission problems.

However, the transmission is not meant to be repeatedly shifted between reverse and the forward gears every 20 seconds, and if you do this, you can be sure that you will soon wear out the transmission's hardware, as well as become very dizzy. The car's transmission allows reverse as a convenience in many situations, but it is not a long-term driving solution. The high costs to repair or replace transmissions can exceed the worth of the car, making such driving patterns unwise.

Such is true of the DV camera as well. And the analogy holds true even for costs, in view of the fact that a really nice DV camera can sell for as much as a used car. Wearing out the tape transport mechanism of a camera is a sad waste of a high-quality imaging device. Most DV decks, on the other hand, are constructed precisely to withstand the constant back-and-forth shuttling that would eventually destroy the transport of even the sturdiest DV camera.

Still, one should plan ahead and think about what a year or two of work will do to your equipment. Another important thing to check for in shopping for a deck is the transport itself. Some lower-end DV decks use a tape transport mechanism that is really not much more solid than the generic camera tape transport. Be careful to check the deck and make sure that you get what you pay for. It is also important to make sure that the various DV formats are compatible with each other and that they get along well with Final Cut Pro.

DV formats: the many faces of DV

Just what do we mean by format? Although we saw that DV uses a standard codec and that Firewire uses a standard communications protocol, there are still proprietary recording formats for DV that will affect your choice of DV device and its relationship with Final Cut Pro. Although the DV codec used and the data rate generated by these devices is the same, some use a different tape recording speed and what is called *track pitch* when recording to tape. The formats to be discussed are DV, DVCAM, and DVCPRO.

The track pitch is the angle at which the record head is positioned on the tape crossing it. The greater the angle and track pitch, the greater the amount of data that can be stored per frame as the tape crosses the head. Of course, the consequence is that the tape speed must be higher if the track pitch is greater. Thus, some material recorded with a higher track pitch might not play back correctly on decks that do not offer multiple playback speeds.

MiniDV

The first and by far the most common DV format is referred to as simply MiniDV. Many cameras use this standard tape speed and format, making it the only format that really isn't the proprietary child of a single corporate parent. The MiniDV cassette size is the most common cassette size for DV cameras, although some DV cameras also take advantage of the greater runtime lengths of larger DV cassettes.

MiniDV runtimes generally top out at one hour. Although many decks and cameras also feature a *long playing* (LP), recording speed that extends the recording length of a tape, it is *highly* recommended that you avoid using the feature. LP record speeds are not only nonstandard among decks and cameras; they increase the risk of tape dropout and other recording errors that could jeopardize your footage.

DVCAM

The next of the formats is DVCAM. DVCAM is Sony's proprietary DV format. Cameras that do not display the term "DVCAM" generally cannot play DVCAM-recorded material correctly, although some do, including many of Sony's own non-DVCAM products. And some non-Sony products not only play Sony DVCAM–recorded tapes, they can record at a comparable track pitch and tape speed to DVCAM.

What is the difference between MiniDV and DVCAM, and why are some non-Sony devices compatible with it? The main difference between MiniDV and DVCAM is that the track pitch of DVCAM is markedly higher. Track pitch is the angle that the record and play head of the deck or camera crosses the tape during the record or playback process. The greater the track pitch, the larger the portion of the video frame crossing the tape head at one time. More video frame crossing the play head at one time results in more information being recorded to tape, and more information recorded is always desirable.

DVCAM decks, without exception, use the same cassettes and tape stock as MiniDV; however, a larger cassette size is available to DVCAM deck users with decks that can use them. Many DVCAM decks and some cameras have a larger tape-loading bay and can accept a special DV cassette size that offers dramatically longer runtimes. There is no difference in the tape stock itself inside the cassette casing. The difference between non-DVCAM MiniDV and DVCAM is the speed and track pitch with which the tapes are played and recorded. In addition to DVCAM, other deck and camera manufacturers are increasingly offering products that use a higher track pitch and tape speed than MiniDV as well. Although some can play back the higher track pitch of DVCAM but can record only at the lower track pitch of MiniDV, higher track pitch will continue to work its way into prosumer camera and deck units until we all have access to it.

Since the track pitch of DVCAM is greater, there is a corresponding required increase in the tape speed. Tape speed/track pitch in a DV deck is usually a transparent function, with the exception of a few Sony decks and cameras that allow the user to select either DV or DVCAM. A deck or camera records at whichever speed it is built to record at. This means that a generic DV tape when used in a DVCAM deck lasts only 66 percent of the advertised runtime length. In other words, a DV tape labeled "60 minutes" will record for only 40 minutes in a DVCAM deck, because the higher speed of the DVCAM deck will run through the length of the tape faster than a DV deck would.

Sony's own DVCAM tape stock is basically standard DV or MiniDV tape, with the exception that the tape speed for DVCAM is correctly labeled on the packaging. Sony MiniDV-sized DVCAM tapes are sold in 40-minute lengths instead of 60-minute lengths, because the amount of tape in the cassette is the same but the recording speed is faster. The larger cassette version of Sony's DVCAM tape stock does extend to 184 minutes, the same runtime length available for DV. Although the tape stock itself is really no different, the intended playback and record speed is. This means that normal DV tapes can be used with a DVCAM deck; simply do the math to figure out how much time your DV tape will lose based on the faster record speed.

You may wonder why Sony would create a different track pitch and tape speed in its product and introduce a potentially incompatible format when there is a popular existing standard that much of the industry agrees on (including Sony, which has quite a few non–DVCAM products). DVCAM is not simply a different format; it includes an improvement on the existing DV format. The quality of the DVCAM format is no different from that of DV, because this is determined entirely by the optics, imaging chips, and hardware codec of the camera.

The added benefit of DVCAM is its robustness and resilience in comparison with DV. DVCAM uses a redundancy technique in recording the DV data. If data is lost, damaged, or corrupted on what are, after all, rather fragile tapes, the redundancy of the data may resolve the problem. You pay a price for this redundancy in the increased track pitch/tape speed/format incompatibility, but if you require absolutely no surprises in your shooting/recording, that price may be a small one to pay for the peace of mind that DVCAM's data redundancy provides.

DVCPRO

The final format of DV is called DVCPRO. Panasonic's DVCPRO is generally more expensive, and is consequently a rarer format in consumer and prosumer markets. It uses a completely different cassette casing and a still-greater track pitch than DVCAM or MiniDV. DVCPRO as a format originated prior to the advent of DV Firewire editing, antedating the Firewire port itself. Originally, it was developed as a high-quality, convenient, and robust format to be used happily along with analog-based systems. With the development of Firewire DV, DVCPRO has shown itself to be valuable on that front as well, bringing lossless editing to a market of users who invested in DVCPRO before such editing possibilities existed. The greater track pitch and robustness of the format is a result of its being designed for rough-and-tumble field acquisition.

A bonus conferred by the greater track pitch of the DVCPRO format is that it can play back any of the three DV formats. Almost a universal format, DVCPRO decks can generally access both DV- and DVCAM-recorded material. DVCPRO has a comparatively long history in professional applications, as reflected in the higher prices DVCPRO decks and cameras fetch. And the DVCPRO format, in another variation of DV, is being used as a tape format for recording HDTV, although such HDTV DVCPRO cameras will be far out of reach for most of us for a very long time.

Is DVCPRO a necessary option for you? The real issue is whether or not you will need to be able to access DVCPRO tapes from other sources. Since the actual video and audio quality is no different for the three formats, the extra expense is likely not to be worth the investment unless you need to be able to use all three DV formats. If this is not the case, and money is an issue, the lower prices of the prosumer DV and DVCAM formats may make for a better investment.

Features to consider in purchasing a DV device

Cassette size and DV format

There are three distinct formats of DV that you may encounter in your adventures. This implies that are incompatibility issues you must take into account, both as you purchase your equipment and afterwards, as you send your finished tapes into the world to be played by other decks and cameras. The issues to keep in mind are (1) cassette size, which means that you need to make sure you will be able to play back any cassette size you want to edit with, and (2) tape speed/track pitch, which means that you know that your deck or camera can play back and record in whichever format that you may need to access.

In general, DVCAM is an excellent choice if you can afford it because it can access the two most widespread formats—DV and DVCAM. DVCPRO, although an excellent format as well, represents a much smaller proportion of the market and may not justify the extra expense, especially for those who know that 99.9 percent of their source footage will be generated by their own non-DVCPRO deck or camera. Do not assume that a DV deck can or cannot play back the DVCAM-recorded tape, however, because many manufacturers are building the ability to play DVCAM track pitch tapes into their own decks. Do the homework, test the decks yourself, and figure out the best deal based on your needs.

Analog video format compatibility

In addition to the format issues already discussed, there are other features that you should consider in your search for the perfect deck for your money. One of the first things you should look for is the analog video inputs and outputs. Even though you may be shooting with a DV camera and mastering to a DV tape in the deck, you will at some point want to share your products with the rest of the world.

The analog video signal leaving the deck should be as high quality as possible, and not all video signals are created equal. If you are combining this DV deck with other video equipment, you should take pains to make sure that the analog format of the output of the DV deck matches that of the equipment you want to record with. There are three main analog video formats that you may output from your DV deck, and you will find that the format offered and the quality of the signal increases proportionally with the price of the deck.

Composite

The most common analog video format is *composite*. Composite video requires only one single RCA or BNC cable for the video signal, and the quality is correspondingly low, though acceptable enough for most consumer-level uses. A composite signal is one that contains all the luma and chroma values in one electrical waveform, thus requiring only one cable. The corresponding compression of the waveform means that the generation loss will be more extreme, resulting in nearly half the resolution of the frame being lost in the transfer. This is why VHS tapes look so poor on the second generation. Although it is a widespread analog format and is used for most of the video we are familiar with, VHS is pretty horrible for serious dubbing purposes.

Y/C or S-Video

It is unlikely that you will even be able to find a DV deck that only offers composite out; almost all DV decks also offer what is referred to as Y/C, or S-Video in common parlance. Y/C is a step up from composite video and offers much higher quality by using two separate wires (though usually both are included in a single cable) to transfer a video signal from one deck to another. The Y in Y/C refers to the luma values, and the C to the chroma. Although the chroma values are still composited into a single electrical wave, they are at least kept discrete from the luma values, thus delivering much higher resolution in a dub. The resolution loss over a generation is much less extreme, resulting in only a quarter of the detail being dissolved.

Although the quality is much higher than that of composite video, Y/C is still considered a prosumer format and is affordable enough to be available to most editors. This is a great analog format for making dubs, and anyone editing video using prosumer-priced equipment should look into purchasing an inexpensive SVHS deck for dubbing their DV products down to a distributable tape. Although the generation loss is still greater than in the component format, the next format to be described, it looks fantastic in comparison to composite video and is priced modestly enough to make it available to most DV editors.

Component

The final format to be considered here is that of *component video*. Component video is the highest fidelity format of analog video. It is something of a universal analog format for applications requiring the cleanest, most high-integrity analog video recording. And it is expensive as analog recording formats go. Even used analog component decks cost at least several thousand dollars, which is between 10 and 20 times more expensive than an S-Video deck that is limited to Y/C and composite connections.

Component decks generate this high level of quality by using three separate cables to transfer the video image. The cables correspond to the three separate electrical waves of the video signal: one for Y, and then one for red and blue chroma each. The Y signal is intact, as it was in the Y/C connection, but the color definition of the chroma is not composited into one cable. The integrity gained by keeping the chroma signals discrete and intact translates into the richest analog video images available. The quality is fantastic, but the cost is prohibitively expensive for most.

This sticker shock goes not only for the analog component video deck, but also for any DV decks that include component inputs and/or outputs. Most affordable DV decks are limited to Y/C and composite in/outs, which are of prosumer quality. To attain all the benefits of component quality video dubbing, both the source and destination deck must have component video connectors. Therefore, in addition to needing an expensive analog deck, you might have to pay much more to acquire a DV deck that inputs and outputs this analog format.

Once again, as with the initial choices to be made about whether a DV deck is necessary as a supplement to your camera and which DV format you should be building your system with, your choice of a destination analog format should inform your purchases. If you restrict your editing to family fun days, you may never miss the resolution you are dropping by using only composite connections and a standard VHS deck for your dubbing purposes. Even so, the price gap between composite VHS and Y/C S-video decks has narrowed dramatically, and even the most miserly would do well to consider the quality increase available for very little expense.

For professionals who intend to make a living (or at least happily pad their present living) with their editing, the investment in component equipment is one that should be considered. If you expect to receive and deliver the highest quality formats, then you have to have the equipment to do it with, and component equipment is the highest standard of analog video there is. Even though you will have to configure things a little differently and invest more capital into your system, the results may justify the cause. When that gold mine client walks in and says, "Can you work from these Betacam tapes?", you want to be able to pop them right in, rather than dubbing them down later when the client leaves and hoping she won't see the generation loss. Just make sure that you also invest the time and forethought in arranging all the pieces of your system so that they serve you best.

But, on the other hand, don't be convinced by reading this section that you won't be taken seriously if you don't have component decks or if you aren't shooting in DVCPRO. I know videographers who make an excellent income using one MiniDV camera and a three-year-old Macintosh. And it isn't always a question of making money either. If your tools make something creative for yourself and the rest of the world to enjoy, then that's professional in my book, too. To put it rather bluntly, "professional" is an attitude you take to the projects you engage in, not the equipment you use.

Film production has been around for 100 years now, and television and video for more than 50. Not having access to the awesome image quality of the video tools of the past few years never stopped any of the great film and video makers of the twentieth century from producing great work, and not having an $8,000 Betacam deck shouldn't stop you either. Firewire DV editing, using inexpensive cameras and decks, is a high-quality gear. Just don't forget that it's what you *do* with the DV that makes a project professional, not what the format inherently is.

Analog/digital conversion boxes

While discussing decks, there is another category of devices that should be addressed, that of *analog/digital DV converter boxes*. Although not a deck, the DV converter is a device that accepts analog video and/or audio on the one side and generates a DV data stream on the other. Because it's a converter box, it can also work in the other direction, accepting a DV data stream and converting this to an output analog signal. These devices run the price range from a few hundred dollars to several thousand, but they all do one thing—convert analog video and audio to or from DV for access through Firewire.

The DV converter can be a really useful piece of equipment, in particular for those who have lots of legacy format tapes to work with. If you have a huge stack of Hi8, 3/4-inch, VHS, or other analog media, a DV converter can provide a very convenient means to access them all in Final Cut Pro. For those who simply need to bring in little bits of media and who are not engaged in deeply complicated editing, this may also be a very cheap and simple answer to your needs. For professionals who need to interface expensive component video equipment and high-quality audio gear with their system, this may be a great option for integrating DV and Firewire-based editing into your system without a major equipment upgrade.

The functionality of any given DV converter is very evident in its price, and you get a lot more features in a converter costing several thousand dollars than one costing a few hundred. Although, as was stated in the last paragraph, all DV converters do the same thing, not all are created equal. At the bottom end of the chain, barebones DV converters like the Sony DVMC-DA2 offer a simple in/out solution. For very little money, you get Y/C and composite in/out and two unbalanced RCA audio in/outs for stereo sound. On the Firewire side, you get a selector for changing the sample rate from 48 KHz to 32 KHz and a Firewire port. The whole thing is the size of a small clock radio, and it runs on an AC plug-in adapter. It's a very simple solution that will get you through the night.

On the other end of the scale, there are the richly featured converters costing thousands of dollars but offering necessary interfaces that professionals often need. Most are rack mountable and carry balanced XLR audio inputs. Many will feature not only component video input/outputs and extra Firewire ports, but also audio level meters and in/outs for other professional digital formats such as *serial digital interface* (SDI). For editors who perform off-line editing, the lower DV resolution in comparison with such high-end formats may not be an issue, but the ability to get any sort of media into your system could be.

Of course, we have to remember that with the use of the DV converter we give up the use of DV timecode. DV timecode is written onto DV tape as video is recorded there. When you stream analog video into a converter box, there is no DV timecode accompanying it, so we lose the ability to recapture should we run into trouble. While we could still recapture the same material through the converter box, we would have to start from scratch and re-edit the entire thing. Using timecode, on the other hand, would put all the pieces back together for us automatically. This may not be an issue for people whose media needs are uncomplicated and who could easily manually recapture and re-edit any media that is lost accidentally. But accidents can be a real heartbreak for the editor engaged in more complicated assignments; in this instance, working with timecode is imperative.

Thankfully, there are solutions for this situation. The most reliable one involves connecting a serial port adapter to the modem in your Macintosh, which can then be connected directly to the analog deck you wish to control and receive timecode from. Thus, instead of accessing DV timecode data through the Firewire cable, you are simply getting it directly from the analog deck via a serial connection. Using such a serial connection directly to the deck would require a few extra settings depending on your deck and converter, since you would now be accessing either LTC or VITC timecode from the analog tape rather than DV.

When it was mentioned earlier in the discussion about timecode that nonlinear editing applications like non–Firewire Final Cut Pro, Media 100, and Avid make use of LTC or VITC, it was through just such a serial connection as is proposed here for Final Cut Pro. This process would allow those who need to use an A/D converter box and non-Firewire formats to use timecode as well. Several good serial converter devices exist on the market that cost very little money. Some converter boxes in the near future will include the ability to convert LTC or VITC timecode into DV timecode and pass that along with the converted video and audio. Do the homework, and you can put together the perfect system based on your needs, your present equipment, and the size of your wallet.

2 The Hardware and Software for Firewire DV Editing with Final Cut Pro

What goes on inside a Macintosh

What you need for Final Cut Pro: the basic Firewire setup

What is necessary for a complete turnkey Final Cut Pro nonlinear editing station? That is a loaded question, and the answer depends primarily on what you want to do with it. But if what is wanted is a base-level editing station that captures, edits, and outputs lossless video through Firewire, then here is the bottom-line configuration that Final Cut Pro version 2.0 requires.

- An Apple Macintosh Computer with a G3 or later processor and built-in Firewire ports, G4 recommended (Beige G3 models not supported)
- Macintosh Operating Software 9.1 or later (OSX not yet supported)
- Apple QuickTime Version 5 or later
- 192 megabytes of RAM or more (256 megabytes is highly recommended)
- Two or more physical hard disk drives
- A Firewire DV device, such as a DV deck or camera or DV converter
- The Apple Final Cut Pro software.

The reader will notice that the words, "or later" recur constantly in the list. This is because, like many applications, Final Cut Pro can function under different versions of the OS or QuickTime and with a huge variety of hardware. The Apple Macintosh is a very adaptable machine that you can customize to suit your needs. When the time comes to upgrade or change your system configuration, you will find that it is much easier and less expensive to do so than to purchase a whole new configuration. Macs last forever. The preceding list gives the minimum requirements.

Welcome to the Macintosh

The Final Cut Pro nonlinear editing station is composed of four principal areas:

1. The Macintosh computer and hardware
2. The Macintosh Operating Software (OS)

3   Apple Final Cut Pro software
4   The external DV device.

Each of these parts has to work together correctly for editing to take place. Although advances in software and hardware development have made much of this configuration transparent, there are still tweaks that you can make for best results.

Final Cut Pro likes its tummy scratched in certain ways that will seem strange or onerous to users with prior experience in other nonlinear editing applications, such as Avid and Media 100. Those systems operate from a hardware and software configuration that is totally closed and proprietary. If you purchase a Media 100, the hardware card that is included with the system is linked directly with its software component and as such is enabled behind the scenes for the tasks it must complete. The result is that you simply start the application and begin editing.

If that sounds great and convenient, it is. But the trade-off is that such a closed system is more expensive in the short term and arguably more limited in the long run. The system you get from Media 100 will work only with its own software and hardware components. This limits users to continually opting for the solutions that Media 100 produces for upgrade or expansion. The Apple Final Cut Pro solution, however, is unlimited in this respect, allowing users to configure their own systems in whichever direction is suitable for their purposes. As new hardware and software options are developed by the teeming third-party industry, the users' options increase many-fold, and Final Cut Pro begins to reveal itself as a wise investment.

This concept is referred to as "hardware-independent." In other words, the Final Cut Pro software and interface can be applied to many different hardware configurations without limitation as to either the hardware or the actual video format itself. Although the hardware requirements must be met, Final Cut Pro will work with similar ease and efficiency whether editing simple Firewire-generated DV media or the super-professional HDTV sources that are just now appearing. After investing in the necessary hardware, users will be able to employ the same low-priced application they are using currently to edit the most up-to-date media that exists now or will come in the future.

A discussion of all the hardware configurations that are possible with the Final Cut Pro editing station would extend far beyond the scope of this book. New solutions are being developed and released on a daily basis, and by the time you read this, new ones that the author could not possibly predict are likely to have hit the market. Such is the nature of the world of technology.

Still, Firewire DV editing with Final Cut Pro is here to stay. The format of DV and Firewire is one that will remain constant for quite some time, so we can focus on its major features to determine the best way to arrange and configure the hardware and software. Let's take the system apart and look at what is involved in the Macintosh's hardware.

The first item, the Macintosh computer, is obviously necessary. A G3 processor or better and Firewire ports are necessary for Final Cut Pro to function correctly. Apple's current Macintosh models sport Altivec-enhanced G4 processors, which are more than fast enough for Final Cut Pro. But a G3 Blue and White Macintosh from yesteryear will function satisfactorily, if somewhat more slowly in more processor-intensive functions like rendering.

Earlier machines with G3-upgrades and the short production run of the Beige G3 Macintosh are not supported by Apple for Final Cut Pro 2.0. Although enterprising hackers will no doubt come up with a way to get them to work, this is not the path for the inexperienced. The recommendation is to invest in a system that is already completely functional.

The four primary areas of the Apple Macintosh

The neatly arranged insides of your Apple Macintosh computer are composed of four major areas: (1) the processor/motherboard, (2) the Macintosh Operating Software (Mac OS), (3) the storage (RAM and hard drives) and (4) the input/output buses (AGP, PCI, Firewire, etc.). Although this description is vastly oversimplified, it is the most general blueprint of the computer. The interaction of these parts results in a working computer capable of running Final Cut Pro.

Although most of the items to be discussed are built into the motherboard and the computer cannot be purchased without them, it really helps to understand how the whole package works together so that you can make more intelligent decisions about your system when the time comes to add new components or troubleshoot existing ones. Although you can sometimes dodge a requirement and still run Final Cut Pro, the requirements are there for a reason, and eventually you may run into behavior that could have been avoided if you started with a machine that fully matches the recommended specifications.

If all of this tech talk seems a little overbearing, relax and just absorb as much as you can. You do not have to be a computer technician to edit with Final Cut Pro, nor should you trade your precious editing time and skills for a fixation on hardware and software issues. But it will make your life easier in the event that problems do crop up to know exactly what everything does, where it goes, and so forth. If you can't get through this discussion, just make sure that your system has the necessary hardware and software requirements and skip on up to the project setup chapter.

The processor and the motherboard of the Macintosh Minitower

The processor located on the Macintosh's *motherboard* (all of that circuit board material inside the Macintosh) can be defined as the brain of the computer. It receives instructions from the Macintosh Operating Software and acquires data resources from input/output buses. As one would expect, the processor must be very fast and efficient to perform all of this at once. Indeed, your Macintosh needs to have a G3 Power PC processor or faster in order to be certified to run Final Cut Pro.

The discussion in this section will assume that the user is working with a Macintosh Desktop Minitower model. Although Final Cut Pro can easily work with properly configured PowerBook laptop and iMac mini-desktop units, the vast majority of editors are likely to be using the Minitower. If you intend to work with a model other than the Minitower, the system requirements are still in effect, and there may be other issues to address, such as hard drive options and RAM. Make sure that you take such issues into account before purchasing a machine that may end up costing you more in the long run.

The Macintosh Minitower is the flagship of Apple's Macintosh line and always sports the latest in Apple's processor development. If you're shopping for a Macintosh from a vendor other than Apple, make sure you know which processor you are purchasing. It is not unusual for one year's model of Power Macintosh to go through three or more major hardware revisions. A machine may have the same processor and the same shell cover, but a different motherboard and faster bus system. As with everything else, a great deal is usually too good to be true. Make sure you get the complete information about the machine that you intend to buy. Do not accept general descriptions. Although an older machine may be fine, you don't want to pay new prices for a machine that is a year old.

The Minitower is also the most easily and thoroughly upgradeable in all respects. Adding new drives, more RAM, and new devices is almost too easy, and there is far more interior space to do so than with any other Macintosh. Any new improvements to the Power PC processor will be included immediately in the Minitower, while for various reasons it may not show up in iMacs, PowerBooks, or iBooks for generations.

Of course, the Minitower is more than a name, it is a descriptive term. This is not a portable computer. It is built to be a stationary workstation that is permanently part of an integrated solution accommodating the hundreds of accessory devices developed for use with it. It weighs in at a good thirty pounds and has a side latch for uncommon ease of access to its interior.

The Macintosh Minitower is currently configured with several different motherboards sporting dual processors and varying speeds of single-processor units. All the different configurations are suitable for use with Final Cut Pro, and your choice of configuration should mirror what you intend to use it for. Dual-processor models may offer vast speed increases in the future depending on software developments, but the present cost of this feature may not be worth it if you do not intend to do effects work that requires rendering. Dual processors can really make a difference in rendering speeds, but they can also make a real difference in the wallet, especially when rendering isn't a burning issue.

Inside the Minitower, you will find room for four extra hard drives in addition to the stock hard drive that the Macintosh ships with. Adding new drives is a snap with the Minitower and takes less than a minute. The standard Macintosh includes two ATA drive buses that can contain a total of two devices each. The upper ATA bus contains the DVD player and an optional Iomega Zip drive. The lower ATA bus contains at least one stock ATA hard drive—the startup drive that ships with your Mac—and can easily take another. The other empty drive bays can be utilized if you invest in an inexpensive expansion card that allows for the inclusion of more hard drive and other devices.

If you order a Power Macintosh with the SCSI option instead of the ATA drives, the system will come with a SCSI card installed in one of the PCI expansion slots described later in this chapter. This is a slightly more expensive option but is completely acceptable and may be a better way to construct your system with a view toward future expansion.

There are either three or four RAM slots, depending on the model, allowing easy installation of additional RAM chips. Your system will come with a certain amount of RAM preinstalled, but you can easily upgrade the RAM to a gigabyte and a half (1,500 megabytes of RAM). How much RAM you should have installed and why will be addressed shortly in the discussion of storage.

Also located on the motherboard are the various input/output buses. They are the pathways from the processor to the rest of the computer's parts through which data is brought in or sent out. The largest of these are the *PCI expansion slots* and *accelerated graphics port*. Your Macintosh will arrive with one stock video card in an AGP slot, which is dedicated to your computer monitor. The PCI expansion slots are for adding additional components such as extra video cards for multiple monitor support, SCSI cards for inclusion of SCSI devices, and beyond this just about anything else you can imagine.

In addition to these PCI and AGP buses and located on the outside and to the rear of the Minitower are the remaining input/output buses. Standard buses on the Macintosh are the Firewire ports, *universal serial bus* (USB) ports, an Ethernet connection, a modem connection, and a speaker jack. In addition to the Firewire connectors on the exterior of the Minitower, another Firewire connector is included directly on the interior of the motherboard for Firewire devices you may want to install inside the Macintosh.

The Macintosh Operating Software (Mac OS)

If the processor can be referred to as the brains of the Macintosh, then the operating software could be called the mind. The Mac OS (logo shown in Figure 2-1) is the complicated set of files and folders that directs all the powerful hardware and makes it available for use to applications like Final Cut Pro. Visitors from the world of Windows may be confused by the seemingly eccentric order of the Macintosh system, but often this is because it is so incredibly straightforward.

The desktop metaphor and the System Folder

The Macintosh OS is organized using the visual metaphor of the physical desktop (Figure 2-2). It keeps its components stored as files that live in folders stored on drives or partitions, which are simply sections of a physical drive that act like a separate physical drive.

Figure 2-1

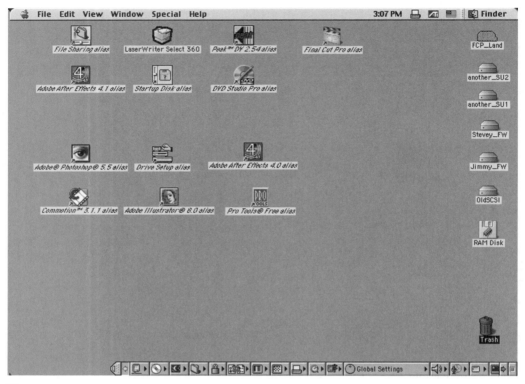

Figure 2-2

You can easily divide a single physical hard drive up into many logical partitions, which the Mac OS will regard as different drives, but which in reality occupy only a specific locked-off area of the same drive. Although this can be useful for setting up different work scenarios as will be discussed later, this does not release the user from the requirement of having two separate physical hard drives for use with Final Cut Pro.

In the Desktop metaphor, every drive or partition acts like a separate desktop/file cabinet with its files and folders stored inside (Figure 2-3). It's like having a lot of filing cabinets in your office. You can have items pulled out of the filing cabinets onto your desktop, but they are still organized and are really regarded as belonging in a proper location in one of the filing cabinets.

At least one of these drives or partitions must contain a *System Folder*. The System Folder is the folder that contains all the software that runs the Macintosh and is necessary to start up the Macintosh itself. We call a drive or partition that has a complete System Folder "bootable" or a *boot partition*, because you can actually "boot" the Macintosh using that drive or partition. You can have more than one bootable drive or partition on your Macintosh at a time. Simply installing a functional System Folder on a drive or partition makes it possible to boot up from that drive or partition.

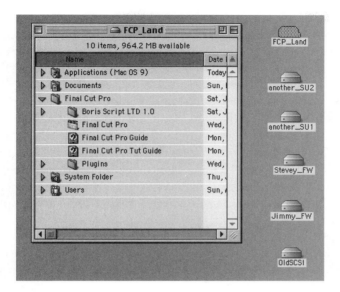

Figure 2-3

This is one flexibility of the Mac OS that is dearly loved by Mac aficionados. It is incredibly easy to set up different boot drives or partitions based on what you want to do. As we will see, sometimes system components that are required for Final Cut Pro or other nonlinear editors do not interact well with other applications such as word processors or Internet applications. So instead of buying a second computer for that, the user can simply set up a different boot partition for each task: one for the editor and one for the word processor. All you have to do to switch between one and the other is to select the partition you want to work from as the boot partition and restart the Macintosh. It is very easy to configure your system to do this and you will find it a very efficient way of working.

One thing that often confuses new users about the Mac OS is the Desktop system and the Finder. The Desktop is essentially what you see on the screen while the Mac is running. The Finder is an application that is always running in the background and allows you to navigate through the Desktop to find and work with the files and folders you want. The Finder never shuts down, and when you are running an application and not using the Desktop, the Finder simply hides in the background until you use it again.

The Finder is a part of the active System Folder though, and this is where people get easily confused. Each drive or partition actually has its own Desktop folder (this is what allows us to install a System Folder on any drive and make it bootable; without a Desktop, how could you see what you are doing?). Whichever drive or partition is the present boot partition and thus has the active System Folder is also the current Desktop and Finder you are working with.

The Desktop itself is actually a hidden folder (not really hidden because you are staring at it the whole time you are working) and you can save things there. When you save to the Desktop, you are actually saving a file into the hidden folder that always displays its contents on the Desktop. New Mac users are often confused by seemingly odd behavior when saving to the Desktop. Every drive or partition has its own Desktop folder, and these Desktop folders are always all visible at once. However, when you save a file to the "Desktop" in an application's dialog box, it is saved only in the Desktop folder of the drive that has the active System Folder, the current boot drive or partition.

You could save something to this one Desktop folder, and then reboot to another partition. The file will still show up on the Desktop, because you are always looking at all the Desktops for all the drives and partitions at once. But you will have saved the file to a different drive than it appears. Just because you can see a file on the Desktop of your Macintosh doesn't mean that it lives in the Desktop of the current boot drive. You can check the actual drive location of a file or folder at any time by selecting the file or folder, then going to the File drop down menu and selecting Get Info. The window that pops up will give the exact directory location of the file and many other details about it.

Although this may seem slightly confusing, it is because the system is so transparent. If you are completely unfamiliar with the Mac OS, you are encouraged to find a book specifically on the subject. We will only go through the most important components here, and you should really become as knowledgeable about your system as possible. The Macintosh OS is filled with little-known tricks and treats that can speed up your workflow immensely, but you'll never see most of them without someone pointing them out.

The System Folder of your Mac OS is composed of files and folders, most of which are constantly working in the background to keep the system running smoothly. Very little needs to be done to keep it running, but there are a few folders that you will probably be accessing periodically to provide regular maintenance or to tweak and upgrade. These components will become familiar as you access them more. One of the fantastic things about the Mac OS and the Apple Macintosh system in general is that it is really infinitely customizable. Rather than call Technical Support when something goes wrong, you simply look under the hood for the probable culprit based on troubleshooting rules-of-thumb and the body of knowledge of the large community using the Mac OS and Final Cut Pro in particular.

If you open the System Folder, you will see many files and folders (Figure 2-4). Let's go through the ones most important to your Final Cut Pro system. The really critical components are the Control Panels, the Extensions, and the Preferences. These are the components that really drive your system and make Final Cut Pro and the hardware do its job. These are also the ones that are most likely to get out of hand and are on the first line of a troubleshooter's defense when the Macintosh's behavior goes south.

The Control Panels

Control Panels are a group of applications that allow you to configure your hardware and software directly (Figure 2-5). Unlike other computer operating systems, the Macintosh OS actually allows you to directly access the hardware of the Mac. When you want to bring audio in from the built-in audio card, you simply turn it on. You don't have to tell each part

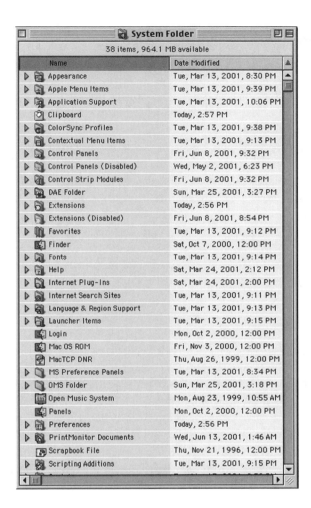

Figure 2-4

Figure 2-5

of the computer where the other part is. This is why you will find that the Macintosh is unusually "plug-n-play" friendly. If you want to install a piece of hardware, you simply connect it and possibly install a software driver. You don't have to tell the system where it is; at most you simply have to tell the system how *you* want to use it.

The Control Panels are the tools that allow you to do this. Some are parts of the Mac OS itself, such as the Startup Disk Control Panel mentioned earlier that allows you to choose which drive or partition to boot from at the next restart of the Macintosh. Others are third-party programs that allow you to customize whichever items they are related to, such as a setup panel for a graphics tablet or the like.

The important thing to remember here is that Control Panels are small applications you run periodically if you need to change something in your system. They are usually linked to another application or piece of hardware, and that item may not work correctly if the proper Control Panel is not installed in the Control Panel folder of the active System Folder. Sometimes your Macintosh may need to be rebooted for a change in a Control Panel to take effect, especially if the change affects hardware components like RAM or your hard drives.

The Extensions

The Extensions folder contains files that act as drivers for hardware and software on your Macintosh (Figure 2-6). The Extensions folder is probably the single most important folder in your System Folder, because it contains the small driver files that actually enable everything else to work. Some of these have names that are fairly obvious, like "QuickTime Firewire DV Enabler." Others have completely unintelligible names that are strings of letters and numbers like "IrDALib." It is not necessary for you to understand the function of each extension in your folder, but it is imperative that you keep track of them and be aware of what extensions get installed and into which System Folder they get installed.

Unlike Control Panels, extensions are not applications per se. They are simply library files that hardware or software items may need to access periodically to function correctly. If your extensions are functioning properly and getting along with each other, you may

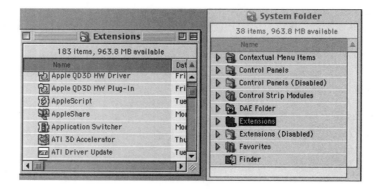

Figure 2-6

never have to bother with them at all. In fact, if you try and engage an extension by double-clicking on it as you would a Control Panel, you will be greeted by a polite message stating that the file is an extension and that to be used, it must live in the Extensions folder. Then it asks, "Would you like me to put it there?"

The Control Panels and extensions are part of a group of files called *inits*. When your Mac is booting up, it loads these files into RAM before anything else. This is important, because most of these inits are actually drivers for hardware and software items you may have installed on your Mac. When your Mac loads these inits into RAM at startup, they stay accessible to the hardware and software all the time, regardless of what you are currently doing. Unlike applications such as Final Cut Pro, which you must open to run, inits are always running invisibly so that if the system needs to use them they will be ready. Most inits must be loaded into RAM at start up to be accessible in this way, thus the origin of the word *init*, or initial.

A problem that can crop up with inits is "extension conflict." Sometimes when you install software, inits may also be installed into either the Extension folder or the Control Panels folder. Unfortunately, sometimes these inits were designed to load into the same address in the RAM that other inits are told to load into. This can result in lockups and crashing and it may even prevent your system from completely booting to begin with. This becomes more likely with third-party products, which don't necessarily know that they may conflict with other third-party inits. The more third-party items you have on your system, the greater the likelihood that this could occur.

There is no absolute way of eliminating the possibility and there is no way of ever creating the perfect extension set that doesn't have conflicts. That's why there are no "magic bullet" lists of the way your extension set should work, short of the original base set that the Mac OS installs. There are ways of troubleshooting and eliminating such problems as they occur. You can be sure that if you experience odd problems soon after installing software or hardware, there may be a conflict somewhere. When you install any software, you should always take a moment to glance through and see if any new extensions were installed so that you can isolate any problems that occur. There is a large community of Macintosh users out there using very similar tools, and any problems such as this are discovered and rooted out quickly. The best strategy is to keep track of your inits and keep your eyes open to Web sites devoted to uncovering such lore.

The Preferences

The third category of system files that you should know about is the Preferences folder (Figure 2-7). Every application on your Macintosh has special settings that allow you to customize things like the workspace, functionality, etc. Whenever you start an application, it looks in the Preferences folder in the System Folder for a Preferences file for that application. When it finds the file, it opens the application based on the preferences laid out in that file. If there is no Preferences file for the application, it creates a new one from scratch based on the original default preferences from the initial software installation.

Figure 2-7

The reason you should know about preferences is that such files are often very old files. Since they are constantly being updated and saved, they are the files most likely to become corrupt. They are accessed and altered not only every time you start an application but frequently during your work. Corrupt preferences are also the most difficult element to find if they cause a problem because they rarely break completely. Instead, a corrupt Preference file can merely exhibit strange behavior that resembles any number of other problems.

Even so, rooting out corrupt preferences is one of the first lines of defense in troubleshooting because it is one of the easiest things to eliminate as a problem. If you remove an application's preferences file from the Preferences folder and restart the application, the application will simply create a new preference file from scratch. If that was the problem, presto—the problem is fixed. If not, you have only wasted five seconds of troubleshooting. Be prepared to reset your preferences with the application though, which can even include needing to reserialize (i.e., to enter the serial number again), since those are sometimes among the contents of the file you are throwing out.

These three file types are the most frequently accessed features in the Mac OS. As you set up and use your editor, it will behoove you to get familiar with and keep tabs on each folder. For simpler purposes using only Final Cut Pro and a few third-party hardware and software items, the preinstalled Mac OS set of inits and preferences will be perfect. On the other hand, beware of becoming too careful and weeding away too much in the search for the perfect extension set. Sometimes, one extension that is obviously necessary may need data from an extension that is less obviously necessary. You don't want to go eliminating problems that don't exist yet.

Data storage: RAM and hard drives

There are two different categories of storage on your Macintosh: *volatile* and *nonvolatile*. Volatile refers to memory that is lost when your Macintosh is powered down. When you shut down your Macintosh, whatever was stored in volatile storage is lost, but whatever was stored in nonvolatile storage is saved until you reboot the computer. RAM, or *random access memory* is volatile storage. Physical hard drives of whatever flavor are nonvolatile storage. Each has its place and function.

RAM

RAM chips are integrated circuit chips that hold electrical charges while the machine is powered up. RAM is incredibly fast, which is why the active parts of the System Folder (e.g., the inits and the system itself) load up into and run from RAM at start up. RAM is also the storage space where applications live and perform their functions while you are running them. When you quit an application, it gives up the space it was using in RAM, making that space available for other uses. You can have only as many applications running as you have RAM to support.

Typically, Macs ship stocked with either 128 megabytes or 256 megabytes of RAM installed, but you can easily install up to 1,500 megabytes, depending on which model of Mac you own. RAM chips are sold in sticks of 64, 128, 256, and 512 megabytes. Since you have only three or four RAM slots, it behooves you to purchase larger sticks so that in future years you may add more as you need them. If you have four 64-megabytes sticks, you will have to throw away 64 megabytes in order to add more RAM.

Obviously, more RAM is better than less; you can't have too much RAM. The average System Folder running Mac OS 9.1 uses anywhere from 60 to 75 megabytes of RAM just for itself. If you had only 128 megabytes installed on your system and you were running Mac OS 9.1, there would not be enough left to run Final Cut Pro 2.0, which requires a minimum of 100 megabytes of RAM and really should have more.

Your applications load up into and run from RAM. How much RAM they use depends on two things: (1) the minimum allowable RAM, which is how much RAM the programmer determined that the application requires for normal operation, often just enough to actually open it but not enough to work effectively; and (2) your user settings for the application, which you set by determining how much RAM you can allocate based on how much physical RAM you have installed and how much is left after the system makes its demands.

There is a feature included with the Mac OS Memory Control Panel called *virtual memory* which you can, and generally should, keep turned off. This feature also exists on the PC platform, although it cannot be disabled there. In the Mac OS, what virtual memory does is simulate the existence of more RAM in your system by using a hard drive as if it were RAM. Using virtual memory from a ten-gigabyte hard drive provides a nearly unlimited amount of RAM. Instead of writing information into the super-fast RAM chip, it writes to a locked-off section of a hard drive.

Why is this a bad idea? As we mentioned earlier, RAM is very fast storage. The processor on the motherboard can access it at nearly instantaneous speeds. But your processor can work only as quickly as it is provided data, so the Macintosh is as slow as the slowest link in the chain. If the processor is receiving data from bona fide RAM chips, your Mac will work at its fastest. If, on the other hand, you are using virtual memory, your processor will receive information only at the dramatically lower speed of the hard drive. The average data rate of a RAM chip is in the neighborhood of 90 megabytes per second. In contrast, the stock hard drive on your Macintosh might keep a sustained data rate of 6–10 megabytes per second at best.

The importance of using real RAM in processor-intensive applications like nonlinear editors should be obvious. Use of virtual memory is highly discouraged in most situations, and it should never be enabled when running Final Cut Pro or any other video or audio application. You will also find that other applications work unbearably slow when it is engaged. Take advantage of the rock-bottom prices on RAM and settle the situation correctly.

Hard drives

In the category of nonvolatile storage, we have the large and sometimes confusing range of hard disk–based drive solutions. As nonvolatile storage, disk-based storage is capable of retaining data saved to it when powered down. Unlike RAM, which is functional only when receiving power in the computer's motherboard, the disk drive is its own physical magnetic medium. When you write information to it, that information is stored as magnetic data and will remain intact when the unit is shut down.

How the computer accesses that data on the drive is what spawns the bewildering diversity of disk-based storage solutions. There are solutions that are appropriate for only one specific purpose, and there are good all-around solutions that function well under many circumstances. Since prices of hard drives can run the gamut from very inexpensive to ruinously expensive, it is important to understand what standards are necessary for Firewire DV editing with Final Cut Pro so that one get the best deal available.

The two most important criteria with which to judge a disk drive are its storage size and its sustained data rate. Your own particular needs with regard to the first issue, the storage capacity, can be determined by reconsidering the data rates we looked at in the preceding section of the book. We know that the data rate of the Apple DV codec we will be using is around 3.6 megabytes per second, which gives us a storage need of roughly five minutes per gigabyte, or 1,000 megabytes, of drive storage (3.6mbs/seconds × 60 seconds × 5 minutes = 1,080 megabytes). Thus a 20-gigabyte drive will hold just under one hundred minutes of DV codec video. If one were using a digitizing card that offered lower compression and higher data rates, that number could shrink to two minutes to the gigabyte, or only forty minutes to a 20-gigabyte drive. The need for large drives should be abundantly obvious.

The second criterion to be evaluated is that of sustained data rates. As we will see, there are a number of different ways for a drive to transfer data back and forth between itself and the motherboard. These differing methods produce limitations on how fast a drive can deliver or accept data. No matter how fast the drive itself is, it can function only as quickly as the interface with the motherboard of the computer. The sustained data rate of a drive depends not only on the drive's internal capabilities, but on the standard bus it uses to communicate with the processor/motherboard. First, we'll look at the types of drive solutions by location, (removable, internal, and external) and then by standard. The major standards appropriate for Final Cut Pro Firewire DV editing are ATA and SCSI, as well as the popular variation of the Firewire drive.

Types of drive by location: removable disk

The first category of hard drives is that of removable disks. The lower-end scale of removable disk drives involves smaller and slower disk types such as the 3.5" floppy and the Iomega Zip disk. These disk types are valuable primarily for their ability to move small amounts of data quickly. With a limit of 250 megabytes of storage and a very slow data transfer rate of around 1.5 megabytes/second, the Iomega Zip disk is clearly not acceptable for use in the capture, playback, and storage of DV footage. Its use is generally limited to saving documents and other smaller relevant files.

Types of drive by location: the internal drive

The next category of hard drive solutions is that of internal drives. The stock hard drive that came as the startup drive for your Macintosh is just such an internal drive. Internal drives may be of either ATA or SCSI, the two major standards we will describe, and their only real limitation is the amount of space inside your Macintosh. There is room on the inside of your Macintosh to add four additional hard drives. Internal drives, sometimes also referred to as "bare," are the least expensive drive solution, as they require only a connection cable, a few screws, and a power supply cable. They do require a little work on the user's part, though, as they must be manually installed. This installation is really quite simple, however, particularly with the new Macintosh body design.

The two major drawbacks to filling your Macintosh with internal hard drives are portability and heat. Because you are actually seating the hard drive in the Macintosh's belly, you are can't easily move it to another Macintosh should you need to do so. Installation and removal are easy, but not that easy. An internal drive is to be considered a fairly permanent item in the Macintosh.

The second drawback is heat. The inside of the Macintosh gets very hot, much hotter than you think. The more items you have installed in it, the hotter it runs, taxing the single cooling fan built in to the system. Overheating systems can exhibit strange behavior or simply melt down completely. Although the built-in fan should keep your internals cool enough, be aware that filling the Macintosh up to the brim and then running it in a 90° Fahrenheit room may cause problems.

Types of drive by location: the external drive

An external drive is simply an internal drive that has been prepackaged in a resilient casing with an input/output interface based on whichever standard it uses. One simply pulls it out of the box and connects it to the Macintosh with no muss or fuss. Of course, the user pays for this ease in price. Expect to see the price for the same amount of storage to double for an external drive solution. The 20-gigabyte internal drive that costs $100 can easily fetch $200 or more once a manufacturer has installed it in an external case for you.

Why choose an external drive rather than an internal? There actually are some benefits to using an external drive that may be less than obvious. As stated before, there is a limit to the number of internals one can fit in the Macintosh. For externals, there is only the limit of your storage standard, and if that is nearly unlimited, so to will be the number of possible drives. Although the ATA drive standard does not exist as an external option, both SCSI and Firewire external drives are in heavy use as external solutions. A Macintosh can technically accept up to 63 Firewire drives at the same time. The larder knows no limit.

External drives are obviously spared the heat issue that can affect congregating a large number of internals inside the Macintosh. Although an external drive also needs cooling resources, generally the casings are designed so that a fan is not necessary, or one is included.

External hard drives also may be considerably more convenient in terms of mobility. Although no hard drive is designed to withstand the heavy shocks of being moved around constantly, it can be done much more easily than doing so with an internal. Moving an external hard drive to another system is as simple as disconnecting then reconnecting a cable. Moving an internal drive will require some surgery, regardless of how simple that surgery can be.

Types of drive by standard: ATA, IDE, or EIDE

There are four primary standards of connectivity with your Macintosh, each of which can be used for communication with drives. These are the input/output buses that will be described in the next section. But since each of these buses determines the standard for the drive we connect through it, it is necessary to look briefly at them. The four, in order of popular usage, are ATA, SCSI, Firewire, and USB.

Each of these standards simply does one thing; it transfers data at a maximum speed or data rate. The drive at the end of this connectivity is limited in its data rate by the actual limitations of the standard. Each standard therefore determines the limitations of our drive choice in terms of maximum data rates. No drive or set of drives ever reaches the limits of its specifications, but some come very close. Our job is to look at what we need for Firewire DV editing with Final Cut Pro and see how far we can make our money stretch.

The first and most popular standard is ATA, also known as IDE or EIDE. This is the most inexpensive drive solution on the market, offering enormous amounts of drive space for the lowest prices. As standards go, ATA is the best deal on the market, and, with some reservations, it is a very suitable drive standard for Firewire DV editing.

The ATA drive interface does, of course, have some limitations that are a reflection of the bargain prices per gigabyte of storage. The first and most serious limitation is that the cables that connect ATA devices must be shorter than 14 inches. This means that there are no external ATA drive solutions. If all ATA drives must be internal, then there is a practical limit to the number of ATA drives one can have in one's system.

ATA drives are further limited by the fact that the standard ATA bus can hold only two devices. As we will see in the chapter on buses, this implies further expense inasmuch as expansion cards are required to expand the number of ATA buses and drives.

Although the specifications for ATA data rates changes with the technology on an almost daily basis, generally speaking, the sustained data rate for an ATA individual drive (given the many different possible variations among drives, Macintosh units, and fragmentation) tops out at around 16 megabytes per second. Since the data rates required for Firewire DV editing are around 3.6 megabytes per second, the ATA standard is well within the range of suitability for use with Final Cut Pro. The low cost and acceptable data rates make the ATA drive a logical choice for most Firewire DV editors.

Types of drive by standard: SCSI

The next popular standard of connectivity for storage is SCSI, which stands for Small Computer Systems Interface. SCSI is a very old standard that has gone through many phases and contains several different levels of data rate and access. A few generations back in the history of the Macintosh, the SCSI interface was the standard for connection with drives of all kinds and came built into the Macintosh's motherboard.

These days, the cost-effectiveness of the ATA standard has eliminated most of the lower-end usage of the SCSI standard. But SCSI has remained in higher-end applications, where its flexibility stands superior to ATA in every way. The SCSI chain is a very wise investment consideration where future expansion will make necessary the ability to adapt to new and more demanding technology.

The acceptable SCSI cable lengths, while still limited to a certain length, are far longer than those of ATA. This means that external drive solutions are possible. In addition, the main SCSI standard allows up to 14 possible devices on a single SCSI bus, as compared with two devices for the ATA bus. And most SCSI cards allow for internal as well as external drives to be used simultaneously, provided that the data rates for each will be limited to the slowest data rate on either side. Add this to the incredible speeds that SCSI card and drive manufacturers currently boast and you have a very impressive standard for connectivity. Depending on the card, SCSI card type, drives, and formatting software system, data rates of 160 megabytes per second are possible.

The catch of course is the price. Even bare internal SCSI drives can cost up to three times as much as a comparable ATA drive. And since the Macintosh does not carry a built-in SCSI bus on its motherboard as it does an ATA bus, a SCSI card must be figured into the price of the system, further driving up the cost of the storage solution. The cost of a 60-gigabyte solution using the SCSI interface could easily amount to $1,500, whereas a sixty-gigabyte ATA drive could be had for around $300. Of course, the savings would appear later, because the SCSI interface easily accepts more and faster-configured drives, internal and external, whereas the ATA bus is completely filled.

SCSI prices tend to fluctuate less and remain at a premium, partially because of their popularity in higher-end solutions and partially because of their special flexibility in terms of expansion and configuration. ATA drives continue to plunge as each new development in drive technology nudges up the data rates and storage capacity of the individual ATA drive. Whether or not SCSI or ATA is a better solution depends on your long-range plans, but ATA is generally the drive of choice for Firewire DV editors, while SCSI is the system of choice for higher-end production stations for which flexibility and the ability to handle extreme data rates are basic requirements.

Types of drive by standard: Firewire

Special mention must be made of an external solution that exists in the Macintosh market today, that of Firewire drives. Although we have discussed Firewire as a system of transferring digital video and audio, the Firewire connectivity is really simply a data transfer system that is fundamentally similar in function to ATA and SCSI. Although radically different in design, the standard is simply a specification of a rate for transferring data between one end of a cable and the other.

Firewire has a specified data rate ceiling of 400 megabits per second. Since there are 8 bits per byte, this delivers a maximum possible data rate of 50 megabytes per second. Of course in practice, this is far higher than the actual possible rates achievable with the present hardware available to use with the Firewire interface. It is only a specification that is used to determine the limits of that particular standard.

High data rates are not the only bonus conferred by the Firewire interface standard. Firewire shares and expands the connectivity gains inherent in the SCSI standard and allows for long cable lengths, enabling it as an external solution. Devices can be daisy-chained, allowing up to 63 possible devices on the Firewire bus. As a bonus, Firewire connectivity confers the ability to hot-plug Firewire devices. Hot-plugging means the user can disconnect a device while the computer is still running, something disastrous for either ATA or SCSI devices. Further still, Firewire is capable of carrying "bus-power," which means that it can deliver the electrical current to run a device that would ordinarily have required a wall socket.

In reality, there is no such thing as a "Firewire drive." Firewire drives are basically external drive boxes containing an ATA drive and a bridge that converts the ATA data to and from Firewire data. Thus, the cost of the Firewire drive is determined by the low price of the ATA drive, the increasingly low price of the ATA-Firewire conversion bridge, and the box that holds it all together. There are several do-it-yourself kits on the market that allow you to purchase and install your own internal drive for use in a Firewire case. Just remember that the drive you install in the case is the weakest link in the chain, and that you shouldn't depend on a drive that you wouldn't want as an internal unit. A slow drive in a fast Firewire chain equals a slow Firewire drive.

Given the benefits of using the Firewire interface for storage, one might wonder why anyone would use anything else for DV editing. The convenience, portability, and rich feature set of Firewire drives make them very attractive, and their reliance on very inexpensive drives makes them a cost-effective solution.

Types of drive by standard: USB

The final standard for connectivity in data transfer is that of the *universal serial bus* (USB). USB delivers a very low standard data rate that does not exceed 1.5 megabytes per second. While this data rate is acceptable for less-intensive applications such as floppy disks and the Iomega Zip drive, it is far below what is needed for editing with the DV codec and Final Cut Pro. The USB standard is not recommended for editing purposes, although it has its uses elsewhere in the Macintosh system as we will see in the discussion of input and output buses.

The Input and Output Buses

The input and output buses of the Macintosh constitute the way in which it receives and delivers data to and from the outside world and other parts of the Macintosh. There are several such buses, and each one is dedicated to a specific task depending on its specification. The primary input/output buses are: the AGP, the PCI, the ATA, the SCSI, the Firewire, and the USB buses. In addition to these, there are other specialized input and output ports (e.g., the Ethernet, the Internet modem, and the external speaker).

The accelerated graphics port

The *accelerated graphics port* (AGP) is a dedicated expansion slot on the motherboard of the Macintosh. It is a very high-speed interface that is designed to communicate with the primary graphics card controlling the computer monitor you use with your Macintosh. The Macintosh comes standard with one such video card in the AGP slot. The AGP card is also the engine that drives the high-intensity graphics requisite for the proper functioning of Final Cut Pro and other applications.

The AGP port is a relatively new feature of the Macintosh, and it is possible that users with an older Macintosh G3 and some G4s will not have an AGP port or card. This is no cause for alarm; the AGP is not required for Final Cut Pro to function correctly, although it may deliver improved performance. Some expansion cards that add real-time or multiple monitor support through one expansion card may require an AGP slot, giving an added benefit to its presence on the motherboard. It should be noted that many video cards are not AGP cards and that such cards will not fit in the AGP slot. Attempts to do so will damage the card, the motherboard, or both. Make sure you are buying the appropriate card for your needs when shopping for a new video card.

The peripheral component interconnect

The second bus is the *peripheral component interconnect* (PCI) bus. These are the expansion slots lined up next to the AGP slot on the motherboard. Different models of the Macintosh include different numbers of PCI slots, from an all-time high of six slots in the old Power Mac 9500/9600 to four slots in the present-day model. PCI slots are one of the most useful items on the Macintosh motherboard, because they allow the user to easily install a completely customizable set of input and output devices such as analog video and audio capture cards, SCSI and ATA expansion cards, extra video cards for multiple monitor support, and a host of other third-party items, the list of which grows by the hour.

One of the most appealing things about the AGP and PCI buses is that, in contrast to their counterparts on the PC side, they are completely plug-and-play compatible. This means that although you may have to install a software driver of some kind in the Extensions folder, you do not have to worry about hardware conflicts and addresses as you would with a PC. The Macintosh hardware and software are able to locate and work with AGP and PCI expansion cards with almost no fuss, provided the proper software drivers are present in the System Folder. PC users who discover this are usually amazed at the ease of installation they encounter in the Macintosh, compared with the nightmare of IRQ settings of other operating softwares.

If the Macintosh was purchased from Apple with a SCSI drive system installed, one of the PCI slots will be occupied by the SCSI card. Since the Macintosh does not come with a SCSI interface built in to the motherboard anymore, a PCI slot and SCSI card are necessary to use SCSI devices. Once again, installing and utilizing a SCSI card is painlessly simple. Most SCSI cards can service SCSI drives inside the Macintosh as well as external SCSI devices simultaneously, making for a uniquely flexible solution. And the SCSI standard is not limited to drives, with every manner of peripheral device offering a SCSI version. Items like scanners, archival data tape backup units, and CD or DVD burners are all available with SCSI connectivity.

ATA expansion cards are also available that allow users to add more than the previously stated limit of two ATA drives to their Macintoshes. An ATA expansion card adds two more ATA buses to the Macintosh, allowing you to install up to four more ATA drives internally. ATA expansion cards are very inexpensive compared with SCSI cards and may prove the best expansion alternative for low-end users looking for more internal drive capacity.

Beyond the SCSI and ATA expansion possibilities lies a range of third-party cards, each of which performs a specific task. Extra video cards for multiple monitor support are inexpensive and painless, most of them not even requiring software drivers beyond the ones automatically installed with the Macintosh OS. No special hardware or software is generally necessary to add a second or even third computer monitor to your editing station. Both the AGP and PCI slots are situated so that the connector side of the card installed in the slot will peek out of the rear of the machine, making connections to computer monitors, external hard drives, and the like a snap.

The ATA Bus

The ATA bus is a dedicated bus that sits directly on the motherboard of the Macintosh. Unlike a SCSI bus, which is a card that requires a PCI slot, the ATA bus is already connected directly to the processor and is hardwired to the motherboard. It allows the connection of two ATA devices, which can be hard drives, CD burners, or anything else that uses an ATA interface. On Macintoshes that were not purchased from Apple with the SCSI option, one of the two devices will already be in place, that being the stock startup ATA drive. SCSI systems are likely to have no ATA drives preinstalled on this bus.

In actuality, there are two ATA buses included with every Macintosh motherboard. One is dedicated to the CD/DVD Player and Iomega Zip drives located in the front panel of the Macintosh. Although it is technically possible to remove these and replace them with hard drives, this is not generally recommended. It would be rather counterproductive to run a Macintosh in the twenty-first century without a CD or DVD drive, since this is how the bulk of software is installed. As for the Iomega Zip drive bay, it is narrow in the extreme, and getting a drive to fit comfortably in that space is difficult even for a professional. For more internal drives, consider an expansion card, or look into the external options.

The ATA bus requires that the two devices on each bus have distinct addresses. One address is called the Master and one the Slave. There is actually no difference in the interaction between the Master and Slave addresses, and it really makes no difference which device is which. This is simply a way for the Macintosh to differentiate between the two devices on the bus. If there is only one drive on the ATA bus, it can be either Master or Slave, but if there are two they cannot both be Master or Slave. The determination of this address is set with small plastic jumpers, based on the drive manufacturer's conventions. These are usually printed on the side of the hard drive itself for convenience.

The Firewire Bus

The third type of bus included in the Macintosh is the Firewire bus, a high-speed input/output port that is the basis for our whole system of Firewire DV editing. It connects the Macintosh with DV cameras and decks, Firewire drives, and CD burners, and potentially any other device that also has a Firewire connection. The Firewire bus has two separate ports to communicate with any external devices you connect to it. A wide range of devices can be connected at once, regardless of their function. You can have a DV deck, a Firewire hard drive, a scanner, and a removable drive all connected at once without fear of failure for any of them. Hubs are also available that allow the user to connect devices to a uniform location rather than stringing them together in a daisy chain.

Firewire devices can be connected in a daisy chain of up to 63 devices, any of which can be connected or disconnected without powering down the computer or device (hot-swappable). They do not require special addresses or IDs and can usually be used by the system without the installation of software drivers. Even so, care must be taken that the

Macintosh is aware that the drive is going to be disconnected to prevent corruption. Simply dismounting the drive by dragging it to the Trash on the Desktop will make it available and safe for hot plugging.

Because there are two different types of Firewire connectors, great care must be taken in selecting the proper cable for use. There is a 6-pin connector, known as a bus-powered connector, and a 4-pin connector. This makes three possible cables: a 6-pin–6-pin cable, a 4-pin–6-pin cable, and a 4-pin–4-pin cable. The connectors on the Macintosh are of the 6-pin variety. Most deck and camera jacks are of the 4-pin variety, so unless your Macintosh comes packaged with the proper 4-pin–6-pin cable you may need to obtain one. Many CD burners and other devices utilize the 6-pin–6-pin connector, so check to make sure you have the correct one before attaching equipment.

The bus-powered 6-pin connector enables you to pass electrical power to Firewire devices that can accept it, meaning that in some cases you may not need to plug a device to a wall AC circuit. This feature is relatively rare and is almost exclusively the domain of pocket-sized portable Firewire drives. Make certain that any device you own can function with bus power before relying on it.

The Universal Serial Bus

The Universal Serial Bus (USB), is a slower data rate connection used for input and output of devices that do not require massive streams of data. Your keyboard, mouse, printers, and graphics tablets are very undemanding devices, and the USB has more than enough bandwidth to handle them. The present standard for the USB maintains a data rate of roughly one and a half megabytes per second.

There are a large number of devices on the market that can utilize the USB connection. Floppy disk drives, lower-end SCSI adapters, and even CD burners and audio and video capture cards have been designed for the USB interface. Like the Firewire bus, USB can carry power to a device. Hubs are also available in both powered and unpowered varieties. Some devices that do not actually power from the USB connection still require a powered connection, so be careful to select a powered hub should you be in the market for one. Like Firewire, USB connections are hot-pluggable as well, giving the standard some excellent flexibility compared with the older serial standard that it replaced on the Macintosh.

Care should be taken in considering the inclusion of far-out third-party USB devices in a system. USB third-party software drivers are notorious for causing extension conflicts with the system software. In general, if you can get a Firewire device for roughly the same price as a similar USB device, you are better off with the Firewire device every time. If you do install a third-party USB device, such as a custom mouse or keyboard, keep an eye on the extensions that get installed as drivers so that in the sudden event of bad behavior, you know where to start investigating for conflicting extensions.

Various other inputs and outputs

Ethernet and internal modem

The Ethernet and modem connections that come stock on the Macintosh allow it to communicate with the outside world of other computers and/or the Internet. The software that operates your Ethernet and modem are built in to the Mac OS, so networking with a Macintosh is a shockingly simple. To network multiple Macintoshes, simply connect them together either directly or through an Ethernet hub. The built-in modem is also a snap to use, and the initial startup of your Macintosh will walk you through the process of either accessing your present Internet account or starting a new one.

Speaker out

Unlike most PCs, a high-quality 16-bit sound card is built into the motherboard of every Macintosh, allowing you to run excellent stereo quality audio out to speakers from your system. As with the Ethernet and modem connections, the sound is controlled entirely by the Mac OS software and requires no extra configuration for audio output.

Checking the pieces: the Apple System Profiler

Although there are an infinite number of other items that you may have installed with your system, the list at the beginning of this section states the bare minimum for satisfactory performance. Although you may have purchased your system preconfigured as an out-of-the-box editing solution, you should still take a moment to check through the process of installation, as detailed hereafter, to make sure that your system is optimized and prepared for editing.

To check on almost everything relating to your Macintosh's hardware and software, access the Apple System Profiler from the Apple dropdown menu on the extreme upper left corner of the Macintosh Desktop (Figure 2-8). The Apple System Profiler is a little application that gives the hardware and software status of your entire system at the moment. It is an excellent tool for quickly checking what version of an extension you are using or whether or not a device is present and doing what it should be doing.

Figure 2-8

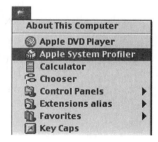

The first tab of the Apple System Profiler shows the General System Profile. It describes the version of the currently loaded system software, the startup drive and its bus and address, the amount of installed RAM and its distribution amongst the RAM slots, etc. It describes the major areas of your system all on one tab.

The second tab, Devices and Volumes, is a detailed map of all hardware components currently connected to your system. Each bus is represented, along with a hierarchical chart of what is connected to each bus and the settings, if any, of the device or volume at the location.

The next two tabs give a detailed listing of all the control panel and extension inits existing in your system. These tabs will display active inits as well as inactive ones you may have disabled. It will also indicate whether or not the init in question is an Apple or third-party init and specify its version and its size. The Applications tab displays every application program on your system. This can include many smaller applications that are included in the installed package of larger application sets. And finally, the System Folder tab displays all the bootable System Folders present on your system.

In addition to being informative to the user, the Apple System Profiler is an invaluable tool for Technical Support or long-distance troubleshooting. You can save a copy of your Apple System Profiler as a report to print out or e-mail to a troubleshooter, eliminating the need for the troubleshooter to be present to look at your system. Simply glancing through the Apple System Profiler and looking for known issues can solve many problems.

The first tab of your System Profiler should show that Mac OS 9.1 or later and Quick-Time 5 are installed (Figure 2-9). Further down the tab, Virtual Memory should be off,

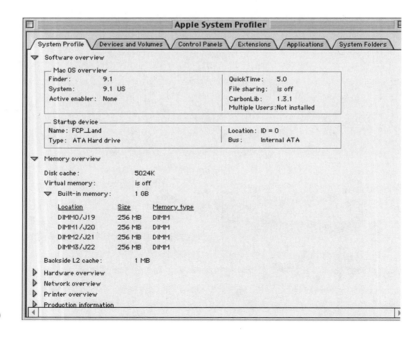

Figure 2-9

and the built-in memory should show at least 256 megabytes or more. If you have less, you should look into purchasing and installing more RAM. The system software can grab as much as 75 megabytes and Final Cut Pro requires more than 100 megabytes for itself. Even so, Final Cut Pro will give unpredictable behavior if it is allocated only the bare minimum to run. Performance will be vastly improved if Final Cut Pro is allocated more than the minimum.

Version 2.0 of Final Cut Pro can actually utilize as much RAM as the user allocates, and performance only improves on doing so. Even if you are running an earlier version, you should make sure that you have enough physical RAM installed to provide over 75 megabytes for the system's use and then at least 120–180 megabytes for Final Cut Pro. Take advantage of cheap RAM prices, and give your system some breathing space.

If you know for a fact that you have more RAM than is actually displayed in the Apple System Profiler, there are several reasons why this could happen. You could have damaged or defective RAM chips. The chips may be installed incorrectly. Some firmware updates to the motherboard have even been known to reject certain makes of RAM chips, even though there is nothing technically wrong with them. In such a case, you will need to look further into the matter and determine the actual cause.

In the Devices and Volumes tab (Figure 2-10), you should see a number of relevant items. If your DV camera or deck is connected to the Firewire port on the Macintosh, you should see it listed as a "DV Audio" device on the Firewire bus. If you have drives located on the native ATA bus, these should be present. If you have a SCSI card and drives installed, the SCSI card should be listed as occupying a slot in the PCI bus, and you should also see a SCSI bus listed that contains your individual SCSI drives.

It is imperative that there be two separate physical disk drives for use with Final Cut Pro. One of these will be the startup drive that currently holds your System Folder and into which you will be installing Final Cut Pro. The other separate physical drive is only for your captured DV footage. It is not recommended that you use the same physical drive for the System Folder/Final Cut Pro and the captured media for reasons that will be detailed in the setup process of your Final Cut Pro project.

Installing Final Cut Pro software

If you have ascertained that all the necessary hardware and software elements are present, it is time to install the Final Cut Pro software. Simply insert the CD, click on the install icon, and let the installer go to work. If you have not upgraded your QuickTime to version 5 or if you are missing any necessary system software, the installation process will stop and allow you to correct the situation. Although Final Cut Pro does not require you to register the full version of QuickTime Pro that accompanies the Final Cut Pro license, it is suggested that you do so to avoid annoying messages about updating as well as to take advantage of some of the more useful features.

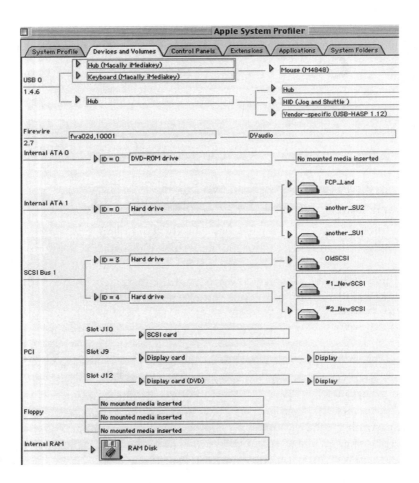

Figure 2-10

At the end of the installation process, Final Cut Pro will ask you to give it some initial preset information about your equipment setup. After entering your serial number, you will be faced with a dialog box entitled "Choose Setup." This is simply Final Cut Pro asking you to establish the initial parameters for a project. You will later be able to change anything you enter at this time. If you are unsure, simply hit OK and pass on to the next window.

On hitting OK, you will have completed the necessary installation steps and will be inside the application itself.

3 The Initial Setup: Optimizing the Mac OS and Final Cut Pro

Final Cut Pro is a fully functional nonlinear digital video editing application. Unlike other software-only nonlinear editing applications, Final Cut Pro stands alone as the most feature-rich application on the market in its price range. With that functionality comes a degree of user responsibility. The user still has to configure the application correctly and organize it for optimal performance and safety. Final Cut Pro will function even if you configure it incorrectly. However, you will pay for such mistakes further down the road when you run across foul-ups that could have been easily avoided had you set up the application and project in an organized and safe fashion at the beginning.

What follows is a tried-and-true path for setting up Final Cut Pro with a Firewire-based camera or deck. It is an especially effective method for multiuser editing stations in which more than one project or editor is working concurrently. If you follow these steps, you will experience few nasty surprises, and the few surprises you do encounter will be much less disastrous than if you simply plug up and begin editing.

The process for setting up your Final Cut Pro preferences begins outside of Final Cut Pro. The first thing you need to do is to take a stroll through your Mac OS system settings and make sure they are optimized for Final Cut Pro. Although we witness only what we are doing at any given time, the Macintosh is constantly doing many things in the background, some of which can have a negative impact on Final Cut Pro's performance. You can always revert the settings after you finish working.

The Control Panels

The Energy Saver Control Panel

STEP 1

Go to the Apple Menu Items in the top left corner of the Desktop. In the Control Panels submenu, select Energy Saver (Figure 3-1). Hit the Show Details button to open the extra custom settings. Set the System and the Hard Disk to Never.

The Energy Saver (Figure 3-2) is a control panel that schedules your Macintosh to go into sleep mode if it is unattended for a predefined period of time. The default setting is for 30 minutes, but you can customize this for your own needs. With the Show Details button

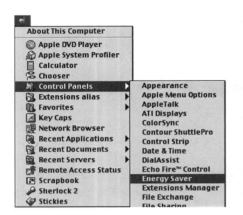

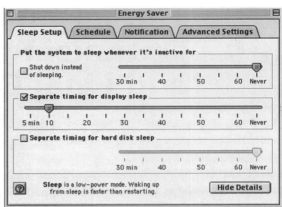

Figure 3-1   **Figure 3-2**

enabled, you will find that there are specific settings for System, Monitor, and Hard Disk Sleep. You will want to set the System and the Hard Disk to Never, which will prevent Energy Saver from putting your Mac to sleep at a clutch moment, such as during a long render. You can also set your monitor to Never, although having your monitor go to sleep will have no impact on Final Cut Pro and may be good for the monitor's life span.

The Memory Control Panel

STEP 2
Go to the Apple menu items in the top left corner of the Desktop. In the Control Panels submenu, select the Memory Control Panel (Figure 3-3). Click Custom Settings on the disk cache, and then change the default value to between 3,000 K and 5,000 K. Make sure that Virtual Memory is disabled.

Disk Cache
At the top of the Memory Control Panel, you will see the Disk Cache setting. Disk Cache is a performance enhancement feature of the Mac OS that allows certain data that is normally accessed from the slower drive but must be accessed very frequently to be written into a protected area of the much faster RAM. The performance boost with Disk Cache enabled is enormous. I highly recommend a healthy amount of Disk Cache unless you have been specifically instructed by an equipment manufacturer to lower it.

The initial size that Disk Cache defaults to is actually determined by the Mac OS and is based on a percentage of your physical installed RAM. The more RAM you have installed, the larger the default setting in the Disk Cache will be. There has been a great deal of disagreement in the Macintosh community about the proper setting for Disk Cache. In the past, many analog capture cards required that Disk Cache be set to its lowest possible setting. This is not the case with Firewire and Final Cut Pro.

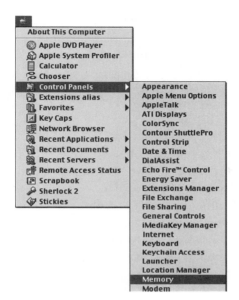

Figure 3-3

The performance gains of Disk Cache seem to roll off at around 5,000 K. If you have a great deal of RAM, Disk Cache may default to more than 8,000 K. Since RAM is more valuable to us allocated to Final Cut Pro than to Disk Cache, you can lower this value. Click Custom Settings, and then lower the default value to between 3,000 K and 5,000 K. You must reboot to free up this RAM, but wait until you have set up the rest of the Memory Control Panel to do so. If you aren't sure, leave the Disk Cache set at its default value.

Virtual Memory

The next thing you need to check is for Virtual Memory (Figure 3-5). This is absolutely critical. Final Cut Pro will not function properly if Virtual Memory is enabled. As described in the last chapter, Virtual Memory uses the dramatically slower hard drive storage as a substitute for super-fast RAM, which has a massively negative effect on the system's performance. Unfortunately, on initial installation the Mac OS defaults this setting to Virtual Memory enabled. Make sure that Virtual Memory is disabled, and remember to check this setting again if you ever reinstall or update your Mac OS. Once again, the Mac must be rebooted for this setting to take effect.

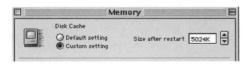

Figure 3-4

Figure 3-5

The RAM disk

The RAM disk, the next option, is less critical (Figure 3-6). A RAM disk is simply a fake volatile-memory partition that consists of a portion of RAM, almost like Virtual Memory in reverse. If you have an enormous amount of RAM and you are working with small documents, you may notice that saving them to a RAM disk gives fantastic performance boosts. Remember that RAM data rates are in the neighborhood of 90 megabytes per second. Many people use a RAM disk for their internet browser's cache because it speeds up page-loading times.

Although RAM is volatile and is wiped when you shut down, the RAM disk can be set to save its content to a physical hard drive on shutdown. This can be a little scary, though, if your Macintosh crashes and has to be rebooted without properly shutting it down. Whether or not you should use a RAM disk on a machine running Final Cut Pro is a trade-off similar to Disk Cache. Although you may experience some performance boosts in certain areas of your Mac, the RAM you allocate to the RAM disk might be more effectively used when allocated to Final Cut Pro. Once again, changes to the RAM disk setting require a reboot to take effect.

If you have a RAM Disk enabled and you begin to experience problematic behavior with Final Cut Pro, you may want to disable the RAM Disk to see whether it is causing RAM allocation problems or any others.

The Startup Memory Tests

In addition, there is a hidden item in the Memory Control Panel. When starting up the Memory Control Panel, hold down the Option and Command keys, and you will be able to access a hidden setting at the bottom of the window called Startup Memory Tests (Figure 3-7). When the Mac boots up, it performs a memory test on all of your RAM chips prior to loading the System software. There is nothing wrong with this; it's meant for your protection. However, RAM chip failures are pretty rare, and if you have 512+ megabytes of RAM installed, as many Final Cut Pro users do, this initial RAM test can seem interminable.

The Startup Memory Tests option allows you to disable this test. Since this is a hidden setting, the default setting is enabled. Setting it to OFF will dramatically speed up the boot process, particularly if you have a large amount of RAM installed. You can always re-enable the setting if you notice behavior that suggests defective RAM. The speed-up in booting your Mac will be noticeable.

Figure 3-6

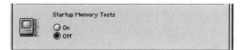

Figure 3-7

STEP 3

The Chooser and AppleTalk

Go back to the Apple Menu and select the Chooser (Figure 3-8). Although not a Control Panel, the Chooser is an important system control application that you may need to adjust. Look to the bottom right corner of the Chooser window, and locate the AppleTalk Active/Inactive switch (Figure 3-9). Here you can enable or disable AppleTalk, a protocol that the Macintosh uses to communicate with printers and networks through the Ethernet port and other connections. If your editing station has no printer and is not connected to a network, turn off AppleTalk. If you are linked to other machines through AppleTalk, you may want to disable it prior to capturing or printing video to tape.

You should generally disable AppleTalk whenever you are preparing to capture video or print it out to tape. Although it could conceivably never cause a problem, particularly if you are not connected to a network or printer, AppleTalk is capable of stealing the processor's attention and degrading performance. AppleTalk connections are notorious for asserting control over the entire Macintosh when there is any activity on a network connection. Final Cut Pro is a very processor-intensive application and, if interrupted by such activity, it will exhibit unpredictable behavior.

Final Cut Pro memory allocation

Finally, the memory allocation for Final Cut Pro should be set. Although you may have plenty of extra RAM installed for Final Cut Pro, it will not access that RAM until you allocate it. To set any application's RAM allocation, first determine the maximum amount of RAM you can give it without choking your system.

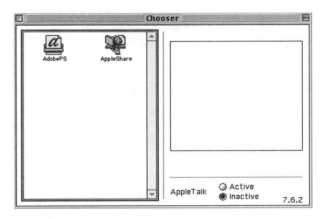

Figure 3-8   **Figure 3-9**

STEP 4

To determine the amount of RAM you can allocate to Final Cut Pro, take into account the actual amount of physical RAM you have installed in the Macintosh. This can be ascertained from either the Apple Menu>About This Computer (Figure 3-10) while in the Finder or the Apple Menu>Apple System Profiler.

STEP 5

Subtract about 80 megabytes from the amount of actual physical RAM, 80 megabytes being roughly the amount of RAM that a healthy System Folder will normally allocate for itself under extreme situations. The number you have left is the amount of RAM you can safely allocate directly to Final Cut Pro. For example, if you have 256 megabytes of RAM installed, subtracting 80 megabytes from this number will give 176 megabytes.

STEP 6

To assign this amount to Final Cut Pro, navigate to the Final Cut Pro application folder on your hard drive. Open the folder, hold the Control key, and single-click the Final Cut Pro icon. Do not double-click or you will actually start up the application, which is not our objective.

STEP 7

In the Contextual Menu that pops up, choose Get Info and then Memory (Figure 3-11). In the next window, change Minimum to 150,000 K and Preferred to 176,000 K. Close the window, and the RAM allocation is set.

The difference between Minimum and Preferred sizes is that Minimum will lock off exactly the amount of RAM you enter into the field, making it available only to the application it is assigned to. If the Preferred size is higher than the Minimum, the Mac OS will dynamically shift RAM so that the amount allocated to the application never goes below the Minimum, but can go higher if more RAM is available.

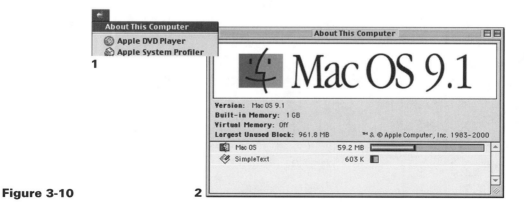

Figure 3-10

This makes it possible to open more applications than should be technically possible. We said that Final Cut Pro requires 106 megabytes of RAM to actually start. We also say that it works much better with 150 megabytes allocated to it and that it will actually use even more if allocated, improving its performance with each extra megabyte. If you set the Minimum at 150,000 K and the Preferred at 180,000 K, then the Mac OS will adjust the amount of RAM that Final Cut Pro is allocated on the fly as RAM is grabbed or freed up by other applications you may be running simultaneously. If nothing else is running, the RAM is available for Final Cut Pro.

The Mac OS has a peculiar way of dealing with RAM, and I generally keep my RAM allocations locked in at a level at which I know they can function well. The problem is that Mac OS 9.1 and earlier don't have what is referred to as "protected memory." Two different applications can try to claim the same RAM, causing a crash of the entire system, as is the case with Extension and Control Panel conflicts. If you allocate too much RAM to an application and don't leave enough for the Mac OS itself, you could choke the system and possibly cause a crash. When your system gets down to only a couple of free megabytes, any large shift in the memory allocations for the system or the applications could cause unbelievably slow performance, a sudden crash, or an error message.

The best bet for the fastest, most reliable performance is to allocate to each RAM-hungry application like Final Cut Pro and Photoshop as much RAM as you can make available to it (assuming that the applications actually can make use of that much—some cannot) while leaving a healthy buffer of at least 75 to 80 megabytes for the system. Set your Minimum and Preferred allocation so that RAM is rarely shifted around by the system without your consent, and then do not open more applications than can be run, based on the amount of physical RAM you have installed.

If Final Cut Pro and Photoshop are allocated 256 megabytes of RAM, they will both run much faster and give much better performance. You may not have enough physical RAM to run them at the same time, so close one and open the other as you need it. You

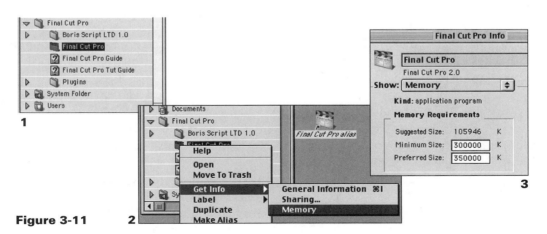

Figure 3-11

may find that the brief delay in restarting an application is a more than acceptable trade-off for being able to allocate it twice as much RAM, particularly with applications like Photoshop and Final Cut Pro that store Undo files in their RAM allocations.

Even though a minimum of 256 megabytes of installed RAM is highly recommended, Final Cut Pro 2.0 will actually run with a minimum memory allocation of 106 megabytes. Running at this minimum allocation may deliver very unpredictable performance (e.g., slowness, crashing, and memory errors). If you begin to receive Type 2 errors, it may be an indication that you are running low on RAM. Working safely and without interruption is much more likely with at least 256 megabytes of installed RAM and 150 megabytes or more allocated to Final Cut Pro. If you aren't sure how much the system is using and think you are running too close to the edge, keep the Apple Menu>About This Computer running on your Desktop somewhere. It will dynamically update, showing you how much RAM each application and the system are using, as well as how much free RAM you have left as a buffer for the system to expand into without crashing. After a while, you'll get an intuitive feel for how much RAM you can effectively give any application and how much the system typically needs, which can vary depending on which applications you are running.

Preferences and Audio/Video settings

Why the settings are required

Next in line is setting up the Final Cut Pro Preferences and the Audio/Video Settings. This step is an absolute necessity. Some other popular applications do not require the user to alter the default preferences at all. Many of the settings in Final Cut Pro, however, directly affect the way it functions and interacts with the Mac OS and your hardware, and as such they must be checked and set correctly.

Preferences are concerned with the toolset and functionality of Final Cut Pro and, generally speaking, won't require changing until your working methods themselves change (the one important exception is that of Scratch Disk assignment). Preferences refine the workflow of Final Cut Pro and allow you to focus the way you approach editing. These settings are generally not application-critical inasmuch as changing them would not cause Final Cut Pro to malfunction.

The Audio/Video Settings (and the Scratch Disk preferences located in the Preferences), on the other hand, are directly related to how the Final Cut Pro application deals with your particular hardware and software situation. Remember that Final Cut Pro can function under numerous hardware and software environments, and the default Audio/Video Settings when you install the application may not reflect your personal setup. These settings are critical to Final Cut Pro's functioning correctly and you must define them.

Unlike the Mac OS System settings you completed in the last section that must be set or checked only initially or when you encounter system misbehavior, the Audio/Video Settings and the Scratch Disk Preferences should be checked and set properly whenever you work with Final Cut Pro. Setting the preferences is a very quick process, and Apple has instituted some great preset tools for streamlining the preferences process in the 2.0 version. However, you still need to check them every time, especially if you are not the sole user or there is more than one active Final Cut Pro project per machine.

The reason for this is that, unlike some professional digital nonlinear video editing applications, Final Cut Pro does not store or link your preference settings with your particular project file. Thus, when you start a Final Cut Pro project, the preferences you encounter will be whatever they were set to the last time Final Cut Pro was used. In a multiuser environment or on a system with multiple projects, this can have disastrous effects if it is allowed to get out of control.

Starting up Final Cut Pro

When you start Final Cut Pro, it usually launches the project that was open when Final Cut Pro was last run. This very well may not be your particular project. Since Final Cut Pro also allows the user to have more than one project open at a time, it is always a good procedure on launching the application to immediately close all open projects. It is not necessary to close an open project just to change the preferences, particularly if it is your own project that is open, but it is a good habit to get into to avoid accidentally altering another user's hard work.

STEP 8

After starting the Final Cut Pro application, go to the File drop-down menu and select Close Project until all the open projects are gone and the Effects tab is the only thing left in the Browser window (Figure 3-12).

There is one exception to the rule that Final Cut Pro usually opens the last open project when started up. If you start Final Cut Pro up by double-clicking your project file rather than the Final Cut Pro application alias, it will start with only your project open (Figure 3-13). If you follow this method of starting Final Cut Pro, you can skip the process of closing all open projects.

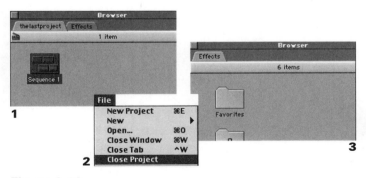 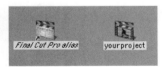

Figure 3-12   **Figure 3-13**

The Preferences

STEP 9

After closing any initially launched projects, go to the Edit drop-down menu and select Preferences (Figure 3-14). A dialog box with five tabs at the top will appear. Each of these tabs needs to be set up for your particular needs before you move on.

The General Preferences tab

The first tab is General settings (Figure 3-15). Each item on this tab has important implications in the workflow of Final Cut Pro, as well as making sure the application warns you in the event of problematic behavior in the system. Be sure to go through each item and configure it according to your need.

Levels of Undo

The first item is the Levels of Undo. Levels of Undo are the number of actions that you can undo in sequential order should you make a mistake while editing. Having a certain number of actions available for Undoing is very handy if you want to try something in your editing that you're not sure you will want. If you don't like the results of any action, select Undo from Edit drop-down menu or simply hit Command-Z, the keyboard shortcut. Final Cut Pro will return your project back to the way it was prior to initiating the last action. You can go backwards in your actions as many times as you have specified in the Levels of Undo in the General tab of the Preferences.

The number of undoable actions can be set up to 99, although be warned that greater levels of undo require more RAM. As you perform actions, they are stored in Final Cut Pro's RAM allocation for instant removal should you choose to undo them. The more actions you store away, the more RAM you burn up. There are more effective ways of

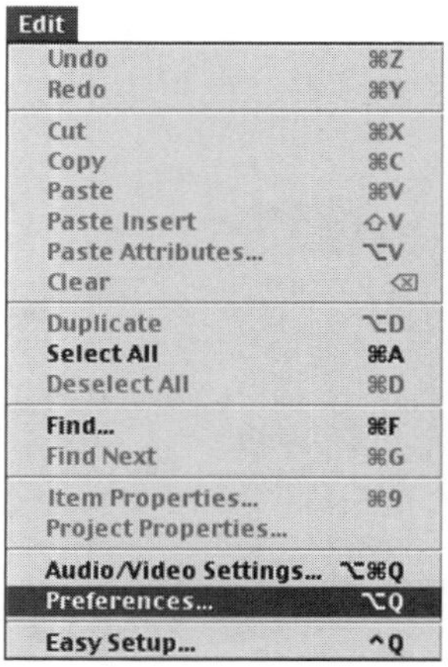

Figure 3-14

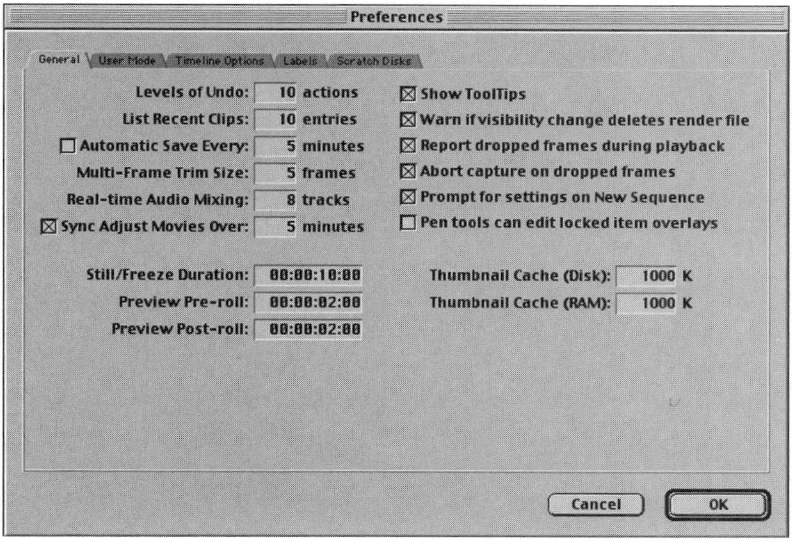

Figure 3-15

returning to earlier versions of your project, and it is recommended that you leave Levels of Undo set to the default value of 10.

List Recent Clips

The List Recent Clips box allows you to specify how many recently used clips are displayed in a certain pop up menu of the Viewer window. This can be a handy way to access clips you want to work with without having to manually drag them across the Desktop. The default value here is 10, although you can set it for up to 20. However, 20 is really too many to choose from in most situations and will create a giant drop-down bar if set that high. Leave this set at the default of 10 unless you find that you need more in your process.

Automatic Save Every

Automatic Save Every, or Auto-save, actually has two parts; there is a check box for enabling it and a box for entering the frequency with which it occurs. The default for Automatic Save Every is to be checked on and Save set for every five minutes.

There are two important things to remember with the Auto-save function. First, Auto-save is an automatically timed function that does not wait until you have completed a task before beginning a timed save. It is enabled whether you are in the middle of a two-hour render or simply doing nothing. As such, it can be a very interruptive function, particularly since a highly complicated project file can take up to 30 seconds to completely save. If you are concentrating on your editing, having the whole application lock up every five minutes to save itself can ruin your concentration.

Second, there have been isolated reports that Auto-save is responsible for the corruption of project files. This has never been officially confirmed, so it is impossible to say for certain that the Auto-save function is the culprit. It is certainly conceivable that if a massive save operation began at a critical moment such as rendering or capturing, corrupt data could make its way into your project file.

The way Auto-save works is that every so often, at the predetermined intervals that you specified in the Preferences, the application saves a copy of your project file in the same folder as your project file. This Auto-save file carries your project file name plus the suffix "-auto,"—for example, "yourprojectfile-auto." Auto-save never actually saves your original project file, and the "-auto" file remains and is continually updated. When you manually save your real project file, the Auto-save copy disappears.

Should your machine crash, you might not be able to save before rebooting. On your return to your project folder, you will see the original project file, which will still be in the state it was when you last manually saved it. You will also see the Auto-saved copy of the project file, which will be as recently updated as the last time the Auto-save was performed. If you open the original project file after a crash, a message box will ask if you wish to open the more recent version of the project (i.e., the Auto-saved version). You may take or pass on this option, because after that point, Final Cut Pro will treat the two files as completely separate projects.

Whether or not you choose to use Automatic Save Every should depend on how careful you are. Most people aren't so diligent and have to learn the hard way that regular saving is a must. Although it is difficult at the time of this writing to say whether Auto-save is responsible for project file corruptions, it is the opinion of the author that the diligent Final Cut Pro editor actually has no need for the Auto-save function. Manually saving backups of your project file on a regular basis is far safer and more responsible behavior than simply running a background saving function all the time. And Auto-save does not allow you the ability to archive your project. In the next chapter, you will be shown a backup saving technique that will make the Auto-save function completely superfluous. Your project will be safer and more flexible, and you will sleep better at night.

Multiframe trim size

Multiframe trim size is a setting for a specialized editing tool with Final Cut Pro. Each editor has an opinion about how many frames should be available for quick trimming in the Trim Edit window, and until you have used it, leave this set for the default five frames. When you get to the editing stage and begin using the Trim window, you will find that this setting not only allows for a very precise adjustment of your edit but also helps you make very quick adjustments of the edit. This is not an application-critical setting, but one that must be set according to your editing style. You can always change it later as need be.

Real-time Audio Mixing

The Real-time Audio Mixing setting allows you to put a limit on the number of audio tracks on the Timeline that Final Cut Pro is able to play back without rendering. You can have up to 99 tracks of audio in any timeline, but your system hardware determines how many of these can be played back simultaneously without having to be mixed together in a rendering process.

Many factors enter into the equation of how many real-time tracks can be mixed and played back without rendering. Processor speed, the amount of RAM and the speed of your hard drives all have a bearing on whether or not your system can deliver the number of tracks you require without rendering. Also, any audio effects you apply to the audio clips will decrease the actual number of real-time mixed audio clips you can muster.

The default value of eight tracks is the maximum for Firewire configurations of Final Cut Pro, provided there are few or no audio filters applied and your hard drive is not overly fragmented. Eight tracks of real-time mixed audio should perform fine in a system that has the recommended amount of RAM and is using a hard drive configuration that isn't below par or overly fragmented. If you use more than eight tracks of audio at a time, simply render them, which will allow them to be played back correctly. Audio rendering is incredibly fast, which may be a minor trade-off compared with having to purchase more hardware for your system.

Sync Adjust Movies Over

This setting is fraught with confusion in the community of Final Cut Pro users. To dispel that confusion, it is necessary to understand what this feature is supposed to address and how it works. The Sync Adjust Movies Over feature (in previous versions referred to as the Auto Sync Compensator) was designed to deal with the issue of standard sample rates in prosumer DV cameras and decks.

As we saw in Chapter 1 of this book, Digital Audio is audio that has been sampled. The quality of the digital audio file and the fidelity to the original source from which it was created depend on the number of times per second that the source audio was sampled. We refer to this as the digital audio sample rate. In DV cameras and decks, the two sample rates used are 32 KHz and 48 KHz. Unfortunately, none of these prosumer cameras and decks are sampling at the exact rate of either 32 KHz or 48 KHz. Although the margin of error may be only a few samples off from this number, most DV cameras and decks actually sample the audio at slightly different rates from what the specification states should be there. Thus an audio sample may be at 48,048 Hz instead of exactly 48,000 Hz.

In ordinary playback this isn't an issue. The problem is that Final Cut Pro at the outset is unaware of the actual sample rate of the audio. In our settings to come, we will see that we can select only these two exact sample rates. Thus, when Final Cut Pro captures this audio, it is playing back the file at a slightly different speed than it was actually recorded at. This would be similar to shooting a film at a nonstandard rate of 30 frames per second, but then viewing it with a film projector that plays back only at the standard of 24 frames per second. The result would be barely perceptible, but it still would be slow motion footage, as the camera and film footage's frame rate is slowed down to the frame of the projector.

Of course, the difference between 48,048 and 48,000 samples per second is very minor and nearly imperceptible. The problem is that the video and audio are recorded together. If the video is playing back at the correct rate, but the audio is playing back slightly too fast or slow, the result over a long period of time is that the two can go progressively out of sync with each other. This is noticeable only over extremely long stretches of footage, since it takes many seconds for those 10 out of 48,000 samples per second to generate a serious sync issue.

The Sync Adjust Movies feature causes Final Cut Pro to calculate the exact number of samples for the captured audio clip rather than assume that it is precisely 32 KHz or 48 KHz. This does not require any more processor or RAM resources and is actually a good feature to have enabled all the time. Certain cameras, such as the Canon XL-1, actually require this to be turned on because of the disparity between the assumed and the true sample rate.

If Sync Adjust Movies is such a great feature, why would Apple give you the option of turning it off? The answer lies in a serious chink in the Sync Adjust Movies armor. If footage is being captured and a timecode break or area of the tape that contains no video is encountered, an incorrect number of audio samples are registered inside the audio file's data. Although this may not be overtly apparent, use of the Sync Adjust Movie feature in this instance can yield a complete lack of sync between video and audio.

Which setting is correct? Under normal circumstances, leaving this feature on and set for five minutes is completely acceptable and even recommended. If there is any aberration in the sample rates, it will be caught and recalculated automatically. Unfortunately, if there is the slightest likelihood of a timecode break on your tapes, it could result in a serious sync issue and the feature should be disabled.

The smartest choice would be for those capturing footage using Timecode and Device Control to leave the feature enabled and set for five minutes. If there is a timecode break in the tape, your capture will be terminated anyway, and you will be alerted to the problem. Those using an analog digital converter box or a pass-through device of some kind and who are therefore not using timecode for capture would be best advised to leave the feature disabled, since it is fairly easy to accidentally capture blank video footage. In addition, remember that if your clips are shorter than five minutes or so, the sample rate irregularities will be largely unnoticeable. Sometimes the best idea is just to keep your clips short and sweet.

Show ToolTips

Returning to the top, the ToolTips setting should be enabled, particularly for new users and those unfamiliar with the Macintosh platform's keystroke conventions. When ToolTips is enabled, Final Cut will actually tell you the function of an on-screen item. Simply leave the mouse pointer over the item for a couple of seconds without clicking and a small text box will appear telling you what the item does. The bonus is that many ToolTips boxes include the keyboard shortcut for the item as well. As we will see, keyboard shortcuts are something all users of Final Cut Pro should master—your wrist will last only so long.

Warn If Visibility Change Deletes Render File

As with all "alarm" switches, this preference should be left enabled. It refers to a situation in Final Cut Pro in which Timeline media can become unrendered simply by turning off track visibility, a timeline feature that is discussed in a later chapter. This can eliminate hours and hours of rendering with the push of one button. This preset simply forces you to go through a dialog box before committing to that action. Leave it enabled.

Report Dropped Frames During Playback

This setting is an "alarm" switch to let you know if, for some reason, playback performance was impaired and Final Cut Pro was unable to play back at the required frame rate. Dropped Frames is a situation in which, for whatever reason, Final Cut Pro was not able to process the video and/or audio data quickly enough to provide the 29.97 or 25 frames per second demanded by NTSC or PAL. Many different things can cause Dropped Frames; this is the primary reason for doing your Preferences and Audio/Video settings! Sometimes they are hardware related (e.g., fragmented drives), but most often they are the result of not having your settings in order.

Dropped Frames are not acceptable. Final Cut Pro is fully capable of playing back NTSC and PAL video at the proper frame rate, given that your hardware and settings are in order. If you are receiving reports of Dropped Frames, go through your settings first and make sure that you haven't missed something. You want to keep the Report Dropped Frames During Playback switch enabled so that you become aware the instant that Final Cut Pro is not performing properly. Disabling the Dropped Frame report is like taking the batteries out of a smoke alarm to stop a raging fire. If you receive a Dropped Frame warning while playing back, eliminate the problem, not the alarm.

Abort Capture on Dropped Frames

This alarm setting is similar to the one previously discussed in that Final Cut Pro will halt what it is doing if it encounters impaired performance and Dropped Frames. The only difference here is that it is occurring during the capture and not the playback process. That distinction is unimportant. Dropped Frames are unacceptable in either location. This setting should always be enabled. If you are not capturing and playing back at the full frame rate, then you are not exercising the true value and potential of the Final Cut Pro editing software.

At the time of the writing of this book, there were reports that erroneous Dropped Frame errors have been generated by Final Cut Pro during the capture process for some individuals. There may or may not be a software issue related to this situation, and the first conclusion of the new user should definitely not be to attribute the behavior to a "Final Cut Pro bug." Of all problems encountered by new Final Cut Pro users, 99 percent are settings or drive related. Make sure that you fully troubleshoot such issues before screaming about software bugs.

If you receive Dropped Frame messages during the capture process, make sure that all of your settings are in order and that your hard drive configuration is not causing the problem. In particular, if you are receiving Dropped Frame messages when you play back clips as well as capture them, your settings or drives are the likely culprits.

Prompt for Sequence Settings on New Sequence

This preset is another setting that should be enabled for safety purposes. Whenever you create a new sequence in the timeline, Final Cut Pro will ask you for its specifics, ensuring that you choose the right ones instead of simply using whichever sequence settings have been previously invoked. Sequence settings are some of the most critical ones, so having Final Cut Pro put this reminder in front of you is a good way of making sure you never forget to do the settings correctly, even in mid-project.

Pen Tools Can Edit Locked Item Overlays

This setting is not application critical, and it's probably not a good idea to use it if you aren't sure why you'd even want to override a locked item on the Timeline. Under certain advanced editing circumstances it could be a useful feature, but for the moment it's best to leave it disabled.

Still/Freeze Duration

This setting allows the user to determine the duration of the clip that results from importing a still image generated by another application such as Adobe Photoshop into Final Cut Pro. The default is 10 seconds. Although such a clip will have to be rendered to play back correctly, it will at least be of a serviceable length. This setting also applies to the length of a still frame generated by Final Cut Pro. Although you can easily change the lengths of either the imported image clip or the frozen frame, this setting allows you to specify its initial duration.

Preview Pre-Roll and Preview Post-Roll

These two settings are preferences that relate to the way the Canvas, Viewer, and Trim Edit windows allow the user to quickly check an edit within a user-defined range of time. That functionality is described later. For now, it's best to leave this set at the default.

Thumbnail Cache, Disk, and RAM

These settings allow you flexibility in dedicating drive space and RAM for thumbnails of the clips you will be working with. Thumbnails are very small icons that represent your captured or imported clips in a project and contain a still image showing what footage is in the clip. If you will be using many clips, you may want to set this as high as 1,000 K each, although the default settings of 512 K and 256 K respectively should be sufficient for normal circumstances. You can always adjust this amount later if you notice sluggish performance in previewing thumbnail clips.

The User Mode tab: Cutting Station vs. Standard

The second tab in the Preferences is for the User Mode (Figure 3-16). This setting allows you to place limitations on the feature set available when working on a project. In the Stan-

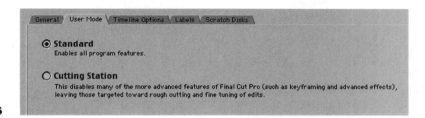

Figure 3-16

dard Mode setting, all of Final Cut Pro's functionality is enabled. In Cutting Station Mode, however, the editor is restricted to the following limitations in the editing process:

- A sequence is limited to only two video tracks and four audio tracks, in contrast to the 99 video or audio tracks normally possible within a sequence.
- There is a limit of seven filters, or effects, per sequence.
- Transitions, or effects that occur in the overlap of two clips in a sequence, are limited to the following four types: Oval Iris, Edge Wipe, Page Peel, and Rectangle Iris.
- Motion Effects are disabled, except where they are applied to every instance of a clip's use throughout the project, otherwise known as Global Use.
- Speed Settings are disabled.

Although it might seem unrealistic to want to limit the options available to the editor, there are actually a few situations in which limiting the possible options might be a good idea.

The most practical situation would be where the finished project will be exported as an Edit Decision List (EDL). An EDL is simply a spreadsheet database file that allows you to carry the timecode information from a project to another editing station. Most digital non-linear editing stations can accept EDLs from other editors, and Final Cut Pro is no exception. However, most EDL formats require that there be a limited number of tracks as well as transition types and effects. Since many editing stations have their own system for defining these things, only trouble can come from trying to include these things in a project that is destined to be exported to another system as an EDL. Cutting Station limits the editor's ability to include such problematic information.

Cutting Station is also valuable for logging and rough-cut stages of an edit. At the earliest stages of editing a project, logging footage and constructing a roughly-cut sequence requires only the most basic tools. Limiting the clutter of the toolset can focus the editor's concentration on the task at hand, which is to get the initial footage in the right order and with the right edit points.

As you begin your first projects with Final Cut Pro, leave the User Mode set for Standard Mode until you have worked with the application a while and are aware of what you are giving up by using the Cutting Station Mode. Further, keep in mind that changing between User Modes requires Final Cut Pro to be restarted to re-enable or disable any features affected by the change in mode.

The Timeline Options tab

The third tab in the Preferences is used to set the Timeline Options (Figure 3-17). In general, you will set these once and not alter them again until your editing needs change, and although they are mostly noncritical, they deserve mention. Since Final Cut Pro contains many separate windows and Desktop real estate is at a premium, the way you set your Timeline Options can optimize your workflow, prioritizing the way you see things in the Timeline as well as its defaults. The Timeline is the only truly linear part of Final Cut Pro. It is a beginning-to-end cutting window that can be customized according to your needs. Each of the following settings should be determined by your needs and the editing style you develop as you become familiar with the application.

Starting Timecode

By default, the Starting Timecode of the Timeline is set to 01:00:00:00. This is the timecode number that the first frame of the sequence will begin counting from. If you set this number to 00:00:00:00, the sequence will begin counting from that timecode value instead. For most users, changing this number will not have much effect other than the ability to organize numbers more efficiently. But for some users, it will provide a very convenient option when outputting finished video to tape. With some serial deck control systems and expensive production decks, you can preset the timecode recorded to tape by using the timecode value of the sequence being output to tape.

Most DV users do not have the option of presetting the timecode their recording deck writes to tape as it records the incoming video. Even though the incoming Firewire DV data stream may contain timecode information, most DV decks are not capable of reading that data and writing it to tape. Some feature-rich DV decks do have the capability of recording user preset timecode values. If you own such a deck, you can preset the timecode number that is recorded to tape—for example, beginning your tape with 00:58:00;00 instead of 00:00:00;00.

This functionality can make a great deal of difference if you are preparing tapes for a production house or broadcast facility that requires a specific beginning timecode value on submitted tapes. Often such facilities demand that tapes start with color bars and a test

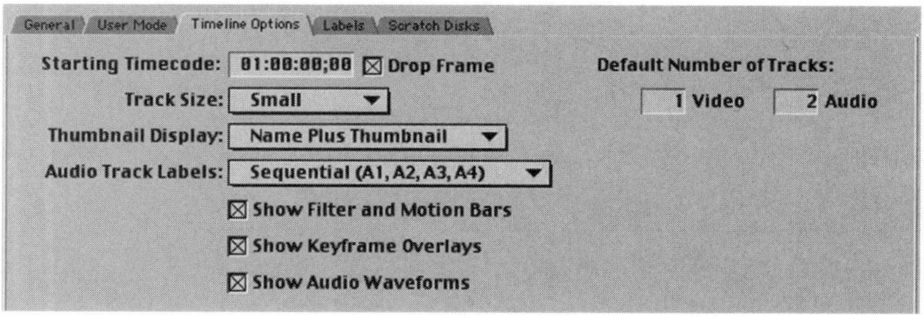

Figure 3-17

tone before the beginning of the actual edited piece, although they require the first frame of the edited piece's timecode to be at the top of the hour. Being able to make something like 00:58:00;00 as the timecode value of the first frame of recorded video can be an important option in professional production. Check the specifications of your DV deck to find out if it has this capability.

Unfortunately, the Starting Timecode information still cannot be used to preset the timecode recorded through Firewire onto DV tape. The ability to automatically preset the timecode value recorded to tape as you record out your video from Final Cut is available only with the use of serial deck control and with a deck that offers that feature. DV decks do not. If you need to produce such a tape with DV, you will have to use the Edit to Tape function described in the final chapter of this book.

Drop Frame Timecode Checkbox

Next to this setting is a checkbox labeled Drop Frame. This refers to whether Drop Frame or Non–Drop Frame Timecode numbers are used in the Timeline. NTSC DV Timecode is always Drop Frame whereas NTSC SMPTE Timecode, which uses the LTC and VITC analog systems, can be either Drop Frame or Non–Drop Frame. Although most Firewire users will never have to worry about the difference, those using the aforementioned DV converter box and a serial adapter for access to an analog deck's timecode might have to. PAL users of course have no need for Drop Frame settings, since PAL frame rates are based on a whole number.

Track Size

The next setting is for Track Size. The default track size of a new sequence can be preset using this option. "Small," the default setting, represents the most efficient trade-off between convenience and accessibility, although you can adjust this according to your needs and available workspace. Like many options in the Preferences, this setting affects only the default settings for new sequences. To change the appearance or features of an existing sequence, you will have to make the change in the Sequence Settings, which are addressed later on in this book. This Timeline option can also be adjusted directly from the Timeline.

Thumbnail Display

Thumbnail Display is an option that displays a thumbnail of the clip as it sits on the Timeline. This can be a very handy way of remembering which clip you are looking at on the Timeline. The options in this field are Name, which displays only the name of the clip; Name Plus Thumbnail, which displays the name of the clip plus one thumbnail image from the clip; and Filmstrip, which displays as many thumbnail images as can be fitted within the clip depending on the current time scale and range of the Timeline itself. The default is Name Plus Thumbnail.

Audio Track Labels

Audio Track Labels defines how the audio tracks are named, and it should reflect your audio sources. The default setting, Sequential, simply names each new audio track with the next largest number in order from the last (i.e., 1, 2, 3, 4) without respect to the track's association with any other track. Tracks are simply created in numerical order. Paired tracks, on the other hand, are labeled in pairs (i.e., 1A, 1B, 2A, 2B) to allow easy association between stereo audio tracks. Paired tracks may offer ease of organization if you use a lot of stereo audio tracks, but they can get confusing if you bring in a lot of monaural material. It does not affect the system's performance at all and should merely reflect your own working style.

Show Filter and Motion Bars

Show Filter and Motion Bars slightly expands each track in the Timeline and includes a small area underneath the clip. If any Effects Filters or Motion Effects are applied to the clip, they will be identified with a colored bar. A green bar denotes that an Effects Filter has been applied to the clip. A blue bar denotes that a Motion Effect has been applied to the clip. After a clip has been rendered, it can sometimes be difficult to remember whether or not it has been manipulated. Enabling this preference can make it easier to recognize such manipulation and avoid accidentally unrendering clips. This Timeline option can also be turned on directly from the Timeline.

Show Keyframe Overlays

Show Keyframe Overlays enables a special tool for each clip on the Timeline. This functionality is commonly known as *Rubberbanding*. When Keyframe Overlays is enabled, you can adjust the opacity, or transparency, of a video clip or the volume level of an audio clip simply by grabbing the line that appears in the clip on the Timeline and dragging it. This is a very easy way to create fades or quickly control audio levels without having to go into a clip window to apply an effect.

The keyframing of these lines is also possible, allowing you to change the level of opacity or volume over time. The term rubberbanding comes from the fact that the straight line created from keyframe to keyframe when rubberbanding resembles a rubber band stretched from nail to nail on a flat surface. Keyframing is thoroughly covered later in the book. This Timeline option can also be turned on directly from the Timeline.

Show Audio Waveforms

When this preference is enabled, clips on the timeline will display their audio levels as a waveform. Audio level, or the clip's loudness, is often displayed as a long continuous series of waves. The higher the peak and the lower the trough of each wave, the higher the level of audio. Since audio recording is not limited to the slow 29.97 or 25 frame rates of video, it's possible to be more precise in editing the beginning or end of a sound based on its waves.

Sometimes, the easiest way of finding footage in a clip on the Timeline is to have the Timeline display the Audio Waveform and look for the waveform spike that identifies your footage. You'd be surprised how easy it is to see the relationship between volume and different types of footage. Audio Waveforms can also be a very convenient marker for finding footage or for re-syncing tracks that have accidentally been unlinked. Unlike other Timeline preferences, this setting cannot be enabled by a button on the Timeline itself, but can be turned on and off using the keyboard shortcut Command-Option-W.

One might wonder why such a useful tool should not be enabled all the time. The simple answer is that the Audio Waveform must draw many thousands of separate Audio Waveforms on the Timeline when enabled. Each time you make a change, it must redraw all of these thousands of waves. The delay caused by redraw can take forever, even using the latest and fastest processors of the Macintosh line. In general, you want to keep this feature disabled unless you have a specific need for it.

Default Number of Tracks

The final setting for the Timeline Options is for the Default Number of Tracks. This will determine how many new tracks of video and audio are created when you create a new sequence. The default is set at one video track and two audio tracks, although you can set either for up to 99. In general, you should leave the default setting. Creating new tracks on a Final Cut Pro Timeline is as easy as dropping a clip where you want the new track to be. There is no reason to start out with extra tracks; simply create extras as they become necessary.

The Labels tab

The fourth tab in the General Preferences is called Labels (Figure 3-18). This tab allows you to customize the color-coding system available in Final Cut Pro. This is not a critical setting, but you may find it useful, especially for quickly organizing clips and sequences according to content.

Using the Label Preference Tab, you can change the color of the icon for clips and even sequences in the project Browser window. This color will also be displayed for the clip in the Timeline window if the Show Filter and Motion Bars box has been enabled for the Timeline. The only limitation of the Labels tab is that the five colors available for color coding cannot be changed, although you can change the text each color represents to any-

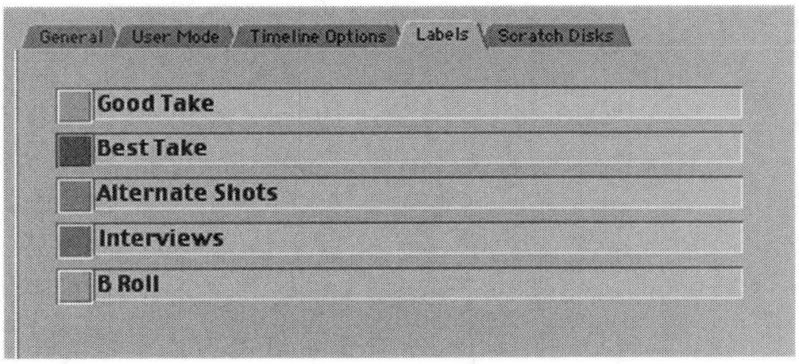

Figure 3-18

thing you want. That text will be displayed in a special contextual menu for applying color coding to clips and sequences.

The Scratch Disk Preferences tab: the issues for scratch disk assignment

The fifth and final tab of the Preferences is perhaps the most critical setting in Final Cut Pro. It is called Scratch Disk Preferences, and its name does not belie its importance to Final Cut Pro's healthy performance (Figure 3-19). It is in the Scratch Disk Preferences that you assign the location to which all the media you capture are written to. It is imperative that this location be properly set up and that it meet the standards we lay down for achieving acceptable performance. There is no getting around the need to visit this setting each time you work to make sure that you aren't jeopardizing your work.

Before describing the proper procedure for setting up the Scratch Disk tab, it is important to take a moment and examine the physical issues at play here and make sure that you have arranged your system in the best possible configuration. From the first section of this book, we recall that the video footage on the DV tape is actually a digital video file rather than an analog signal. Since the footage is already digital, our capture process through Firewire will be simply a matter of copying the digital video file from the camera or deck to the hard drive of the Macintosh.

We should also remember that our footage is already compressed in the camera with the DV codec and that its data rate is just over 3.6 megabytes per second. For the data to be

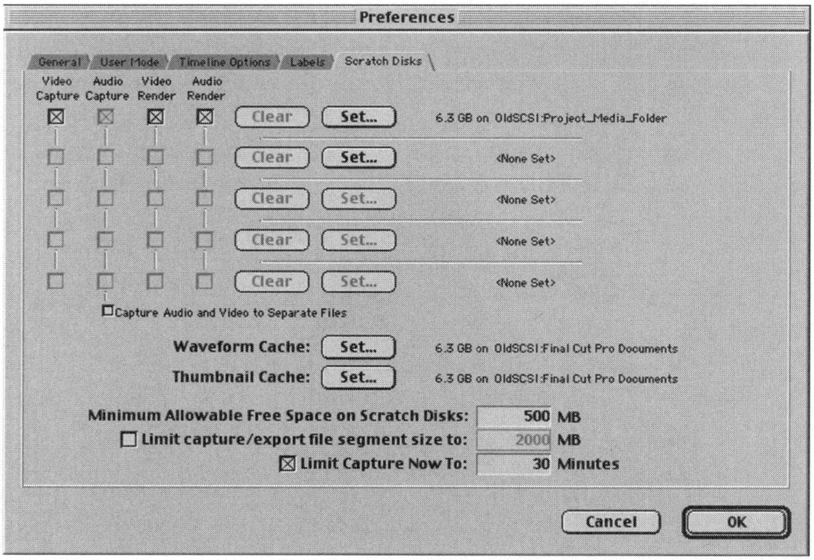

Figure 3-19

written to the drive correctly or played from the drive correctly after the capture process, the hardware configuration of our Macintosh will need to meet or exceed this 3.6 megabytes per second data rate.

In practice, this is relatively easy. The standard hard drives and buses on today's Macintosh far exceed the 3.6 megabytes per second data rates required. There's no need to spend lots of money on extremely fast drives to work with broadcast-quality video any more. However, the inexpensive drives we are talking about must be configured correctly, and you must access and assign them in Final Cut Pro appropriately if you expect them to deliver satisfactory performance. Like anything else, the devil is in the details, and if you pay attention to the details, you will find that you never run into data rate–related issues.

The first issue to be addressed is that it is highly recommended that you use at least two separate physical hard drives with Final Cut Pro. One of these hard drives is your startup drive. This is the drive that contains your System Folder, the Applications Folder, Final Cut Pro, and any other software you need to install on your system.

This hard drive is dedicated to software use only. Accordingly, there is no reason for this drive to be enormous. Large drives have no real inherent value over smaller ones other than the amount of data they can hold. For use as a startup drive, a 10-gigabyte hard drive will function just as well as a 40-gigabyte drive. For the purpose of efficiently organizing your drive resources, it is recommended that you choose the smallest drive on your system for this startup drive. If all of your drives are of equal capacity, simply use any of them for this dedicated startup drive.

The rest of the drives on your system will be dedicated as media drives. These media drives will be used only to capture and play back the digital files you bring in from the deck or the camera. You will also use these media drives for the render files, which are generated whenever you render an effect in Final Cut Pro.

Separating your drives into a startup drive and media drives will keep Final Cut Pro from fighting itself to maintain an acceptable data rate; consequently, it will play back your video files correctly. Although 3.6 megabytes per second is not an astonishingly high data rate, it does require that Final Cut Pro have unfettered access to the media files on the drive. That access can be compromised if a drive is busy accessing system files or application files at the same time it is trying to access your media files. Any interruption in the steady data rate of 3.6 megabytes per second can result in unacceptable performance and a break in capture or playback.

Another problem with using a startup drive as a media drive is that it may result in fragmentation of the drive and thus poor performance. A startup drive containing the System folder and applications contains thousands of small files that are used constantly by the system. As you work, you will also be saving files all the time that relate to your work. These files range in size from 1 kilobyte to a few megabytes or perhaps even 10 megabytes. Even though 10 megabytes seems large, in the scheme of things this is not an enormous file size. Since files are constantly being created, saved, thrown away, and so forth, the hard drive can quickly become littered with files scattered across a large relatively empty disk.

A frequent result of mixing a lot of these small files and large files on a single hard drive is that sometimes the larger files must be broken up into smaller pieces to fit on the drive. Although there is plenty of free space on the drive, it may not be contiguous (i.e., unbroken) space. For example, if you have 2 gigabytes of space left on a drive but that drive has a few small files written randomly around its area, you may have only a few much smaller contiguous spaces left on the drive. The space left may equal 2 gigabytes, but it is actually 1 gigabyte plus 250 megabytes plus 500 megabytes plus 250 megabytes.

In order to write or read at the highest level of efficiency, which is a good idea when you are trying to maintain high data rates like our 3.6 megabytes per second, you want to be able to write large files as contiguous files that are not broken up into smaller places to fit available drive space. In other words, you want to avoid fragmentation. Fragmentation is not merely a malady for startup drives. It occurs anytime you begin to reach the limits of capacity for a hard drive. Fragmentation can also occur on a dedicated media drive, even though there are few small files there. Whenever you are constantly writing and throwing away files, especially as you approach a drive's capacity, you risk disk fragmentation.

Consider this analogy. You are a parking lot attendant, and you have a certain number of parking spaces to use. At the beginning of the day, there are a lot of free parking spaces for you to choose from to park the incoming cars. It takes you almost no time to find a place to park a car. But as the lot fills up, the empty spaces get further and further apart and harder to find until it takes you several minutes to find a free space to put the next car. And the cars leaving the lot do not leave in an orderly fashion, but just free up spaces based on where they were originally parked. As the cars back up waiting to be parked in your lot, your performance as an attendant decreases.

Essentially, this metaphor applies to drives as well. As the drive fills up from capturing, the free spaces get harder and harder to piece together to write large video files to. As the drive reaches full capacity, it takes longer and longer to write or read the file. As a result, our possible data rate dips down below the necessary 3.6 megabytes per second. Correct playback and capture become impossible, and you start to get Dropped Frames warning messages.

One method that some people use to try to wring more drive space from their Macintoshes is to partition their single hard drive. Partitioning involves creating two apparent logical drives from a single physical hard drive. The single physical hard drive is formatted into two distinct sections that are treated by the system as two completely different hard drives. If you had a 40-gigabyte startup drive, this might seem like a clever way to reclaim much of the wasted startup drive space for media purposes, since any given startup drive uses at the absolute most 10 gigabytes.

The problem is that this method does not really increase data rate performance, since you are still dealing with a single physical hard drive, however dissected. While you may eliminate fragmentation as an issue, you will encounter spotty performance as your startup partition and the media partition fight for dominance. Using two separate physical hard drives on the other hand ensures that if the system or application files need to be accessed while media files are in use, there will be no competition between the two and there will be a lower probability of fragmentation resulting from the mixture of system or application and media files. There's nothing wrong with partitioning your startup drive or partitioning your media drive either, as long as you don't use the same physical drive for your startup drive and for your Scratch Disk media drive.

Learning to manage your media drives through the Scratch Disk Preferences will keep this from happening to you. You must learn to organize your media and not simply blindly accept the default media locations that appear when you start up Final Cut Pro. Although you may not experience problems initially when you begin working or even after many edit sessions, the time will come when sloppiness in maintaining your media will come back to haunt you. Those situations can be easily avoided if you begin work with your media resources organized and under control.

If you do not have a second physical hard drive as you begin work, this does not mean that you cannot work. Indeed, Final Cut Pro will happily accept whatever Scratch Disk Preferences you give it. But be warned that you can expect problematic performance if you do not use a dedicated media drive and a dedicated startup/system drive. Although a few hundred dollars for a second drive may seem like a big expense, a dysfunctional system that does not live up to its potential can be more expensive, particularly if you are having a go at professional video production for pay.

To properly set your Scratch Disk Preferences, take a look at the top of the Scratch Disk Preferences tab. From the left of the tab you will see four check boxes, two buttons, and a bit of text. The four check boxes are labeled Video Capture, Audio Capture, Video Render, and Audio Render. The two buttons are labeled Clear and Set. The information to the right of these items is a number describing the amount of free space that currently exists on the drive, followed by the directory path to that drive. The default drive assignment when you install Final Cut Pro is the startup drive, and it should be changed before you begin work. You will see that there are a total of five lines of check boxes and buttons, which means that you can assign up to five different media drives or partitions.

When you capture or render media, the check boxes on the left send the designated media files to the drive location you specify when setting the Scratch Disk. If you are using Firewire and DV, you should always have all four enabled for each Scratch Disk location you assign. You should be using the same drive and folder for not only the video and audio files you capture through Firewire, but also the video and audio files that you generate when rendering footage in Final Cut Pro. If you assign more than one drive or folder for your media, make sure to enable each check box for each assigned location.

The Clear and Set buttons allow you to enter and change the media locations when necessary. Clicking Clear will remove the current media location from the right and replace it with the message <None Set>. The only exception to this is that there must always be at least one media location assigned, so if you have only one line assigned, the Clear button will be grayed out and unavailable.

Scratch disk assignment: Follow the next steps precisely

STEP 10
Click the Set button (Figure 3-20). You will be immediately be confronted by a dialog box asking you to select the appropriate drive or folder for designation as the Scratch Disk. Click on the Desktop button in the box to move your location to the Desktop level.

STEP 11
Once you are at the Desktop level in the dialog box, all the available drives and partitions should be listed. Find the drive you intend to use as your media drive and double-click the name.

STEP 12
The location menu bar at the top of the dialog box should now show that you are inside the media drive. Now hit the New Folder button and name the folder "Project_Media_Folder" or something equally project specific and distinctive. Click the Create button.

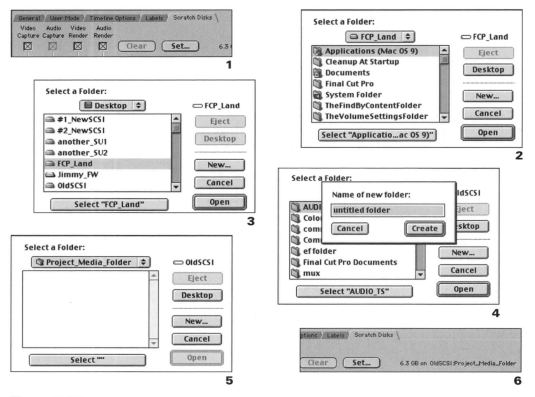

Figure 3-20

STEP 13

After clicking the Create button, you will find that the location bar has moved inside of the "Project_Media_Folder" which is, in turn, inside the media drive on the Desktop of your system. To see this graphically, click on the location bar drop-down menu at the top of the dialog box and you should see the directory order from Desktop all the way down to "Project_Media_Folder."

STEP 14

From inside the "Project_Media_Folder," hit the Select " " button at the bottom of the dialog box. Immediately after hitting this button, you will be returned to the Scratch Disk Preferences.

Take a look at the information next to the Set button, and you should see the amount of free space on your media drive displayed in gigabytes (GB), followed by the exact directory path to the place in which you saved the folder.

When you name your media drive and your project folder, make sure to give them distinctive names. Final Cut Pro can miss the difference between "Media1" and "Media2," for example. A better system is to actually number the drives, using a number at the beginning of the drive's name (e.g., "1Media" and "2Media"). If you have more than one type of drive on your system, such as both ATA and SCSI drives, you may want to include such information in the drive name (e.g., "1_SCSI_10GB"), where 1 is the drive name, SCSI is the drive type, and 10 gigabytes is the capacity. Then if something goes wrong or a drive is missing from the Desktop, you know right away which one you're dealing with.

To rename a drive or partition, simply single click on its present name. When the name becomes highlighted, type in the new name, up to 31 characters. You cannot rename a drive or partition that is being shared. If the "single click and rename" method doesn't work, go to the Apple Menu, select the Control Panels submenu and choose the File Sharing Control Panel. If File Sharing is enabled, disable it. This will allow you to rename any drives or partitions. It will also close any network connections to other machines on your network if you have one, which is probably a good idea if you plan to capture or print video to tape.

Similarly, you will want to be careful in naming the folder that is at the end of the directory chain. "Project_Media _Folder" may be a good idea for the folder you establish on your first media drive, but you will want to name the next one something distinctly different. Once again, the sequential numbering system is probably the wisest and easiest to keep under control. Thus your media folders might be named "1ProjectMedia," "2ProjectMedia," etc. "Fred," "Sam," and "Jim" might be even better choices if you can keep up with your system.

Each drive that you have dedicated as a media drive will have one such project media folder on it. It may seem a little strange to create a folder on the media drive, but understanding how Final Cut Pro organizes its captured media resources will demonstrate how quickly media can get out of control and clarify how important it is to have media folders set up uniformly and consistently.

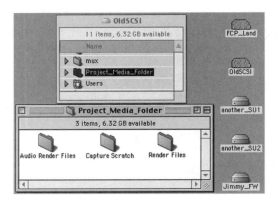

Figure 3-21

Take a look at the project media folders you just created on your media drive. When you assign a drive or folder as the Scratch Disk, Final Cut Pro creates three folders inside of it: a Capture folder (for Video and Audio, which Firewire captures together and considers as one file), a Render folder (for rendered video from the project), and an Audio Render folder (Figure 3-21). These Final Cut Pro–generated folders cannot be renamed without unlinking everything in your project.

When you go to capture video or audio, Final Cut Pro looks into the Scratch Disk folder or drive you assigned (or which has been left there by default if you did not set your preferences). If it does not find a folder whose name matches the name of the current project you are capturing media for, it creates a new folder for this purpose. Every time you capture media using a new project name, you will create a new media folder in the Capture folder in the disk or folder you have assigned as the Scratch Disk (Figure 3-22).

It is easy for this process to get out of hand, particularly if you are sharing your editing station with other users. The problem can be compounded by the fact that, when started up from the application alias, Final Cut Pro always opens the last project that was open. This means that unwary users can start up another person's project, capture a lot of media to the wrong drive, and then walk away without realizing the damage they have done to both themselves and their colleague.

These issues highlight the care that must be taken in the proper setup and maintenance of media resources in a Final Cut Pro editing station. Setting up the Scratch Disk is not difficult, and if you follow the preceding method for setting it up, you may never have to find out why the process is so necessary.

After you have set up a Scratch Disk folder for each media drive you want to use, you can get to work. As you return to Final Cut Pro to continue editing your project, select the media folder you created the first time rather than setting up a new folder each time you edit. Make sure you do so, though. One of Final Cut Pro's few serious flaws is that it cannot store individual project preferences (all those settings we just went through) within the project file. Any changes to the project preferences remain active until you change them back. This means that the previous editor's Scratch Disk assignment is still active when you begin work. If you skip changing the preferences, you could accidentally fill the other

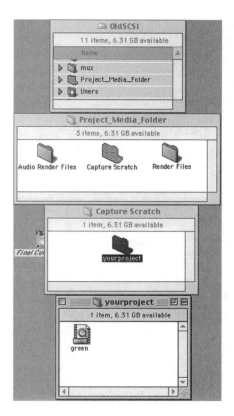

Figure 3-22

editor's Scratch Disk drive completely or accidentally assign the startup drive, creating a serious mess. Be diligent and avoid such problems altogether.

To wrap up the settings on the Scratch Disk tab, look underneath the Scratch Disk assignment columns. You will find six items that need to be addressed.

STEP 15

Capture Video and Audio to Separate Files

Disable the check box "Capture Audio and Video to Separate Files." For Firewire DV capture, audio and video should never be separated on capture (Figure 3-23). Separating audio and video is necessary only for capture cards with much higher data rates than Firewire DV capture. Separating audio and video files also requires a high-performance ATA or SCSI RAID disk array system to function correctly. Checking this box while capturing and playing back through Firewire will only result in problems.

☐ Capture Audio and Video to Separate Files

Figure 3-23

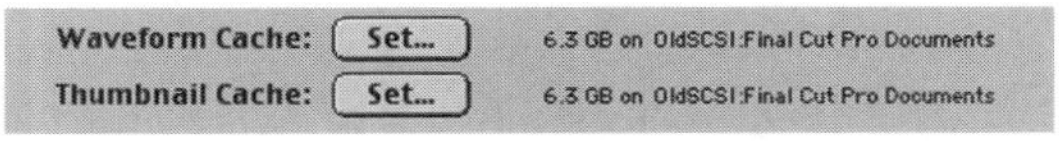

Figure 3-24

Waveform and Thumbnail Cache

Waveform and Thumbnail Cache locations may be set on this tab (Figure 3-24). Thumbnail and Waveform caches store thumbnail and waveform data for faster screen redraw performance. Final Cut Pro will function quite well with these caches defaulted to the startup drive, although choosing a fast media disk can dramatically improve the rubberbanding performance described earlier in the chapter. Some users have reported performance boosts through using high-speed disk solutions or RAM disks for these caches.

STEP 16

Minimum Allowable Free Space on Scratch Disks

Adjust the Minimum Allowable Free Space on Scratch Disks to 500 megabytes or 5 percent of your smallest media drive capacity, whichever is less. Minimum Allowable Free Space On Scratch Disks is a setting that helps to eliminate the problem of disk fragmentation mentioned earlier in the chapter (Figure 3-25). The field next to it allows you to enter an amount in megabytes. When a Scratch Disk fills up and has only the assigned value of remaining storage space, Final Cut Pro either switches to your next assigned Scratch Disk, or, if another Scratch Disk has not been assigned, aborts the capture. We can set a healthy buffer here to avoid reaching a situation in which an overcrowded drive might deliver compromised drive performance.

The default setting for this field is 10 megabytes, but this is far too low to provide any measure of security. There is much room for argument regarding the best Minimum Allowable setting, and it really depends on the size of the media drives in question. The best rule of thumb is that the Minimum Allowable should be set at either five percent of the drive's capacity or 500 megabytes, whichever is smaller. This means a setting of 500 megabytes for any drive larger than 10 gigabytes. A buffer space of 500 megabytes is large enough to protect a drive from fragmentation but doesn't waste more than about two minutes of drive space.

Limit Capture/Export File Segment Size To

This setting contains a couple of options, a couple of implications, and a bit of history. The options of the setting are to disable it completely or to enable it with a specified limitation to the size of the file either captured or exported using Final Cut Pro (Figure 3-25). If enabled, the default file size limitation is set for 2000 megabytes.

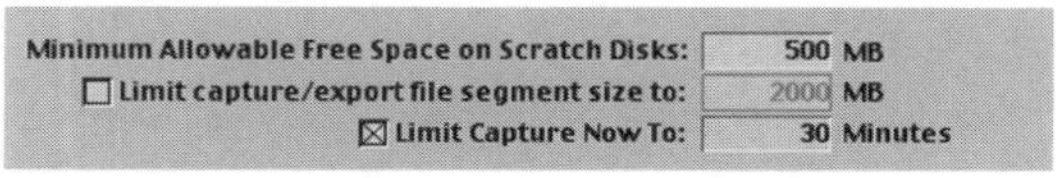

Figure 3-25

In the old days of pre–Mac OS 9 disk formatting, file sizes larger than 2 gigabytes were illegal and would result in a file error. In order for Final Cut Pro to capture DV clips that were longer than about ten minutes, the solution was to segment the capture files so that no individual file size reached this 2 gigabyte file size limit. The segments were linked together for correct playback in Final Cut Pro, but they were individual files on their own. This was not an issue only for Final Cut Pro; all video editing applications had to deal with the file size limit of 2 gigabytes in some proprietary way.

With the arrival Mac OS 9, the 2-gigabyte file size limitation was eliminated, allowing the capture or export of file sizes up to 2 terabytes. File segmentation is no longer necessary, and this option may be safely disabled, with an important rare exception: Files larger than 2 gigabytes, or >2 gig as they are sometimes referred to, require the drive to which they are written to be formatted with the Mac OS Extended Format. If you have a disk in your system that is formatted using the original Macintosh OS Standard Format, you cannot write files larger than 2 gigabytes to it. If you are receiving file errors whenever you capture or attempt to export DV-codec files longer than ten minutes, this may be the cause. Since the default disk format type in Erase Disk, Apple's quick format option, is set to Standard, this can be an easy mistake to make.

If you are not sure about the type of format of a disk, select the disk at the Desktop level, go to the File drop-down menu and select Get Info, or hit Command-I for the shortcut. The General Information will describe the format type as either Standard or Extended. If the drive shows Standard, you will want to reformat it to Extended at your earliest convenience using Apple's Drive Setup application or the Erase Disk command found in the Special drop-down menu in the Finder. Be aware that this reformatting will wipe the drive and irrevocably delete any data on it.

Limit Capture Now To

There are two basic ways to capture video and audio in a Final Cut Pro project. The preferred method, called Log and Capture, involves setting in and out points for the footage on the tape based on the Timecode. This is preferable because it limits what you capture to only the footage you actually need to work with. This confers a few benefits, not the least of which is the amount of disk space saved by not capturing minute after minute of footage you will not need.

There are times when logging clips before capture will be unnecessary or even impossible. If you only need to capture quick clips for instant use or you need entire tapes captured for multicamera syncing, logging tapes may not be the most efficient way to work. Those using A/D converter boxes that have no timecode in the DV data will find that they have no way of logging clips anyway, since such Firewire data streams contain no timecode information. Such users will want to opt for Capture Now. With Capture Now, Final Cut Pro begins capturing whatever is being fed to it from the Firewire input. Later on we will investigate that process thoroughly.

The setting Limit Capture Now To has bearing here (Figure 3-25). It can be disabled or set to limit the length of the capture to a specified number of minutes. You can use this setting to capture long clips without having to sit and watch your machine go through the process. Since there are issues with capturing clips that have timecode breaks, you can avoid the problem by starting a Capture Now process and setting a limit to the length of the capture such that the capture process ends before the footage on the tape does without your being present to stop it.

Another issue of relevance is that this avoids long hang-ups between the time you hit the Capture Now button and the time the capturing process actually begins. When Final Cut Pro captures a clip, it has to prepare a space for the clip on the Scratch Disk drive. If you are using the Log and Capture process, this is quick and simple. Final Cut Pro knows exactly how large a space will be necessary for the clip to be captured, since it knows both the data rate of the capture and the length of the capture.

If you are using Capture Now, Final Cut Pro has not been informed of the ultimate length of the capture and assumes that the resultant clip will use up all the available disk space on the Scratch Disk. It can take time to arrange this, which can lead to a rather long hang-up as you wait for Final Cut Pro to actually begin capturing. If you have very large Scratch Disks, you might have to wait for as long as 30 seconds, all the while watching your footage pass by with Final Cut Pro still not beginning the capture. Timing the beginning of the capture so that it begins where you want it to begin can be frustrating as well.

If you cannot log your clips and must rely on Capture Now, the Limit Capture Now To value can correct this problem. If you set it to, say, 5 minutes, Final Cut Pro will know that the captured clip will need only so much Scratch Disk space, and the capture process will begin almost instantly. The default time, if enabled, is 30 minutes, but you can choose a limit that suits your purposes.

The audio/video settings

Issues involved in the audio/video settings

Having completed the Preferences for the application, we now move on to the actual hardware/software configuration. Many of these settings are critical, and you should visit them at each edit session and make sure that they are appropriate for your situation. As with the Preferences, these settings are not stored in the project file and will simply be the same as the last time they were used.

Although the hardware on your Final Cut Pro station will not change much over time, you must inform Final Cut Pro correctly about such things as the type of data it will be receiving, whether or not timecode will be present, where you want Final Cut Pro to send the video to, and the like. The beauty of Final Cut Pro is that it is what we call *hardware agnostic*. This means that it will function more or less the same whether we have a $6000 capture card or are simply using the included Firewire ports for video capture. When you get ready to move to new hardware, Final Cut Pro is still there working for you.

The only catch is that we must accurately inform Final Cut Pro as to what hardware it is working with. Mistakes made here can have negative effects on system performance as well as cause impish problems that might not crop up until well into the editing process. Such late discoveries can mean tons of lost work, time, and frustration. An ounce of preventative medicine, good work habits, and respect for the process of setting up correctly can eliminate this altogether.

With the 2.0 version of the Final Cut Pro software, Apple has instituted a couple of new features that can streamline the process, assuming they are initially configured correctly. The most important inclusion is the development of the Easy Setup in the Audio/ Video Settings. Easy Setups are simply presets of all the Audio/Video Settings that are to be used at a given edit session. The Final Cut Pro software installs four presets that are a pretty good generic guess at the Audio/Video Settings for Firewire users.

Unfortunately, even within the closely standardized world of Final Cut Pro Firewire DV editing, there are subtle variations in the Audio/Video Settings that can have a major impact on the progress of your project. The presets that Final Cut Pro installs may not be accurate for your DV source footage, camera, deck or converter box, timeline needs, or video and audio monitoring situation. Clearly you need to correctly configure based on your specific needs. The rule of thumb is "what's on your tape must match your capture settings, and your capture settings must match your sequence settings."

Fortunately, this is an easy process. After configuring the Audio/Video Settings correctly, you will be able to create Easy Setups that reflect your configuration. At the initial configuration of the Audio/Video Settings, you will go through and customize for your machine. At the end of the process, you will take these customized settings and create a unique Easy Setup for future one-click configuration of your Final Cut Pro system.

You cannot inform Final Cut Pro about hardware that is not present. Upon launch, Final Cut Pro is able to detect video and audio hardware such as camera and decks at each of the Macintosh's ports and buses. Before going through the process of setting up your Audio/Video Settings, make sure that your camera, deck, or converter is plugged in, turned on and connected to the Macintosh via Firewire. If you do this and still do not see the device as an option as you step through the settings, quit Final Cut Pro, reboot your Macintosh, and then restart Final Cut Pro. That should cause Final Cut Pro to recognize and initialize the device.

The Summary tab

STEP 17

To set up the initial configuration, first go to the Edit drop-down menu and choose Audio/ Video Settings (Option-Command-Q). A window will pop up with five tabs (Figure 3-26). The first tab in this window is simply a summary of all the present settings in the other four tabs. Each of these displayed settings is actually a template that you can configure on

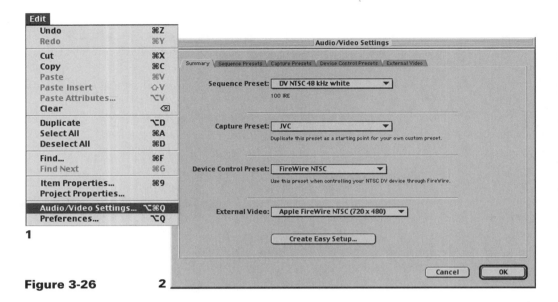

Figure 3-26   **2**

the following tabs as you cycle through them. Each section of the Summary tab is a drop-down bar that allows you to quickly select a preset from each of the tabs following the first. For the moment, skip the Summary tab until we have correctly set presets on the other four.

The Sequence Presets tab

Select the tab labeled Sequence Presets (Figure 3-27). Sequence is the term Final Cut Pro uses for individual editing timelines, as opposed to the Timeline window that sequences are used in. But in addition to being a linear timeline you edit items into, the sequence is also like a frame or window through which you watch what you have edited in the Canvas window. As you edit items onto the sequence Timeline, you'll see them appear in the Canvas window and out the Firewire tube to the video monitor.

For Final Cut Pro, the preview of the sequence in the Canvas window is the video rectangle, or frame, of the medium the sequence is intended to be used in—the Web, a television, or even printed out to film and run through a projector. The size of the frame is defined by the width and height of the medium you are preparing it for, based on the number of pixels, as well as by the shape of those pixels.

The Sequence Presets tab lets you specify how Final Cut Pro treats the items in a sequence. In the Sequence Presets, we are defining what a Sequence is for Final Cut Pro. Any clips you later place in the sequence will be forced to fit into that definition. In order to be processed correctly, the dimensions of the media clip must match the Sequence dimensions you insert it in. If the dimensions do not match, there will be two problems. First, the clip will have to be rendered in order to play back correctly. Next, the clip will not

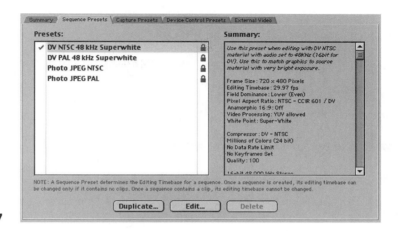

Figure 3-27

fit the canvas correctly, being either too small or too large, depending on the mismatch between sequence and clip.

In addition to this, if the audio sample rates for your clip do not match those of the sequence or are just incorrectly assigned, you will experience crackling distortions, speed variations, and/or progressive audio/video sync issues. If you experience any of these problems while editing, you can bet that you made an error along the way in establishing the sample rate for the Sequence or Capture Settings. Unlike video settings, you will not receive a red render bar warning that your audio is set incorrectly.

Size and Sample Rate are not the only ways in which settings can be mismatched. There are many factors involved in proper playback of material in a Sequence. Information such as the codec being used, the frame rate, and letterbox settings have to be included here. Settings must match precisely to ensure trouble-free performance. If you are told you need to render after inserting a clip on the Timeline prior to having applied any effects to it, then you can be pretty sure that you got the settings wrong somewhere along the way.

Why are the Sequence Presets, which have to do with the Timeline, grouped in with what are obviously hardware settings? This is because the sequence is directly in line with what goes out through Firewire to the camera or deck. If the sequence is configured incorrectly, then the device at the other end of the Firewire chain will not be able to operate correctly, resulting in poor performance or failure.

That said, it is easy to set up the Sequence Presets correctly for Firewire. On the left of the tab, you will see a white window with four preset names: DV NTSC 48 K Superwhite, DV PAL 48 K Superwhite, Photo JPEG NTSC, and Photo JPEG PAL. These are the pre-installed presets. If you click on them one by one, you will see the summary information of the preset in the box on the right update to reflect the differences between the presets. The information in the box on the right is all the information contained in the preset you select on the left.

It is possible that you might never need a preset differing from either the DV NTSC 48 KHz Superwhite or DV PAL 48 KHz Superwhite settings. These are the most generic settings. Since most DV cameras record audio at a standard 48 K audio sample rate, chances are that you will never have to use another setting. But, as the rule of thumb states, our audio and video settings must match the source tape. We will step through the process and set up a sample preset that uses the exact same settings except for audio that was recorded at 32 KHz in the camera. We will make this a new preset that we can select if we ever encounter a situation in which we must work with 32 KHz audio instead of 48 KHz.

Using Duplicate for a settings preset

STEP 18
Select either the DV NTSC 48 KHz Superwhite or DV PAL 48 KHz Superwhite preset. At the bottom of the window, click on the Duplicate button. Instead of starting from scratch with a new preset, we will use a preset very similar to the one we need and change the few differing settings. Be careful to hit Duplicate rather than Edit. We want to create a new preset using the DV NTSC 48 KHz Superwhite or DV PAL 48 KHz Superwhite preset as a template, not change the existing preset.

STEP 19
After clicking the Duplicate button, a Sequence Preset Editor window will appear (Figure 3-28). This window will allow us to specify the unique settings that our new 32 KHz sequence demands. The first field will show that the Sequence Preset is currently named "DV NTSC 48 KHz Superwhite copy" or "DV PAL 48 KHz Superwhite copy," depending on which video standard you have chosen. You will want to change this to "DV NTSC 32

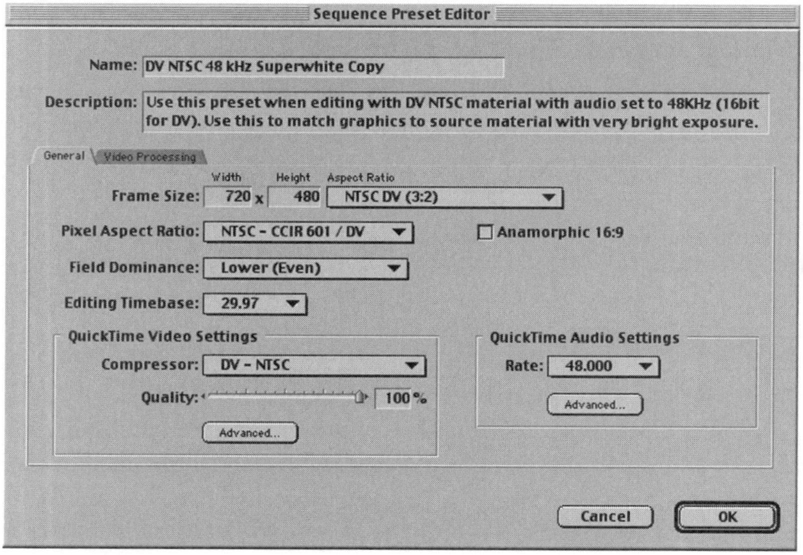

Figure 3-28

KHz Superwhite" or "DV PAL 32 KHz Superwhite," since in this example we will be setting up a preset for use with 32 KHz audio.

STEP 20

In the field immediately below this, type in a description of the preset. The description already present in the box outlines itself as "Use this preset when editing with DV (NTSC or PAL) material with audio set to 48 KHz (16bit for DV). Use this to match graphics to source material with very bright exposure." Simply change the words "48 KHz (16bit for DV)" to "32 KHz (12bit for DV)." The description field doesn't really affect anything in Final Cut Pro; it exists simply as a convenient description of the preset function.

Even though the rest of the information is set correctly with the exception of the audio sample rate, let's step through all the settings and make sure they conform to our Firewire DV editing needs.

STEP 21

Frame Size

The Frame Size setting for Firewire DV should be 720 and 480 for NTSC or 720 and 576 for PAL for width and height and DV NTSC (3:2) or CCIR601/DV PAL (5:4). This is measured by width and height in pixels or by its *aspect ratio*. You will see that you can specify an exact size by entering numbers in the fields, or you can change the aspect ratio using a drop-down bar. You should always set the Frame Size by using the drop-down bar and setting the aspect ratio, rather than simply typing in the numbers, since these Frame Sizes are standard for NTSC or PAL.

Aspect Ratio and Pixel Aspect Ratio

The aspect ratio describes the shape of the screen based on the relative comparison of width to height in pixels. Aspect ratios are most commonly used to differentiate between the wider rectangle of the film screen and the tighter rectangle of video. They are also used to describe the actual shape of the pixels themselves, since, in the digital realm, pixels may be square or non-square, a subject we will explore later in this book. There are other common ratios for larger High Definition Television and smaller Web multimedia frame sizes, and you will find them all available in this drop-down bar.

Firewire DV always uses non-square pixels. For NTSC, this means that the Frame Size should be set for DV NTSC (3:2). For PAL, this should be set for CCIR601/DV PAL (5:4). The pixel dimensions will be set automatically when you choose the correct Aspect Ratio from its drop-down bar.

STEP 22

For Firewire DV, Pixel Aspect Ratio should be set to NTSC-CCIR 601/DV or PAL-CCIR, depending on the video standard you use. Pixel Aspect Ratio may seem a little redundant after having set the Frame Size, since we already told Final Cut Pro what the shape of the frame is which is based on the pixel's shape. But we must also set this to determine the actual individual pixel shape as well. For Firewire DV, this should be set to NTSC-CCIR 601/DV or PAL-CCIR, depending on the video standard you use. Square should come into play only when the video is not to be sent out to Firewire video again.

STEP 23

Field Dominance

Anything using DV compression is always Lower Field Dominant. Field Dominance has to do with the way the lines are scanned in video. As we saw in the first section of this book, each frame of interlaced video is actually composed of two separate fields of video, each of which contains half of the lines of the complete video frame. The term Field Dominance refers to which of the two fields is scanned first. Although it would seem normal for the first field to always be scanned first, this is sometimes not the case. In fact, with Firewire DV systems and anywhere that the DV codec is used, the second field is actually scanned first. We call this Lower Field Dominant as opposed to Upper Field Dominant. This is also referred to as Second Field versus First Field and Even Field versus Odd Field. Either appellation is acceptable; different applications use differing nomenclature, but the implications are the same. For your Firewire DV setting, you always want to set this to Lower (Even).

STEP 24

Editing Timebase

For Firewire DV NTSC, the Editing Timebase will always be 29.97, and for Firewire DV PAL, it will always be 25. The Editing Timebase is the basic rate in frames per second that your sequences will be divided into. If you click on the drop-down menu, you will see choices from 15 FPS to 60 FPS. This must be set correctly. Final Cut Pro must know how many divisions of a second there are.

Because Final Cut Pro is not limited to any specific hardware configuration, it offers much flexibility in the specific frame rate. One can edit using 24 FPS, the native frame rate of film, just as easily as using 29.97 FPS, the native rate of NTSC video, or 25 FPS, the native rate of PAL video. Beyond this, your sequence can be divided up into lower rates for multimedia destinations or much higher ones for other specialized uses. The main thing to be aware of is that the frame rate must match the frame rate of the media in use. For Firewire DV NTSC, this will always be 29.97, and for Firewire DV PAL, this will always be 25.

STEP 25

Anamorphic 16x9 Checkbox

If you shot your footage in Anamorphic 16x9, you will want to enable this setting; if not, you will want to disable it. If at any point your footage appears to be uniformly stretched such that objects appear thinner than they actually are, you probably have set this incorrectly.

To the right of these settings is a check box for Anamorphic 16x9. Many cameras allow the shooter to use a different window size, referred to as widescreen. Although not exactly the same dimensions as widescreen cinema, Anamorphic 16x9 gives the feel of cinematic, letter-boxed framing, which many find aesthetically pleasing.

Unlike the idea of square and non-square pixels, Anamorphic 16x9 does not actually change the shape of the pixel to alter the screen shape. The footage is stretched in the camera, either through a lens or a digital process, and is then squeezed when brought into Final Cut Pro. The result is no change in the amount of information in the frame; the composition of the frame will be exactly the same whether16x9 is on or off. However, if the footage was shot in 16x9 and the sequence is not set to reflect this, the footage will appear stretched and distorted.

The QuickTime Video Settings

STEP 26

For Firewire DV media, the QuickTime Video Settings should be set for either DV-NTSC or DV-PAL, depending on the system in use. Quality should be set to 100 percent. This is where you define the codec to be used in a sequence. Hitting the Advanced button brings up the standard Compressor codec window (Figure 3-29).

With most codecs, the Compressor window is the location in which the settings for any particular codec can be tweaked for optimum efficiency. With the Apple DV codec, however, the compression is uniform, allowing no adjustments. Opening the Compressor window just shows us the same codec-choice drop-down bar and the quality slider. In a later chapter on exporting, we explore some of the other codecs and their settings, but for the Sequence and Capture Settings, we want to stick with the Apple DV codec.

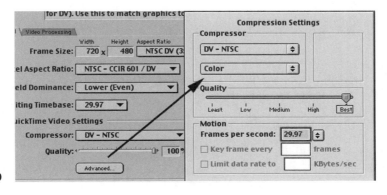

Figure 3-29

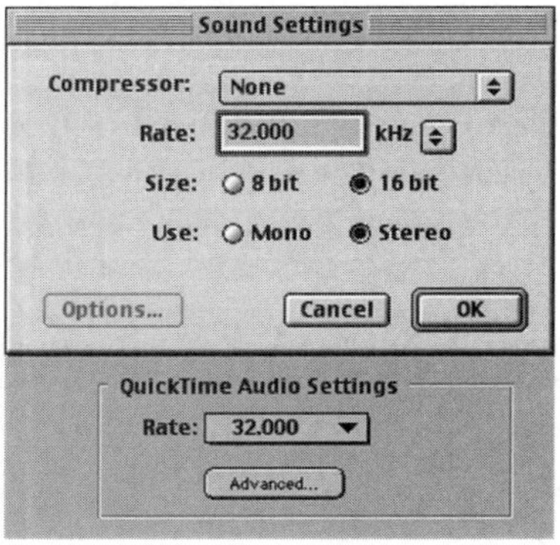

Figure 3-30

The QuickTime Audio Settings

Finally, the QuickTime Audio Settings must be addressed. This setting depends on the media you will be using in the sequence.

STEP 27

Since we are setting up a Sequence Preset for Firewire DV using 32 KHz sampled audio, change 48.000 to 32.000.

There are actually many more choices available than the three visible in the initial drop-down bar, which offers only the principal rates of 32 K, 44.1 K, and 48 K. The other rates are behind the Advanced button and allow many lower sample rates that can be useful when working with media destined for the Web or CD-ROM distribution (Figure 3-30). But remember that the sequence settings must match the media. If you want to work with lower sample rates for some reason, you will want to change the sample rate elsewhere, such as the Export function we will experiment with later. As long as you are working with Firewire DV, this sample rate should match the media.

Return to the Sequence Presets tab

STEP 28

After completing all the settings for the 32 K Sequence Preset copy, hit OK and go back to the original Sequence Preset window. Our new 32 K preset now appears in the presets window. Put the little gray check mark to the left of whichever preset you will need to use (Figure 3-31).

Remember that setting up the 32 KHz preset was only an exercise. If your audio on the DV tape was not recorded at 32 KHz, this would be the wrong preset. Only you can know which is appropriate based on what you have recorded on your tapes. If you aren't sure, play the tapes back in your camera or deck and check before you capture them.

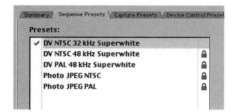

Figure 3-31

Capture Presets tab

Now move on to the Capture Presets tab (Figure 3-32). Since the Capture is the link between the media on your tapes and the media being available for editing in the sequence, capture settings really determine whether the media works correctly once in your system. If they are not set correctly, the media in your projects will misbehave, causing you no end of grief.

The Capture Preset setup window procedure is very similar to the Sequence Preset window. There are initial presets for generic DV Firewire capture with a sample rate settings for 48 K for NTSC and PAL. In addition, there is a Generic Capture Template for setting up presets that vary from the pre-installed DV Firewire presets. We will use the Generic Capture Template for setting up a DV Firewire Capture Preset at 32 KHz.

STEP 29

Using Duplicate for a Template

Select the Generic Capture Template, and hit the Duplicate button at the bottom of the window. This will open up the Capture Preset editor window, where you will define every setting for the new DV Firewire 32 KHz Capture Preset (Figure 3-33).

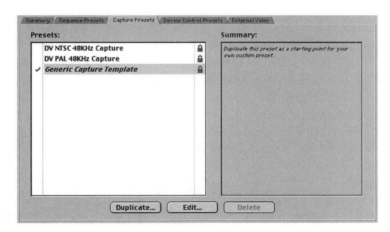

Figure 3-32

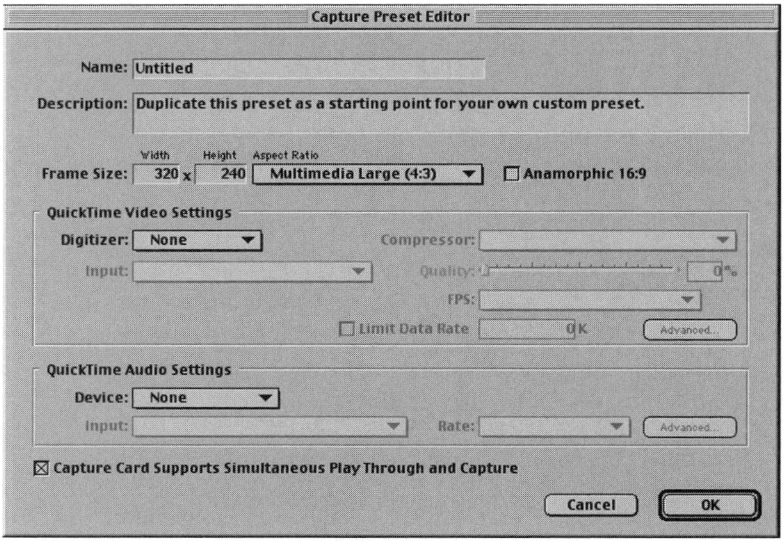

Figure 3-33

STEP 30

In the first field at the top of this window, name the preset "DV Firewire 32 KHz." In the Description field, type in a description identifying this preset as the DV Firewire Capture Settings for DV using 32 KHz audio.

STEP 31

Frame Size and Aspect Ratio

The next field and drop-down bar are for the Frame Size and Aspect Ratio of the media to be captured. Since this will be standard DV Firewire, select "NTSC DV (3:2)" or "CCIR 601/PAL (5:4)," depending on your video standard.

Anamorphic 16x9 Checkbox

Next to the Frame Size and Aspect Ratio setting is a checkbox for the use of Anamorphic 16:9. There are two different situations for the Anamorphic 16x9 setting in the Capture Preset. If you shot your footage in a DV camera that had a widescreen mode, Final Cut Pro can automatically detect this based on data in the DV stream coming from the DV tape. You do not have to tell Final Cut Pro that this sort of footage is 16x9 in the Capture settings.

On the other hand, if your Anamorphic 16x9 footage was created using a special lens on the camera, an option that potentially offers higher resolution, you must have this box checked when capturing the footage. Since such an Anamorphic stretch would be a physical lens effect and not a digital one, Final Cut Pro cannot see that it is stretched, so you have to inform it of this in the Capture Preset. You may want to create yet another Capture Setting preset named DV Firewire Widescreen that has this checked for convenient use when using such lens-distorted footage. If you do not have Anamorphic 16x9 checked when using such footage, or if you have it checked when not using such footage, you will end up with distorted results on capture.

STEP 32

If your footage is not Anamorphic 16x9, disable the box. If your Anamorphic 16x9 footage was created using a special lens on the camera, you must have this box checked when capturing the footage. If you shot your widescreen 16x9 footage in a DV camera using a digital widescreen mode setting, you do not have to enable the Anamorphic 16x9 checkbox.

The QuickTime Video Settings

Next, you will see a group of settings named QuickTime Video Settings (Figure 3-34). This is where you will define what the video source, or Digitizer, will be for the Capture Setting preset.

STEP 33

Unless you have other hardware on your system, you will be able to choose only between None and DV Video. If your DV camera or deck is not turned on and connected to the Firewire port of the Macintosh, you will not be given the option of DV Video, because Final Cut Pro can configure only hardware that it can communicate with.

After you have selected DV Video, you will see that other boxes in the QuickTime Video Settings have become available. Final Cut Pro can identify the brand name of the device it will capture from, so underneath DV Video it will likely display the DV deck's manufacturer (e.g., Sony DV or JVC DV).

To the right of these fields, you will find a Compressor area. You must identify for the Capture Setting Preset the codec you intend to use, as well as the quality setting and its frame rate. Since we are using DV Firewire capture, these settings are pretty easy.

STEP 34

The Compressor should be set for either DV NTSC or DV PAL, depending on the standard in use. The Frame Rate should be set for 29.97 or 25, depending on whether you are using NTSC or PAL. Quality should be set to 100 percent.

The Advanced button underneath the Compressor settings simply takes you to the standard codec settings dialog box we just visited in the Sequence Preset QuickTime Video

Figure 3-34

Settings. In most cases, you will never have to use this Advanced box until you add additional capture hardware or software to your system. Such hardware or software may have additional settings that must be configured for proper use. Firewire DV, however, requires no more configuration than simply choosing it as your Compressor.

The QuickTime Audio Settings

Underneath the QuickTime Video Settings, you will find the QuickTime Audio Settings (Figure 3-35). As with video, the audio capture settings must be the same as the media that is being captured to avoid problems later. The Template we are using here defaults to None for audio, and we will have to change it.

Unlike the QuickTime Video Settings, if your Firewire device is not plugged in and turned on, you will still see options for capturing audio. If you click on the Source drop-down bar, you will see two other options for capturing audio—Built-in and Internal CD.

"Built-in" refers to the fact that your Macintosh has a high quality 16-bit audio card built in to the motherboard. Not only can the Macintosh send high quality audio out to speakers for your listening pleasure, it can also capture audio through a properly configured USB microphone connection. This can be handy for getting sounds into Final Cut Pro very quickly without resorting to recording to tape and then capturing. The highest sample rate that the built-in sound card can capture audio with is 44.1 K, which is quite good, although somewhat lower in quality than the standard DV camera and deck sample rate of 48 K. You should not mix sample rates when working in Final Cut Pro sequences, but you will be provided a method later in the book for safely changing the sample rate of a file you capture or otherwise introduce to the project at 44.1 KHz.

"Internal CD "refers to the CD- or DVD-ROM drive in the Macintosh. Since the Macintosh is aware that the CD-ROM is potentially a source of digital audio files, it includes this CD or DVD-ROM drive as an option anywhere that QuickTime Audio is addressed in the system. However, since Final Cut Pro captures audio in a strictly linear fashion similar to a tape recorder, the Internal CD is more or less useless for our purposes as a capture setting. We will see later that it is much easier and faster to manually import audio tracks from a CD than to capture them through the built-in audio card.

STEP 35

For our Firewire DV capture, we want the QuickTime Audio Settings to be DV Audio. Once DV Audio has been selected, the other panels become active in the QuickTime Audio Settings. Immediately beneath the Device drop-down menu, you may choose an

Figure 3-35

input arrangement. The choices are First 2 Channels, Second 2 Channels, and Mix 4 Channels. This allows you to select the proper tracks to capture if you recorded the optional extra two tracks using 12-bit 32 KHz audio described in the first chapter in the section on digital audio. Select First 2 Channels if you are using 32 KHz but have not recorded extra tracks to the DV tape.

Next, you need to set the Rate, which refers to the audio sample rate of the material you will be capturing. You could easily set this at 32.000 and be done with it, but the Advanced button for the QuickTime Audio Settings actually deserves some examination. Since you can also set the sample rate in the Advanced settings, we will do so there.

STEP 36

Click on the Advanced button and you will be greeted by the QuickTime Audio dialog box. In the top left-hand corner, you will see a drop-down bar that reads either Sample or Source. Set it to Sample for the moment. When set to Sample, you have the option of setting the capture sample rate to 32.000, 44.100, or 48.000. Since we are creating a preset for 32 KHz audio, we will set this for 32.000.

Beware that because Final Cut Pro regards CDs as a digital audio source, 44.100 is selectable as a sample rate, even though most, if not all, DV cameras and decks limit the sample rates to 32 KHz or 48 KHz. It should be noted that some DV decks can in fact record audio using 44.100 sample rate, although this involves a nonstandard system of juggling the sample rate as the audio is played back, a process referred to as using unlocked audio. Under most circumstances, though, if the sample rate is set for 44.1 KHz and your media was not recorded at that sample rate, you will end up with video and audio synchronization problems.

Changing the drop-down bar from Sample to Source yields the same choices we had available in the Device drop-down bar in the QuickTime Audio Settings initially, so there is no reason to change them here. Looking to the right of the box in either Sample or Source, however, shows speaker settings (Figure 3-36). These settings control how Quick-Time interacts with the Macintosh's built-in audio and the speakers you have connected to the Macintosh.

Although it may seem strange, you will not be using the speakers that are connected directly to the Macintosh to monitor audio as you edit. When editing using Final Cut Pro

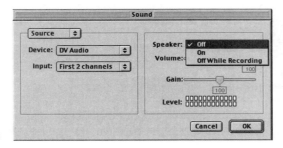

Figure 3-36

and Firewire DV, the video and audio are both sent out through Firewire to the deck or camera simultaneously. You will be listening to the audio through whatever speaker arrangement you have attached to the deck or camera—an audio mixer, an amplifier, or simply the same television monitor you are sending the video to. Unless you disable sending video out to Firewire, you will always be sending audio out to Firewire.

There is one exception to this rule. When you are actually capturing video and audio, it is possible to have the audio streaming to the Macintosh's speakers as the video and audio plays in the Capture Window. In all likelihood, you may hear an echo effect caused by the fact that you hear the Macintosh speakers play the same thing the deck or camera just played, delayed by a few thousand samples or a fraction of a second. This occurs because it takes a short time for the audio to pass from the deck through the Firewire bus and get processed on the motherboard before playing through the speakers. This delay is referred to as *latency*. A more critical issue is that capturing audio with this speaker setting set to On has in the past resulted in crackling audio in some cases. In any event, there is no reason to have it enabled, and the Speaker setting should be set to Off.

Capture Card Supports Simultaneous Play Through and Capture

Finally, at the bottom of the Capture Settings Preset window, there is a checkbox labeled Capture Card Supports Simultaneous Play Through and Capture. This setting is irrelevant for users of Firewire DV, because we are not using a capture card. Some third-party capture cards are able to capture video into Final Cut Pro and send out another video signal to a monitor simultaneously. Firewire DV users can watch their video as it plays back on the deck or camera it is being captured from, making the setting irrelevant. Enabling or disabling the box will make no difference.

STEP 37

Return to the Capture Presets Tab

After entering all of these settings, hit OK and return to the Capture Setting Preset window (Figure 3-37). You should now see your newly created DV Firewire 32 KHz Preset. If you wish to use this new 32 KHz preset, make sure that the little gray checkmark is next to its name. Otherwise, place it next to the DV 48 KHz preset for NTSC or PAL, depending on the video standard you use.

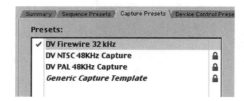

Figure 3-37

Device Control Presets

The next tab that you must set is for the Device Control Presets (Figure 3-38). This tab is concerned with Final Cut Pro's ability to remotely control the deck or camera you are capturing from. As we saw earlier, DV Firewire carries timecode and Device Control data along with the video and audio. Correctly configuring the Device Control Presets allows you to take advantage of this feature, adding both security and convenience to your working process.

As with other Preset tabs, Final Cut Pro preinstalls several default settings that work with many generic editing setups. The preinstalled presets are Firewire NTSC, Firewire PAL, and Non-Controllable Device. The Firewire NTSC and Firewire PAL presets will work with nearly any DV camera or deck that is approved for use with Final Cut Pro, although some devices may need a little tweaking for perfect operation.

Non-Controllable Device

Non-Controllable Device is a preset for working with a DV device that does not generate timecode, or is used in situations in which timecode could hamper your capture operations. DV converter boxes convert analog video and audio into DV data that can be captured using the Firewire connection and Final Cut Pro. However, most do not generate DV timecode or Device Control data along with that DV data stream. Unless you tell it not to in the Device Control Presets tab, Final Cut Pro always looks for timecode and Device Control data in the Firewire data stream prior to a capture. In order to capture video through Firewire from a source that does not generate timecode and Device Control data, you must tell Final Cut Pro to stop looking for it.

Suppose you are working with a deck or camera that does generate timecode and Device Control data but has lots of timecode breaks on the tape. When there is a space between two recorded segments on a tape in which nothing is recorded, we say that there is

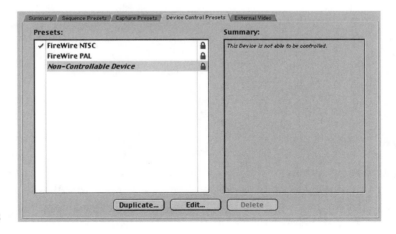

Figure 3-38

a timecode break on the tape. This is easy to identify on most consumer and prosumer DV cameras and decks, because when the device begins recording on a tape where there is no timecode already present, the deck or camera sets the timecode of the first frame recorded to 00:00:00:00. If recording is begun where timecode already exists, the device will pick up the timecode number that is present and continue from that number, a process called *regenerating timecode*.

You can therefore easily determine whether your timecode is clean (i.e., has no breaks) by watching the numbers as the tape plays. As long as the displayed timecode number never resets to 00:00:00:00 (or skips a large number of frames), you can be reasonably sure that your tape has no timecode breaks. In the final chapter of this book you will be shown a nearly foolproof way to avoid such timecode breaks on your tapes completely.

If you have timecode breaks that keep you from being able to capture specific areas of your material, turning off Device Control may provide some flexibility in overcoming the problem. In general, though, the best medicine is always prevention. Avoid timecode breaks, and always work with Device Control enabled. Taking advantage of timecode will make your editing experience much safer and more efficient. Remember that capturing footage with timecode breaks can have a consequence with the audio sample rate and the Sync Adjust Movie feature of the Preferences. Timecode breaks are just no good.

Using Non-Controllable Device as a Template

We will create a special preset based on a few of these tweaks. We will use the Non-Controllable Device as a template, since it will start us with a clean slate.

Although the presets for Firewire NTSC and PAL should work quite well for most users of Firewire in most situations, there are a few tweaks that could eliminate rare problems. Instead of using the generic default presets, we will set a bullet-proof preset based on the defaults that Final Cut Pro preinstalls.

STEP 38

Select the Non-Controllable Device preset and click the Duplicate button to create a copy for our specialized preset (Figure 3-39). In the Name field, type in "My DV Device" as the name of your preset. If you have more than one camera or deck, you can make up a preset for each one—for example, a "Sony Camera" and a "JVC Deck." In the Description field, type in a short description of which device the preset will be used with.

Protocol

The first choice you must make is Protocol. This drop-down bar contains a long list of Device Control Protocols that Final Cut Pro can access. Only the first two choices, Apple Firewire and Apple Firewire Basic, are Protocols that are accessed through Firewire. Although most DV decks and cameras are compatible with the Apple Firewire Protocol, some older cameras are more stable with the Apple Firewire Basic Protocol. Check the Apple Final Cut Pro Web site for details about which to use for your particular DV device.

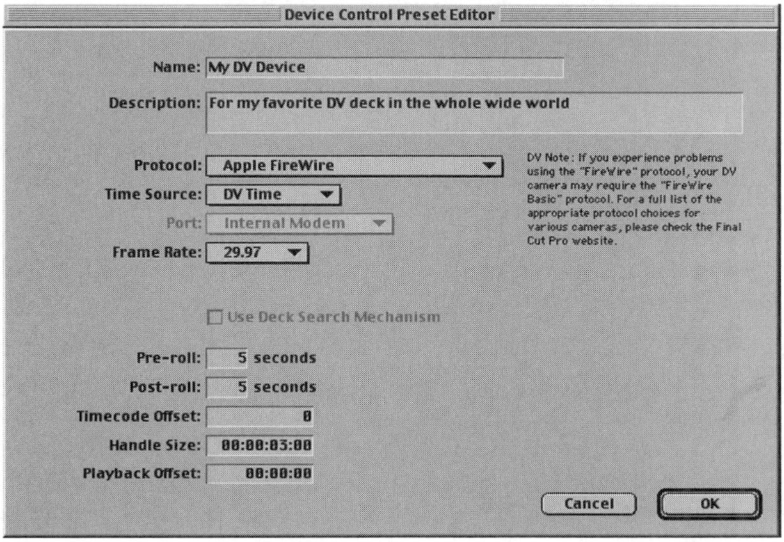

Figure 3-39

The remaining items on the drop-down bar are Device Control Protocols to control a deck through a serial connection. Before Firewire (and still, in many cases), nonlinear editors performed deck control by communicating with the device through a connection that is very similar to the way a computer connects with a phone modem or a printer. In fact, most expensive edit decks still require this sort of Protocol for deck control.

Although your Firewire DV device does not need a serial connection for Device Control, if you have other analog or digital equipment (e.g., Betacam or other professional formats that you want to use with your Final Cut Pro station), you may want to look into this option. If the deck has a serial connection and is controllable, a serial controller adapter could put that reliable, though Firewire-deficient, deck into action. To figure out which serial Protocol would be correct in such a situation, consult the user manual for the deck and serial adapter.

Time Source

The next drop-down bar you must configure is the Time Source. This refers to the type of timecode to be used by the Protocol you have selected. If you have chosen Apple Firewire or Apple Firewire Basic, your only choices will be Longitudinal Timecode (LTC) and DV Time. If your timecode was generated onto the tape by a DV camera or deck, select DV Time here.

If you chose Apple Firewire or Apple Firewire Basic for your Protocol, the Port drop-down menu should be grayed out. This is because the Port defines where the timecode data is to arrive in the Macintosh. Since Firewire Protocol works only with the Firewire ports, there is no optional port here. If you have selected one of the serial connection Protocols, you will see that there are two other ways that timecode and Device Control data can be communicated to the Macintosh: (1) the Internal Modem, which can be converted into a serial device connection, and (2) the optional Infrared Port.

Frame Rate

The next field is the Frame Rate. As we are working with timecode data as well as Device Control, it is imperative that this be set correctly. For NTSC, this should be set for 29.97. For PAL, it should be set for 25.

Pre-Roll and Post-Roll

The first two fields in the lower section of the Device Control Preset tab are the Pre-Roll and the Post-Roll. You can enter values in seconds in these fields. The default values are three seconds, but you will generally want to set these slightly longer. Pre-Roll and Post-Roll are required to make sure that the tape in the deck is crossing the tape head at the proper speed. This is referred to as *locking the deck servo*, because after a few moments of getting up to speed, the servo motors are locked in to the precise tape speed required for proper playback. We don't have to tell it what that speed is, but we do have to allow it the time to get there.

Whenever we are dealing with tapes that have timecode, whether analog or digital, it is imperative that the tape crosses the playhead at the proper speed. A certain amount of Pre-Roll allows the deck to make sure that it is playing back at the proper speed before any capturing begins. Thus, all decks utilizing timecode require a certain amount of Pre-Roll to function correctly. Unfortunately, different decks require differing amounts of time to reach servo lock, so your deck may need more than the three seconds of Pre-Roll that are applied in the default Preset. A value of five seconds is generally acceptable for most decks. If you receive "unable to lock deck servo" error messages later on when capturing, chances are that your Pre-Roll was not set long enough. Post-Roll should generally match Pre-Roll settings, although the setting is less critical since it refers to how long the tape continues to roll after a capture.

One common problem that Pre-Roll presents some users is that the Pre-Roll amount determines how close to the beginning of a tape you can capture from. If the beginning of the tape has the timecode value of 00:00:00:00 and your Pre-Roll value is set to five seconds, the earliest footage you can capture with Device Control and timecode is 00:00:05:00. That means that the first five seconds of the tape are inaccessible, because there wouldn't be enough time for a five-second Pre-Roll before capturing begins.

The problem can be further compounded by timecode breaks resulting from sloppy recording techniques. As we said earlier, timecode breaks reset the timecode clock such that a break in the footage on a tape gives the new footage a timecode value of 00:00:00:00. Important footage existing 30 minutes into your tape could end up as inaccessible as the first five seconds of your tape. The solution, as always, is preventative medicine. Always put a 5- to 10-second head and tail on the end of each and every shot to make sure that you always have a little buffer for your five seconds of Pre-Roll. Always be careful to begin recording on the tail of the last shot to avoid ending up with timecode breaks.

Timecode Offset

Timecode Offset is a very important though often ignored feature. If you care about the frame accuracy of your captures and edits, you should make sure that the Timecode Offset is set properly for your deck or camera. Although video and timecode data both stream into the Macintosh through the Firewire connection, there can be a slight difference between the times the video and the timecode data for a frame are passed through and written to disk. Although in many cases the difference is less than one frame, in some cases it could be off by several. This could result in the timecode numbers for clips that you want to capture actually being wrong even though you log them correctly.

Determining the Timecode Offset for your equipment is actually pretty easy, although it requires capturing a clip, which we have yet to do. After completing the lessons in this book, consult the Apple Final Cut Pro manual for a detailed description of the process. If, upon using the method described there, you determine that a Timecode Offset is necessary, simply return to your special Preset and enter one.

Handle Size

The next field, Handle Size, is actually a preference for having the system grab a little extra video during a capture than what you asked for when you logged timecode numbers for the clip. If you set a value for handles of 15 frames, Final Cut Pro will always capture 15 extra frames at the beginning and end of every clip you capture. Although it may seem like a waste of drive space, it is standard practice to capture digital video with at least 15-frame handles, and many editors use much higher settings of up to two or three seconds.

If you need a few extra frames once you have finished capturing and are starting to edit, you'll be able to access them without having to go back to your tape and recapture. This happens far more frequently than most of us are willing to admit. Despite the fact that the extra footage is available for editing purposes, it remains hidden, and the clip in question will always show the in and out points we logged until we change them. In addition, a transition requires handles when being inserted between two clips. Adding valid handles when you capture can make your editing possibilities a lot more flexible when the time comes.

Playback Offset

The final field is the Playback Offset. When you are finished editing your piece, you will Firewire it back out to DV tape to record it. There are a few different ways to do this, one of which uses a Device Control function called Edit To Tape. Edit To Tape allows you to simultaneously play back your edited program from the Timeline and remotely begin recording on to DV tape. Since this requires Device Control, any inconsistency between Deck and Final Cut Pro must be addressed.

Similar to the Timecode Offset, Playback Offset compensates for the fact that sometimes there may be a slight difference in when the playback of your edited piece begins playing in Final Cut Pro and when the Device-Controlled deck or camera begins recording.

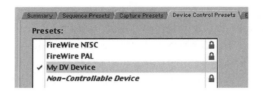

Figure 3-40

As with the Timecode Offset, you will need to determine whether you have an offset to compensate for and, if so, you will enter that value in this field.

Under some circumstances, Final Cut Pro can accidentally duplicate the first frame of video being sent out to Firewire during an Edit-to-Tape operation. At the completion of the lessons in this book, you will have tested Edit to Tape. If you detect this behavior, use the Playback Offset to negate the duplicated frames.

After you have entered these settings, hit OK and return to the Device Control Preset tab (Figure 3-40). If you have your new Preset selected on the left side of the tab, the box in the right side of the tab will show the settings you established for it. Make sure that you have clicked the gray checkmark for the new preset and move to the next tab.

External Video

The last tab in the Audio/Video Settings is the External Video tab (Figure 3-41). This tab lets you determine how you will view your video as you edit in Final Cut Pro. There are two drop-down bars for this tab: View During Playback Using and View During Recording Using.

View During Playback Using

When you are editing in Final Cut Pro, you have several preview options: you can watch your progress on the computer screen, watch it on a video monitor connected to your DV deck/camera, or watch on both computer screen and video monitor simultaneously. Many

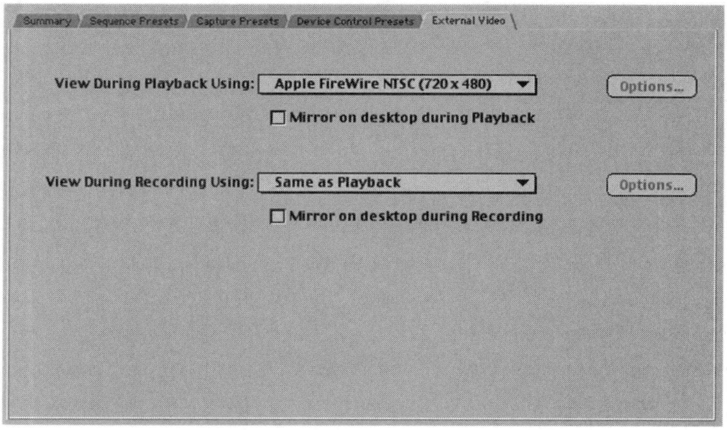

Figure 3-41

Figure 3-42

people find it easier to evaluate their edited work on a full-screen video monitor rather than on a small box on the computer screen. If you have your Firewire DV device connected to a video monitor, you can watch the results as you work there by selecting Apple Firewire NTSC or PAL.

There are other available options in this drop-down bar. If you have no Firewire device connected to the Macintosh, you will want the View During Playback Using set to None. If you do not, Final Cut Pro will constantly search for the Firewire device that it is supposed to be previewing video out to. On not finding a DV device, it will ask you where the device is, which can be very disruptive. Setting the bar to None eliminates this behavior.

If you have a second computer monitor and video card connected to your Macintosh, you can also select Desktop 2 as the preview screen (Figure 3-42). Selecting Desktop 2 on the bar replaces the second computer monitor's regular Desktop display with a low-resolution playback of the video being edited. This really pushes the processor hard, and the display has very low resolution, but it will work in a pinch.

Mirror Onto Desktop During Playback

Clicking on the checkbox labeled Mirror Onto Desktop During Playback has the system play back video on both computer monitor and out to Firewire simultaneously. Be aware that on some older and slower Macintosh models such as the Blue and White G3, you may take a performance hit by attempting to do this, resulting in a Dropped Frames error message. This is not the end of the world; it just means that your system cannot process and deliver video to both places simultaneously. Most new Macintoshes are powerful enough to deliver this performance though, and if you are receiving this message, you may need to review and make sure the rest of your system configuration and settings are optimized for best performance.

View During Recording Using

The View During Recording Using drop-down bar is very similar to the previous drop-down bar with the exception that it concerns where video is sent while outputting your Timeline to DV tape. If you have other hardware in your system such as a real-time card or a capture card, you can choose to send video to that while you are simultaneously recording your finished product to tape. The choices are Same as Playback (which simply uses the same setting as View During Playback Using), None, Apple Firewire NTSC, and PAL and Rendered Frames.

If you are using your DV deck or camera for editing and recording to tape purposes, you can simply leave this set for either Same as Playback or Apple Firewire NTSC or PAL. As with View During Playback Using, you may set the Mirror to Desktop during Recording to display on the computer screen as it records out to tape. Once again, be aware that on some systems performance may be compromised by Mirror to Desktop. Mirror to Desktop is somewhat gratuitous anyway when you are standing next to a full-size video monitor playing back the exact same thing.

Creating the Easy Setup: After this initial configuration

Return to the Summary tab

After completing the last tab, return to the Summary tab (Figure 3-43). Now that you have set each tab for a preset that you want to use, your selections should be reflected in each of the drop-down bars on the tab. If you have accidentally enabled a preset you don't want to use, such as activating the DV 32 K Sequence Preset or Capture Preset that we created as an exercise, you can change it here quickly without having to go through the tabs again. Simply select the preset you want enabled by clicking on the drop-down bar for the preset type you need to change.

Create an Easy Setup

To make this even easier, we will now create an Easy Setup that automates the whole process so that when you start an editing session, you need choose only one drop-down bar item rather than going through the whole Audio/Video Settings list again. Although you still need to check your Scratch Disk Preferences each time you begin working, you will have to set only one menu bar item to get all the Audio/Video Settings in order.

In the Summary tab, make sure that you have selected the presets you want to use for your general working situation. You can make as many Easy Setups as you want, so go through one at a time and create new ones based on every variable you can think of—for example, using Sony camera, using a JVC DV deck, or using no DV device at all. When you go to edit, you'll be able to easily switch between them.

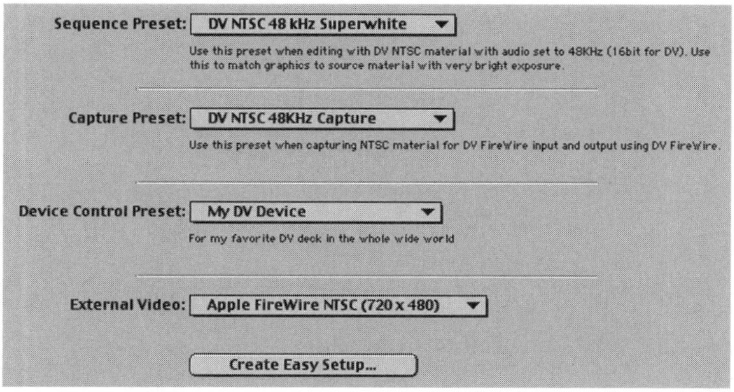

Figure 3-43

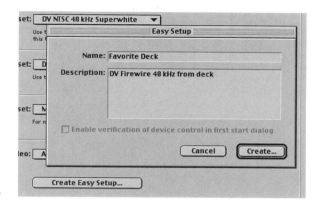

Figure 3-44

Look to the bottom of the Summary tab and locate the Create Easy Setup button. Click this button to bring up the Easy Setup dialog box (Figure 3-44). Type in a distinct name for the Easy Setup as well as a thorough description of it, including the particular hardware or software issues that make it necessary.

Click the Create button and you will be taken to a dialog box that requires you to save the Easy Setup (Figure 3-45). Make sure that this dialog box is set to save the Easy Setup file in the proper location, which is actually a folder inside the Final Cut Pro Data folder in the Preferences folder of the System Folder (Figure 3-46). If the file is not located in the proper folder, Final Cut Pro will not be able to find it when you wish to invoke it. Save the Easy Setup file using the same name you used in creating it and click Save.

To see your handiwork, go to the Edit drop-down menu and select Easy Setup or simply hit Control-Q for the shortcut (Figure 3-47). The dialog box that pops up will have a drop-down bar that includes your new custom Easy Setup. Changing your entire Audio/Video Setting configuration is now as simple as one Control-Q keystroke and one drop-down bar choice. Continue creating Easy Setups until you have every possible configuration covered.

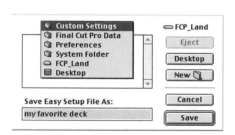

Figure 3-45

Figure 3-46

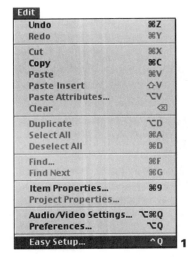

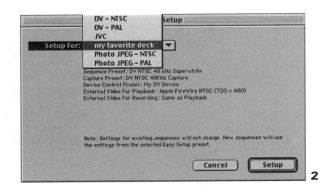

Figure 3-47

After the initial configuration—What you do and don't have to do every time

After you have completed the initial configuration of all the Preferences and Audio/Video Settings, you needn't go through the whole painstaking process again every time you begin editing, but obviously there are a few things you really need to revisit each time. What follows is a quick checklist to make sure that you get everything right before beginning work.

The checklist

STEP 1

When you begin an editing session, whenever possible start your project by double-clicking your project file rather than the application alias or icon. This keeps Final Cut Pro from potentially opening someone else's project. If you do start from the application icon, be sure to close projects you don't want to work with.

STEP 2

Go to the Preferences in the Edit drop-down menu. Glance through them to make sure that they are set the way you want them. Then take the time to set the Scratch Disk tab correctly, as described in this section.

STEP 3

Go to Easy Setup in the Edit drop-down menu and choose the Easy Setup that reflects your current working configuration; then move on to Project Setup.

4 Project Setup: Do It Right the First Time and Every Time

Why we do it right every time

Final Cut Pro is a software application. Applications run from inside of RAM, which is volatile memory. Thus when you shut down the application, everything you have done up to that point disappears. Of course if that were the limits of technology, computers would not be worth much to us. In order to work over the course of many different editing sessions, we must save our projects to a hard drive. This allows us to return to the project at a later time without having lost our previous work.

The act of saving files in anything from word processors to video games really isn't difficult. But it is imperative to understand the necessity of consistency and organization in the saving process. One of the most common mistakes new users make with nonlinear video editors is being inconsistent in project backup strategies. Either they do not back up their projects at all, or they do so in a way that becomes counterproductive in the event that they actually need to recover their work.

Don't let this happen to you. Properly saving your project and then backing it up on a regular basis takes no more effort than not doing so. However, the benefits are that you can always return to exactly where you want to work in the history of your project, no matter whether it's the last time you were editing or 12 o'clock noon, two weeks and four days ago before your client informed you of the correct spelling of that corporate vice-president's name. Not archiving your projects is a mistake you tend to make only once.

The proper method for creating and saving your new project is a step-by-step process, similar to the process of completing the Scratch Disk Preferences, and similarly, the process is very fast once you get the hang of it. You should build your editing session startup procedures based on the following step-by-step process. Doing so will consume no more than a minute of your time and will give you the security of a bullet-proof hardware, software, and backup strategy. You'll be able to forget about the annoying configuration pitfalls and spend your time doing the fun part: editing.

The process

Starting from the application alias

STEP 1

Start up Final Cut Pro from the application alias. If you already have a project you are returning to, you can start Final Cut Pro by double-clicking on the project file itself, which will cause Final Cut Pro to open only your project, not all others that were open when it was shut down last.

The first step in creating a new Final Cut Pro project is to start the application by doubling-clicking the application icon. Of course, if you have already worked through the preceding section of this book, then you have already opened the application at least once.

Closing all open projects

There is a reason for attention to this step in the project startup procedure. Unlike other applications, Final Cut Pro, when started from the application alias, always automatically opens all projects that were open when it was last shut down. This means that if you are working on more than one project at a time or if you share a workstation with other users, when you start up Final Cut Pro from the application alias, some project other than the one you want to work with could open up.

This is further complicated by the fact that Final Cut Pro allows for more than one open project at a time. It is easy to accidentally begin working with someone else's project open in the background. From there it is only a matter of time before you make accidental changes to a project you weren't even aware was open. The easiest way to avoid this is to develop the habit of closing all open projects when starting from the application alias.

STEP 2

Immediately go to the File drop-down menu and select Close Project. Then take a look at the Browser window and make sure that the only tab in this window is named Effects. If there are other Project tabs remaining, continue to select Close Project until you have closed all open projects.

It is important to remember that if there was no open project the last time that Final Cut Pro was shut down, it will still create a new open project named Untitled Project 1. Although this untitled project is very likely a blank slate, Final Cut Pro will allow the user to save using this name. For this reason, you cannot assume that anything named Untitled Project 1 is not someone else's hard work. Be safe; always immediately close any project opened at the startup of the application.

Complete the necessary Preferences and Audio/Video Settings

STEP 3

Go through the Preferences and Audio/Video Settings. While there is no project currently open, go through and set up the Preferences and Audio/Video Settings for your project, using the procedures laid out in the previous section of this book.

Remember that these settings were not saved with your project file and will simply be set to whatever they were the last time that the application was open. If someone else was working with a different configuration, it will still be there, regardless of the open project.

Creating a new project and saving it correctly

After completely checking your Preferences and Audio/Video Settings, you are ready to create, name, and save your project.

STEP 4

Go to the File drop-down menu and select New Project (Figure 4-1). On doing so, a couple of things will happen. First, if you set Prompt for Settings on New Sequence in the General Preferences, a Select Sequence Preset box will appear (Figure 4-2). Since all new projects are created with one new sequence, Final Cut Pro is asking you to specify its preset. Choose the appropriate preset based on your media.

After selecting a sequence preset, a few windows will pop up on the screen. We will explore all of these windows momentarily. Second, a new tab will appear in the Browser window named Untitled Project, followed by a number (Figure 4-3). The number is derived from Final Cut Pro's automatic naming convention. If you create multiple projects with the same name during the same editing session, Final Cut Pro will automatically add a sequential number to the end of the name in an effort to differentiate between them.

Figure 4-1

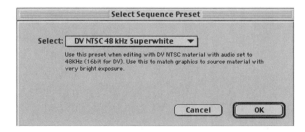

Figure 4-2

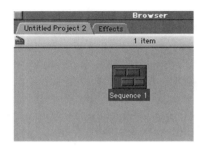

Figure 4-3

STEP 5

Now that there is a new open project, go back to the File drop-down menu and select Save Project As (Figure 4-4). In the dialog box that follows, click on the Desktop button to move the file location bar to the Desktop.

Unlike captured media files, the project file you save as you work should live on the startup drive. Hitting the Desktop button moves the dialog box save location to the Desktop of the startup drive. Your project file, which is not a media file, should be located on the startup drive in its own folder. For the same reasons that we don't want to capture our media files to the system and application drive, we don't want to save a project file to the media drive. There's nothing wrong with backing up extra archive copies of your project file to the media drive, but don't run your project from there.

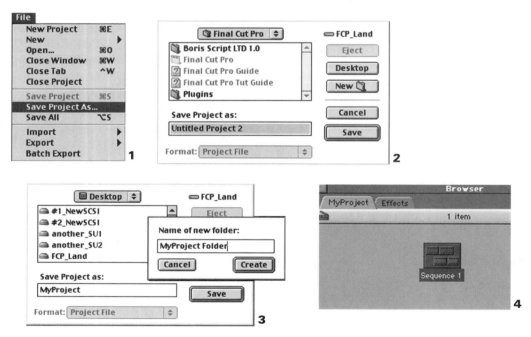

Figure 4-4

STEP 6

When at the Desktop level, click on the New Folder button. This will bring up another dialog box asking you to name and create a folder. Name the folder MyProject Folder and click Create. This will move the file location bar of the first dialog box into the MyProject Folder you just created on the Desktop. This will be the folder containing your actual project file. Although it is not required that you create such a container folder, you will find that it offers archiving benefits a little later in the process.

STEP 7

Now enter the name MyProject in the field and click Save. Your project file is now saved in its own folder on the Desktop of the startup drive.

Hiding Final Cut Pro/Showing the Desktop

Because there are so many windows in Final Cut Pro, sometimes it can be difficult to see items on the Desktop, especially if you are working with only one computer monitor. The Mac OS has a feature called Hide that allows you to make all the parts of an application invisible temporarily so that you can see the Desktop.

STEP 8

To Hide an application, make sure that it is active, meaning that its name is displayed in the Finder bar in the top right-hand corner of the screen (Figure 4-5). The Finder bar always displays which application is active. When Final Cut Pro is the active application, click this bar and select Hide Final Cut Pro.

STEP 9

You can also use the keyboard shortcut for hiding an application. Make sure the application is active, and then Option-click anywhere on the Desktop's background. All of Final Cut Pro's windows will disappear when you hide it. The application is still running; it is just out of sight to allow you an unobstructed view of the Desktop. To return to the application

Figure 4-5

Figure 4-6

Figure 4-7

that has been hidden, go back to the Application bar and find Final Cut Pro on the list (Figure 4-6). It will be slightly grayed out because it is hidden. Selecting it in the Finder bar will bring it back to the front again.

If you hide Final Cut Pro at this point, you will see that a new folder named MyProject Folder has appeared on the Desktop (Figure 4-7). Opening that folder will reveal the project file that you have just named and saved. Each time that you select Save from the File drop-down menu, this is the only file that will be updated. This method of saving your project will prevent you from committing the rookie mistake of accidentally saving your projects all over your multiple hard drives. It's easy to end up with different versions of the same project littering your system. Using this system will eliminate this possibility, as well as simplify the process of backing up.

Backing up; Archiving your project

How to do it

Finally, you will want to back up this folder containing your project to another disk. You can back this up to any disk other than your startup drive, although it is recommended that you back up to a removable format such as an Iomega Zip Disk. The purpose is to spread a couple of copies around so that when a system goes down, it doesn't take all of your project files with it.

STEP 10

Pop in an Iomega Zip disk. When the disk appears on the Desktop, double-click the Zip disk to open it up on the Desktop. Pick up the MyProject Folder and drag it over into the disk window and it will be copied onto the Zip disk (Figure 4-8).

STEP 11

After the short copying process, locate the MyProject Folder you have just copied over to it. Once you locate the folder, single-click the name of the folder.

STEP 12

A moment after being single-clicked, the folder's name will become highlighted, meaning that it can be renamed. Rename it with the date, time, and any other relevant information you want to include, using up to 31 characters. Remember that you are changing the name of the project *folder* and not the name of the archived project file itself.

Now, each time you finish with an edit session, pop in the archive Zip disk and repeat the process, each time changing the name of the newly copied project folder to reflect the time and date that it was archived. Eventually, you will have a Zip disk full of versions of your project at each stage of its development. If you want to go back to the way your project was on such and such a date, just go into your project archive Zip disk and copy the folder from that date to your Desktop. You may have to re-render a few things, or even

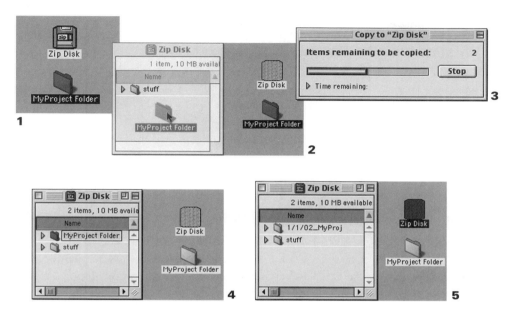

Figure 4-8

recapture if you've thrown away media, but at least you won't have to rebuild your project from scratch.

Its a good idea to do this at least every few hours, since the process only takes a moment, and three or four hours is enough time to radically change the entire nature of your project. Remember to remove your Zip disk after using it for this purpose; a mounted Zip disk can deliver a performance hit on your system even if you are not accessing it. Mount a Zip disk only for the purposes of backing up files, and then get it out of there.

Why we do it

There are two benefits to backing up this way. First and most obvious, such a backup will ensure that no matter what happens to your project files, you have a backed-up copy of your project. This is, after all, a computer, and files tend to corrupt at the most inopportune times. The more frequently you back up your project, the less you lose when your system crashes and takes your project file with it. Such things happen rarely, but they do happen.

A second benefit is less obvious. Backing up your project to another disk on a regular basis also allows you to step back into your editing process in a manner limited only by the frequency with which you back up to disk. As with most applications, you can perform Undos of actions while the application is running. But once you quit, you lose the ability to go back and change things to the way you had them previously. Unless you have saved prior versions of the project elsewhere, you can never return to an earlier state.

Imagine that you have edited an entire program. After finishing and copying it out to tape, you watch it and realize that there are serious problems with it that cannot be fixed easily by adjusting the sequence recorded to tape. It would be easier if there were a copy of the project that dated back to when you first made your mistake.

Archival backup procedures provide this option. The process is also referred to as Sequential Backup, because as you periodically archive copies of your project, they will be identified and organized by their date and other information you include in your naming system. If you ever need to step back into your editing progression, simply go back into your archived project files and locate the folder that is nearest to the date you need to return to.

The reason for Sequential Folder Saving instead of Sequential File Saving

You may be wondering why we created a folder for the project file and renamed it rather than simply renaming the project file itself. There is another method of Sequential Backup in which users save a new copy of their project with a different name each time they back up their project. Instead of copying the project file and folder to the Zip, this process saves a new copy of the project file. This means that every time you archive your project file, you would change the name of it, usually including the same information that we put in the archived project folder: name, date, time, etc.

There is nothing intrinsically wrong with this other method of archiving project files for most editing applications. In fact, in some ways the process is a little more robust than archiving the folder. The problem is that it's ultimately a bad idea for Final Cut Pro 2.0, because of the new way that Final Cut Pro manages its Scratch Disk resources.

The reason has to do with the way Final Cut Pro organizes its media after you capture it. When you set your Scratch Disk Preferences, you gave Final Cut Pro a location to save media files to: a Scratch Disk folder. When you did this, Final Cut Pro created three folders inside of the Scratch Disk folder: Capture Scratch, Render Files, and Audio Render Files. (See Figure 4-9.)

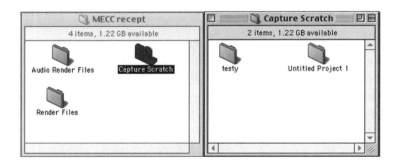

Figure 4-9

The wrinkle is that Final Cut Pro does not simply stream files into these folders. Inside these folders, Final Cut Pro creates yet another new folder for each differently named project that is opened while that location is set in the Scratch Disk Preferences. This means that if you open five different projects while the Scratch Disk is set for one folder, you will end up with five differently named folders inside that Scratch Disk's Capture Scratch folder.

Thus if we sequentially back up our project by renaming it each time, we would create a new project capture folder in our Scratch Disk folder each time we Saved As with a new name. If you are archiving regularly, as you should, this could lead to four or five new capture folders per editing session. If your project runs for a couple of weeks, you could end up with a real mess in your Scratch Disk folder.

Don't blame Final Cut Pro for this. It arranges things this way so that you will have more control over your media from inside of the application, which is a feature you will discover is well worth the project setup confusion. But it is important that you organize your Scratch Disk folders and utilize an archival backup strategy so that you can make sense of your media folders outside the application as well as from inside. Nothing is more painful than looking at a mess of files and folders and wondering what goes where and what can be safely thrown away. Use this archival system, and you will be spared such trauma.

The Windows of Final Cut Pro

After starting the new project and saving it correctly as described above, several windows will have opened up. These windows, in clockwise order from the top left, are the Viewer, the Canvas, the Audio Meters, the Toolbar, the Timeline, and the Browser windows (Figure 4-10). The arrangement you see when you start the project is the default arrangement, which you can customize any way you like. If you ever get mixed up and are missing windows that have somehow gone off screen, look under the Windows drop-down menu, select the Arrange sub-menu, and then choose Standard, which will return you to this exact state (Figure 4-11). If any of your windows are closed and you need to access them, they are available from the Windows drop-down menu as well.

Before beginning the discussion of Final Cut Pro windows, it is important to note that most of the windows utilize what is referred to as *tab architecture*. This means that one window may hold more than one item, just as one filing cabinet can hold more than one folder. At the top of each window, you will see the tabs. To switch back and forth between the different tabs in a window, simply click on the tab you want to bring up front.

Tabs can also be "broken away" from a window to create a new window occupied only by the "torn-away" tab. If you need to look at two tabs from the same window, simply grab the tab and drag it away from the window to create a new one. Most tabs can be restored to the windows from which they originated by being dragged back into place.

Figure 4-10

The Browser window

The Project tab

The Browser window is the top-level window of the application (Figure 4-12). It contains the Project tab that represents your entire Final Cut Pro project and all its contents. The three objects associated within the Project tab are the Clip, the Sequence, and the Bin. As you capture or import video, audio, and graphics clips, they will appear in this Project tab. When you create a new sequence to edit in the Timeline window, it will also appear in the

Figure 4-11

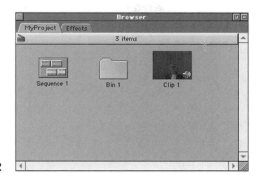

Figure 4-12

Project tab. There should initially be one sequence in the Project Tab, because all new projects are created with one sequence. Finally, anything that goes into the Project tab can be further organized into bins, which are simply organizational folders that can be created in the Project tab.

The two primary resources of a Final Cut Pro project are the Clip, which is the individual icon for the media we edit into the sequence Timeline, and the Sequence, which is the linear timeline into which we are editing the Clips. They are available here in the Project tab. They are readily organized through the Bin, the third object of the Project tab, a sort of folder that can store and organize Clips and Sequences according to whichever categorization you find convenient.

The Effects tab

There is another tab in the Project window that is always open, even when there are no open projects. This is the Effects tab (Figure 4-13). The Effects tab contains many different built-in effects such as transitions, color bars, special effects, audio filters, and titles. When you have clips available, you will be able to apply Effects to them from this tab simply by dragging and dropping the Effect you want to use on the Clip. We will return to the Effects tab later on in the book.

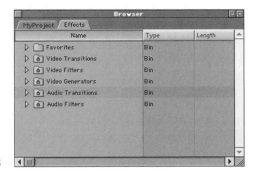

Figure 4-13

Figure 4-14

The Viewer window

The Viewer window is associated with clips (Figure 4-14). When you load a clip into the Viewer window, you can manipulate it in many ways. You can play it back and review the footage. You can assign new In and Out, or Edit, Points, defining how much of the clip is used when you edit it into a sequence, no matter how long the clip originally was. The tabs at the top allow you not only to adjust features of the video and audio, but also to apply special effects. These effects range from the motion-based suite, which alter size, position, scale, and many other attributes, to the effects filters, an enormous choice of special effects plug-ins that you can apply to your work.

The Canvas window

The next window is the Canvas window, which corresponds to the Sequence (Figure 4-15). Whereas the Viewer window allows you to load, prepare, and preview clips for editing into a sequence, the Canvas window allows you to preview sequence Timelines. Once again, there is a tab architecture to the Canvas, allowing you to switch back and forth between the different sequence Timelines in your project as you work.

Both the Viewer and the Canvas have video windows through which you view your progress as you edit. You can—and should—adjust these windows for optimum perfor-

Figure 4-15

mance. Although some machines may have no trouble with other display settings, you should generally set the frame size to 50 percent when playing back video. Having the video windows play at odd percentages can yield quirky performance, and 100 percent is usually too much to ask of the processor and absorbs too much screen real estate. The ideal video window size is 50 percent.

You should take care to make sure that no objects or other windows ever overlap a video window when attempting to play back video. Also make sure that the entire video window is visible and that there are no scroll bars along either the side or bottom of the window. If there is any cramping of the Viewer or Canvas window's style, you will get compromised performance or no performance at all from them. If you get Dropped Frames messages during playback, overlapping windows is one of the first things you should check for.

Because you can have more than one project open at a time, it is easy to edit using resources from many different projects. Clips or sequences from one project can easily be edited into another one by simply dragging them into place from one open project to another. Although this ability may seem minor, Final Cut Pro is the only video editing application that offers this very handy timesaving feature. Rather than spending time shutting down and starting up new projects or exporting or importing huge files, you simply open whichever projects you want to access, move materials to or from, and get right to work.

The Audio Meters

Going clockwise, the next window is the Audio Meters (Figure 4-16). These meters show rising or falling bars in the left and right channels based on the level of audio. This allows you to keep an eye on the decibel level of your audio, as well as troubleshoot your system's audio. If you don't hear audio, but you see levels, or if you hear nasty distortion but your Audio Meter levels are acceptable, it may be time to go in for a little system maintenance.

Those with a background in analog audio should take special notice of these Audio Meters, which are based on digital audio. With digital audio, 0dB is the highest possible audio level before clipping occurs. Unity, or the optimum reference level generally used by professionals, does not occur at zero as with common VU meters, but registers at around –12dB. Thus, we will use a –12 dB reference tone when mastering to tape later in the book.

At the time of this writing, Apple acknowledged that the Audio Meters may be creating performance problems with other areas of the application. If you experience mysterious problems in capture and playback that cannot be attributed to any other issues, try closing the Audio Meters and shutting down and restarting Final Cut Pro. It is possible that the Audio Meters are a source of trouble. Many users have reported that closing the Audio Meters completely eliminated their problematic performance. This obviously is not necessary if you are not yet experiencing problems. Don't go looking for trouble you don't yet have.

The Toolbar

The next window is the Toolbar (Figure 4-17). This strip contains all the useful little tools that come into play when editing. Selecting an icon on the Toolbar changes the mouse pointer and makes the tool available. Some of the tools have several different possible variations, or Toolsets, which you can enable by clicking on a tool and holding it down for an extra moment. These tools will be covered in the sections on editing and compositing later on in the book.

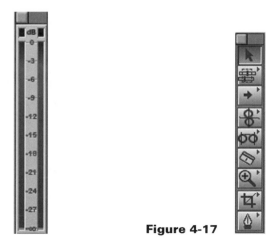

Figure 4-16   **Figure 4-17**

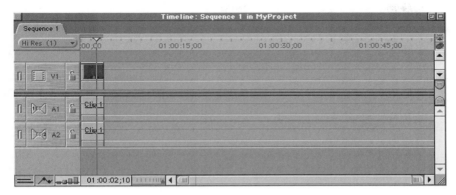

Figure 4-18

The Timeline

Finally, the last window is the Timeline (Figure 4-18). The Timeline is not a sequence in and of itself; it is merely a window in which you work with sequences. As with the Project, Viewer, and Canvas windows, the Timeline is based on a tab architecture so that you can have more than one sequence available in the Timeline window. To switch from one sequence to the next, simply click on the tab.

You will notice rather quickly that each of these windows is linked directly to the others. Adjustments made in one window will change the project *globally*, meaning that whenever you make a change in one window of the project that affects other windows, the change will be automatically reflected elsewhere. For instance, when you switch sequences in the Timeline window, the Canvas window tabs switch around to show the same sequence.

Moving a clip around the windows

The Color Bars and Tone Generator clip from the Effects tab

As an example of this, and to exhibit the way the windows work, let's utilize a little generator from the Effects tab that behaves like a clip and pass it back and forth between the windows. As we don't yet have any captured clips to work with, we'll use the Bars and Tone Clip that comes preinstalled with Final Cut Pro. Color Bars and Audio Test Tones are used professionally with any video production to allow technicians to calibrate their equipment for the best possible playback. Final Cut Pro includes a clip of them for use in your finished projects. They will come into play in the final chapter of the book concerning output to tape. Since the preinstalled Bars and Tone clip acts just like a regular captured clip, we'll use it for example footage.

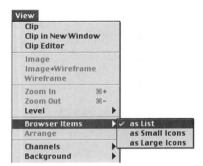

Figure 4-19

STEP 13

First, make sure that the Project window is set to be viewed as a list, rather than as icons. With the Project Tab in the Browser window active, go to the View drop-down menu (Figure 4-19). Select the Browser Items submenu and then choose As List.

If any of your Project or Effects tab items appear as icons or spread-out objects rather than a vertical list with information columns to the right, perform this action for each separate tab or bin of the Browser. For this exercise it is important to see the items as a list. Later on in the book, we will explore the functionality of the Large and Small Icons viewing modes.

Moving the clip from the Effects tab to the Project tab

STEP 14

Go to the Project window and click on the Effects tab (Figure 4-20). Grab the tab, drag it away from the Project window, and drop it so that it becomes its own window. A little way down the Effects tab, locate the Video Generators bin and double-click it. The Video Generators bin will open up in a new window.

STEP 15

In the Video Generators bin, locate the Bars and Tone clip. There will be two Bars and Tone Clips available—one for NTSC and one for PAL. Select the one that is correct for your video standard and drag it into your Project tab in the Project window.

Creating a bin and using it as an organizing tool

STEP 16

With the Project Tab active, go to the File drop-down menu and select New, followed by Bin in the submenu that appears (Figure 4-21). A new bin will appear in the Project tab.

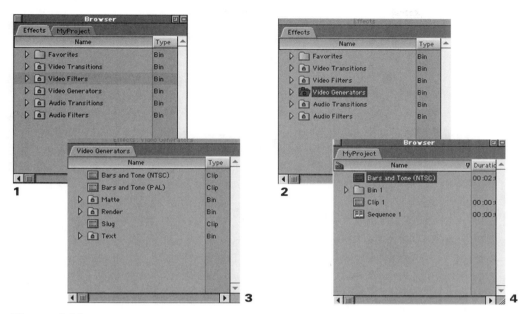

Figure 4-20

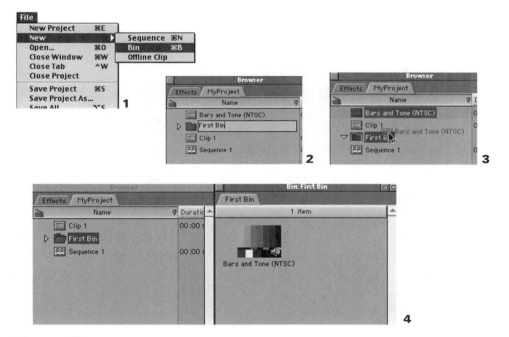

Figure 4-21

STEP 17

When a new bin is created, the name of the bin will be highlighted. Immediately type in the name First Bin. If the bin's name is not highlighted, single-click the default name Bin 1 and the name will become highlighted.

STEP 18

After renaming the bin, grab the Bars and Tone clip you just dragged from the Video Generators bin and drop it onto the First Bin. It will disappear inside. If you double-click the First Bin, it will open into its own window, showing you the Bars and Tone clip inside. You've just completed your first act of project organization!

Moving the clip from the Bin to the Viewer

Now let's open the Bars and Tone clip in the Viewer. This can be done in either of two ways.

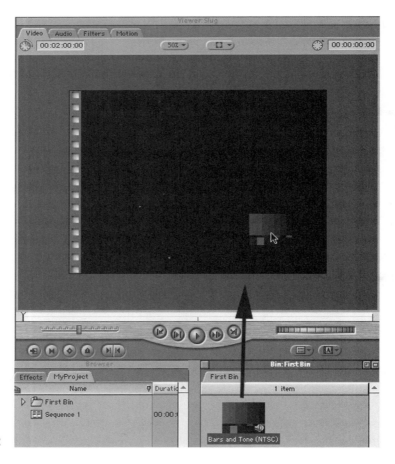

Figure 4-22

STEP 19

You can drag the clip physically from the First Bin into the Viewer window (Figure 4-22), or simply double-click the clip in the bin. When you do either, the title of the Viewer window will change, revealing that the clip loaded in the Viewer is the Bars and Tone clip from your project.

STEP 20

Hitting the play button in the Viewer window will begin playback of the clip. You will hear the 1 KHz audio test tone that accompanies the video color bars in the Bars and Tone Clip. A quick glance at the Audio Meter window will show that the 1 KHz audio test tone registers at −12 dB, exactly as it should.

Although the color bars should not and do not move, the little yellow indicator near the play button at the bottom of the Viewer window moves along as the clip plays. You can click anywhere in the current time indicator bar, or Playhead, and drag to move back and forth in the clip (Figure 4-23).

Clicking on the tabs at the top of the window reveals other aspects of the clip (Figure 4-24). The Audio tab shows a solid gray mass for levels, because the clip is a test tone with no variations in audio level. In other video clips you will work with, you will see a *wave-form*, or a graphic representation of how loud or soft a clip's audio is. You will also be able to adjust those levels on this tab.

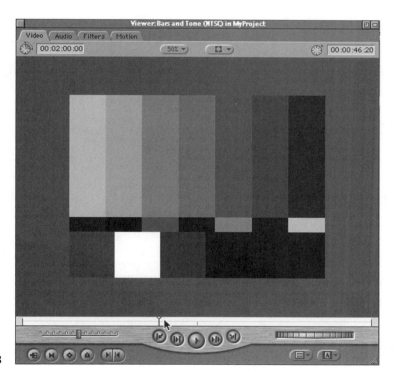

Figure 4-23

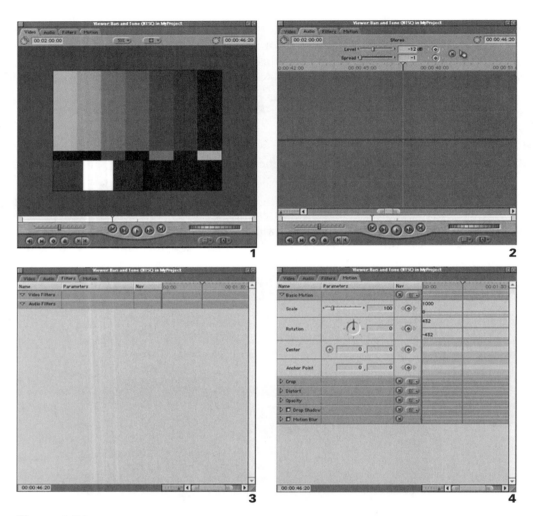

Figure 4-24

The Filters tab will currently have no Effects Filters listed, because we have not yet applied any. These will be covered later on in the chapter on Compositing and Special Effects.

The Motion tab will present many options, including the ability to change the size, position, rotation, and scale of the clip. It will also allow us to change these over time so that we can create professional-looking effects easily and naturally. We will also cover these in the Compositing and Special Effects chapter of this book; for now, leave this and the Filter tab alone.

Moving the clip from the Viewer Window to the Sequence

To move the Clip to the next window, we can once again choose from a few different techniques, each of which will produce the same result. The Canvas window always shows what is happening in the Timeline window, and vice versa. You can therefore move a clip into a sequence by one of the two following methods.

STEP 21

Click anywhere in the video window of the Viewer that has the Bars and Tone clip loaded and drag down into the beginning of the sequence on the Timeline window (Figure 4-25).

Figure 4-25

It will appear as a long linear clip there and stay positioned wherever you drop it. You could also:

STEP 22

Grab the Bars and Tone clip in the Viewer as in the previous example, but drag it instead into the Canvas window, holding the mouse button down momentarily (Figure 4-26). Doing so will reveal a graphic overlay of boxes proposing several different types of Edits. Final Cut Pro is asking in what way you wish to put the clip into the sequence. We will cover these in detail in the next chapter on Editing techniques, but for now, drop the clip onto the Overwrite box. The clip will appear in the sequence, once again, as a long linear strip in the Sequence. It will appear starting from wherever the sequence playhead was positioned right before the edit took place.

Having completed either of these actions, you will find that there is now a Bars and Tone clip in the sequence in the Timeline window and that the Canvas window displays the Bars and Tone Clip (Figure 4-27). Hitting play with either the Canvas Window or the Timeline active will show the current time indicator moving forward and, once again, we will hear the 1 KHz audio test tone.

These are not the limits of our ability to move clips around Final Cut Pro, merely the basic methods meant to show how the windows work together in a project. For instance, we could have simply dragged a clip directly from the bin to the sequence or the Canvas if we had no need to work with it in the Viewer. Similarly, we could actually open a sequence in the Viewer in order to apply effects to the whole thing rather than to the individual clips inside it. There's really very little limit on how you can work with clips inside your project.

But we get ahead of ourselves. The main thing to remember as you work through Final Cut Pro is that there is always more than one way to get something done. The options are

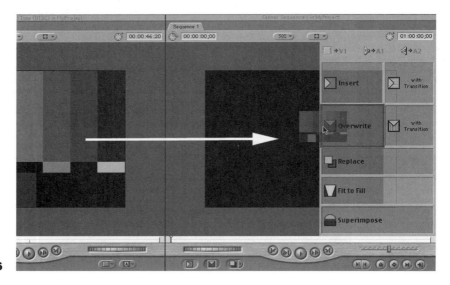

Figure 4-26

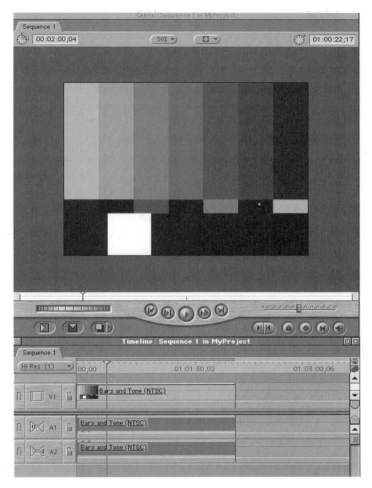

Figure 4-27

there to help you increase productivity and speed, based on the individual nature of your project and media. The more you work, the clearer this will become and the better judge you will be of which method is best for your needs.

What is Media and how do I get it into Final Cut Pro?

What is Media?

In order to edit, you need media to work with. Media is much more than just video or audio that you capture from Firewire. The term *media* refers to any content that you use in a project. This can range from video and audio captured from DV tape to audio imported from an audio CD. You can also use image files created in such applications as Adobe Photoshop or image sequences from compositing applications such as Adobe After Effects or Pinnacle Systems Commotion. You can even import special vector-based files such as the Flash file format. In short, you can use nearly anything you can create elsewhere.

Anything that you bring into a Final Cut Pro project to work with is media. This seemingly simple concept is an important one. There is a tendency among new users of nonlinear editing applications to think of the objects in use in the Timeline as the actual media files. This is an intuitive conclusion that unfortunately oversimplifies what is actually going on when you capture or import.

When you bring media into Final Cut Pro, you are merely allowing Final Cut Pro to access a file that exists somewhere on your hard drives. The clips that you see in the Project, Viewer and on the sequence Timeline are really indexes or references; they are simply icons that point to a media file on the hard drive. When you edit, you are telling Final Cut Pro how much of the original media file to display and where.

This icon is just textual data and is a part of your project file. It is used to determine how much of the indexed media file is accessed and the way it is to be displayed. When you use an effect or you trim a clip for use in a sequence, you are not really affecting the media file on your hard drive. You are simply telling Final Cut Pro how to display the media file in your project. Because the icon's text bits can direct the media to appear shortened or lengthened and arranged in any order or repeated endlessly, they are said to be nonlinear, hence the name nonlinear editor for the systems. This concept is also referred to as Non-Destructive Editing, because you are not making changes to the actual media. You are only changing the way that Final Cut Pro displays the media when it plays back.

This is to be contrasted with the venerable system of linear tape editing, in which one edits with the original tape media and must put each piece in the order it is to occur in playback. The convenience of nonlinear editing should be obvious compared to this. In linear editing, if you aren't happy with an edit you have made, you must go back and rerecord a new edit the way you want it using the original tapes. In addition, all the edits you have made following the edit you are changing usually must be re-edited as well.

With a nonlinear editor, on the other hand, we can simply arrange our clips on the sequence in any order that we like. Not happy with an edit? Just rearrange the clips. Since the clip is just referring back to a media file somewhere on your Scratch Disk, it makes no difference where it is displayed on the sequence; it will play back every time in just the way you desire it.

The reason for belaboring this point is that the safe and secure capturing process depends on our understanding of what media is, the distinction between our project clips and our actual media files and where those media files go when we capture or save them. We have already set our Scratch Disk Preference, which tells Final Cut Pro where to save the media files we capture. We know that we can take advantage of timecode and Device Control, which accompany the Firewire DV data stream. We will now explore how to take advantage of both of these to gather media to work with in a project.

How do we get it inside Final Cut Pro?

We have not actually brought any media into Final Cut Pro yet, so let's get right to it. There are three basic ways to bring media in: Log and Capture, Capture Now, and Import. These three methods have different requirements and come into play based on differing situations. Log and Capture requires not only that you use a deck or camera that Final Cut Pro can control, but also that you use timecode. Capture Now, on the other hand, simply captures whatever is currently streaming through the Firewire connection regardless of the presence or lack of timecode data or Device Control. Finally, Import allows you to bring media into your project that already exists on your system in some format that Final Cut Pro can understand.

Log and Capture

What is Log and Capture?

Logging and Capturing is, without exception, the safest and most reliable system of capture available to you, regardless of hardware or software configuration. When you Log clips using timecode, you create a placeholder for a yet-to-be-captured media file you will want to use. You are simply telling Final Cut Pro that a clip exists out there on timecode tape that corresponds to the timecode numbers you enter as you log.

Since the Final Cut Pro clip is a reference to timecode numbers as well as the actual captured media file, clips can exist as what we call Offline Clips. In other words, the offline clip is just a placeholder for media that you may not have captured yet, or that has disappeared and you want to capture again. A clip is just a malleable little icon that refers to a media file that existed before, exists now, or will exist once you have captured it.

Remembering our discussion of timecode will make it clear how powerful this concept is. If a clip contains all the data about the tape the timecode numbers originated from, referred to as its Reel Name, and the actual timecode numbers of a portion of that tape, there is no limit to how secure and easily contoured our project can be. If we want to give our project to another editor so that that editor can work with it, we don't need to save all those huge media files on a big hard drive to give to the editor. We can simply give the editor the project file and the original tapes and the editor can quickly and easily recapture the same media on his or her system. If we have lost our media for some reason, or if we want to revisit a project we worked on in the past but we have eliminated the media for, we can simply recapture it and get back to work. Nothing could be easier.

There are even more benefits to using this system, which you will discover as you get more involved in the editing process. The main thing to remember is that with Logging and Capturing, you are at the mercy of two things—your Project file and your original source tapes. Always maintain a strong consistent system of Project file backups, and be careful that you do not end up with timecode breaks on your source tapes.

Pre–Log and Capture techniques: The paper log and the Timecode Window burn

Before starting the Log and Capture process, it is important to mention that there are actually a couple of different ways to Log and Capture. The first is referred to as a Paper Log. This old but reliable system involves simply watching your tapes and writing down the timecode numbers of the clips you are going to want to use when editing.

Most shooters produce source tapes that include far more material than will be used by the editor. The amount of material shot on tape compared to the amount of material actually used in the final edited project is referred to as a shooting ratio. Because tape is so incredibly cheap, many shooters are inclined to let the tape run rather than stop and start for the important parts. Common shooting ratios for video production are often higher than 6 to 1, and produce tons of material that will only waste space on your drives if captured en masse.

As we saw when discussing digital video data rates, video files are huge, even when compressed down to the respectable 3.6 megabytes per second of DV. Five minutes of footage takes up over a gigabyte of hard drive storage space. If you were shooting at a roughly 5 to 1 ratio and you just captured an entire tape, this would mean that you stored 25 minutes, or 5 gigabytes of video on your drive when you only needed, at most, to store 1 gigabyte.

Clearly, this is not the way to work, even with today's enormous and inexpensive hard drive solutions. You will find that you still run out of drive space far too quickly. As with RAM, there is no such thing as too much hard drive space. Previewing your tapes and taking loose notes can easily eliminate the problem. Simply jot down the beginning and ending timecode numbers for each section of footage you know you need to capture and edit. Never fear that you might need something you forgot; if you ever need more footage, simply go back in and get that as well. In the final chapter of this book, you will be shown several methods of speeding up this process enormously using Batch Lists for generating paper logs and automating the Logging process.

What if you are collaborating with someone in a project and you both need access to the tapes to create paper logs? You have only one set of source tapes, and moving them around makes you a little nervous, not to mention the fact that constantly playing back and shuttling around on the source tapes increases the likelihood of damage to the tapes themselves.

The solution is another time-tested one. Create what is referred to as a Timecode Window Burn Dub. Most DV cameras will allow you to output the video and audio to analog VCRs. Indeed, your Final Cut Pro editing station will be based on this ability so that you can preview the editing on an NTSC or PAL video monitor. Most DV decks and cameras will also let you display the timecode numbers of the footage on screen in the bottom corner as the video plays.

To create a Timecode Window Burn Dub, simply record a copy of your source tapes to VHS tape with the timecode display turned on in the deck/camera. Then you can use this VHS dub to get the numbers you want by pulling them right off the television screen without having to jeopardize your precious source DV tapes. Since the displayed timecode numbers will exactly match the frame address numbers on the source tape, you will have an accurate paper log of the footage with no excess playing of the original DV source tapes.

Do you have to use paper logs to Log and Capture, though? Not at all. You can easily manually log your clips from right inside Final Cut Pro. Either system is good, although you will find that once again, your particular source material decides which method is most appropriate. Paper logs are most useful when the shooting ratio is very high and when your necessary footage is spread over many source tapes. Manual Logging from inside of Final Cut Pro is more useful when very precise logging is necessary over a limited amount of source material, and, of course, when paper logs do not exist for whatever reason.

Manual Logging is a very simple act. Simply shuttle back and forth around your footage in the Log and Capture window, setting In and Out points for each of the clips you intend to create. After setting an In and Out Point for a clip, you create the offline clip, and it appears inside your project. It will remain offline, signified by a red slash line through its icon, until you capture it, but it is safe as long as your project itself is saved and backed up. If your system should crash, the Logged Clip will still be waiting when you return, provided you saved the project prior to crashing.

Let's go through and log a clip. As you might expect, there is a recommended process for this. Although Final Cut Pro will allow you to log clips in a sloppy, haphazard manner, doing so really negates the safety, security, and reliability you are gaining through the process of logging and capturing. Setting up and organizing your project will save headaches later on.

Create a Logging Bin

First we will create a Logging Bin. A Logging Bin is the specific bin that all clips appear in as they are logged. When you log a clip and create it, it appears offline in the Logging Bin. However, if you have not established a Logging Bin yet, Final Cut Pro assumes that the Project tab itself is the Logging Bin. This might be OK for the first few clips you create, but very quickly, you will have masses and masses of clips cluttering your Project tab, making it difficult to find anything at all.

Don't make that mistake. Start out with good organizational habits. Final Cut Pro offers really strong tools for organizing your Project, so take advantage of them. Start by creating a separate Logging Bin for each source tape you Log and Capture from. As you get the hang of Logging, you will find that you learn to organize for the project itself and not just for the source tapes.

STEP 23

To create a new Logging Bin, activate the Project tab in the Browser window, then go to the File drop-down menu, and select New and then Bin in the submenu (Figure 4-28). When the bin appears in your Project tab, change the name of the bin to #1 Tape_Logging_Bin and hit Enter.

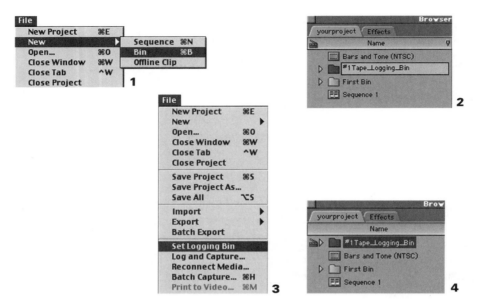

Figure 4-28

STEP 24

After changing the name of the Bin, make sure it is still highlighted, or selected; then go back to the File drop-down menu, this time selecting Set Logging Bin. When you do this, a little film slate clapper should appear next to the bin's icon in the Project tab. This clapper identifies the bin as the current Logging Bin. Until you change that status, any clips you log will be automatically sent to that bin for storage.

Opening the Log and Capture window

STEP 25

Make sure that your deck or camera is connected through Firewire to the Macintosh and is powered on. Make sure that there is no DV tape inserted in the deck or camera. Although it is not technically necessary to remove the tape before going through the next few steps, in a moment you will see the rationale for doing so.

STEP 26

Go back to the File drop-down menu and select Log and Capture (Figure 4-29). On doing so, a window will appear that bears the name Log and Capture. From this window, you will not only log your clips manually or from paper logs, but you will also capture those clips, either as a batch group or individually.

You may notice that when the Log and Capture window appears, the video signal from your deck or camera jumps a little and switches from whatever was showing before to a

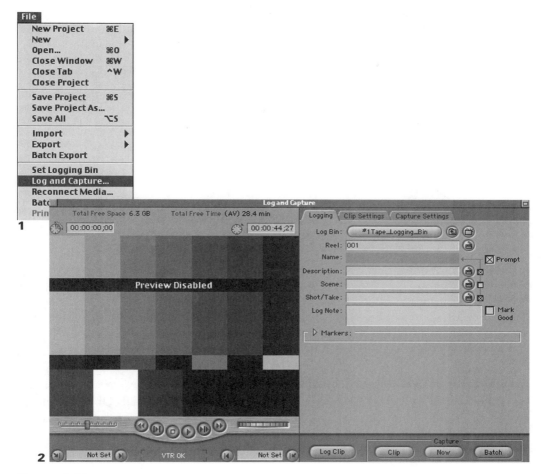

Figure 4-29

blank screen. This is because, despite Firewire's amazing flexibility, it can still send data in only one direction at a time. When you initialize Log and Capture, Final Cut Pro reverses the direction of this data from going out of the Macintosh and into the DV device, to going out of the DV device and into the Macintosh. For this reason, you have to remember to close the Log and Capture window when you are finished working in it. Otherwise, you will not be able to switch back to viewing your editing progress.

The Log and Capture tabs
The Log and Capture window is divided into two halves. On the left is the Video Preview window, which initially displays color bars when not being fed a video source (Figure 4-30). Ignore the Preview Disabled message. It will be enabled as soon as you feed it some video. More important, below the preview window and underneath the Play transport button, you should see the message "Not Threaded," which means that the DV device is

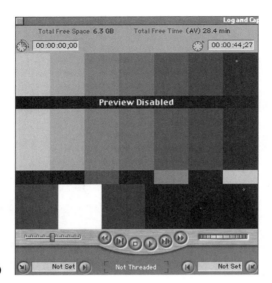

Figure 4-30

present, but no tape is inserted, or "VTR OK," meaning that a deck or camera is connected, it currently has a tape loaded, and Final Cut Pro sees it. Either of these messages indicates that Final Cut Pro is in fact communicating with the deck or camera.

Any other messages in this location when opening the Log and Capture window could be an indication of misconfigured settings or even hardware problems. Setting your Device Control preset to Non-Controllable Device, for example, will yield a "No Communication" message, as if you had a Firewire device connected but turned off. Of course, if you are using a converter box without timecode or Device Control, you will have to get used to the "No Communication" message.

In addition, do not be concerned that the video quality in this window is rather low. DV Firewire capture is simply a data transfer, and your footage will look exactly the same in Final Cut Pro as it does on the tape. The low quality in this window is simply a result of Final Cut Pro's saving a little processing power for the capture process. If you need to see higher quality as you Log your Clips, you can watch the footage on a video monitor coming from the video outputs on your DV device. The frames you choose will be the same ones whether you select them by viewing the monitor outputs from the DV device or from the Log and Capture window.

The right half of the Log and Capture window is divided into three separate tabs: the Clip Settings tab, the Capture Settings tab, and the Logging tab.

STEP 27

The Clips Settings tab
Click on the Clip Settings tab on the right side of the Log and Capture window (Figure 4-31). The Clips Settings tab is mostly available for editors using analog capture cards and

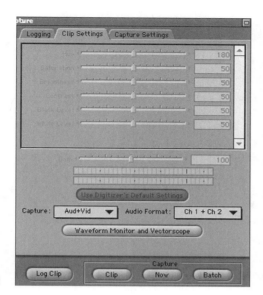

Figure 4-31

has tools for calibrating Final Cut Pro to deliver the proper luma and chroma values to work correctly with them. There are, however, two settings that apply to Firewire DV users. There are two little drop-down bars that allow you to refine what you capture in regards to audio and video.

The drop-down bar on the left allows you to select Video Only, Audio Only or Video + Audio from the DV data coming through the Firewire tube. The drop-down bar on the right allows you to specify how Final Cut Pro will regard the audio channels on your tape when you capture them. This can be set for: Channel 1 and Channel 2 separate; Channel 1 only; Channel 2 only; Channel 1 and Channel 2 as a stereo, or panned, mix; or Channel 1 and Channel 2 as a mono mix. If you are not sure, leave this set at the default of Channel 1 and Channel 2 Stereo.

STEP 28

The Capture Settings tab

Click on the Capture Settings tab on the right side of the Log and Capture window (Figure 4-32). The Capture Settings tab is actually a quick shortcut to your Scratch Disk Preferences and Audio/Video Settings. If you realize that you need to capture your clips to another hard drive right before you capture, there's no need to leave the Log and Capture window to do so. Simply click the Scratch Disk button, and the Scratch Disk Preference window will appear. Make your changes, close the window, and carry on with your capture.

Similarly, the Device Control and Capture Settings Presets drop-down bars from the Audio/Video Settings can be accessed here as well, although you can only choose between the presets you established in the Audio/Video Settings tab. If you need to access some

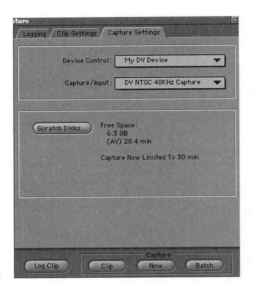

Figure 4-32

other hardware or software configuration that is not yet available as a preset in the Audio/ Video Settings, you will not be able to select it here and will need to quit Log and Capture and return to the true Audio/Video Settings window.

If you initially set your Preferences and Audio/Video Settings when you started up your project as recommended, you would need to visit this tab only when your situation changed. If you are running Final Cut Pro as a dedicated DV Firewire editing station, your hardware and software configuration will likely never change. Once again, a quick check at the beginning of your edit session will keep you from having to remember to check these windows as you work.

STEP 29

The Logging tab

Click on the Logging tab on the right side of the Log and Capture window (Figure 4-33). The final tab, which incidentally is the front tab in this window, is the most important. The Logging tab contains information fields that we need to fill out for each clip as we log it. The information you include in this tab when you log a clip will accompany it in your project, even after you have captured media and converted the clip to Online status. While most of the information you enter here is basically gratuitous and is necessary only for your own organizational purposes, some is critical to your ability to later recapture the media you associate with the clip.

You can never have too much information about a clip. Be as thorough as possible when filling out the fields. You might not think some items are important, but keep in mind that you will be able to organize or perform searches for clips based on the information you put in these fields. For instance, you could stratify a bin based on which clips are marked Good, or alphabetically by name, description, date, or Reel Name. The more infor-

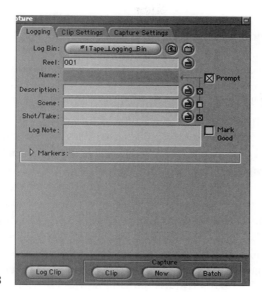

Figure 4-33

mation, the better. Because most of this sort of logging work is often repetitive, Final Cut Pro will automatically repeat information in some fields, such as the Reel Name and the Clip Name, from one clip into the next as you Log.

The Reel Name

The most critical field for you to address here is the Reel Name. The Reel Name refers to the particular tape that relates to the timecode numbers you assign with the clip you are logging. The Reel Name is the special link between the real-world tape and the offline clip you are logging. Many different source tapes could and do have the exact same timecode numbers you are associating with this clip. The only way to differentiate one clip's time-code numbers from another clip's identical timecode numbers is giving them different Reel Names based on the different tapes they were logged from. If you wanted to capture a clip based on its timecode numbers and then were unable to remember which particular tape the clip was originally logged from, you'd be in a lot of trouble very quickly.

It is imperative, therefore, that you develop a system of individually naming your tapes and labeling them in a durable and easily recognizable fashion, then consistently changing the Reel Names as you log clips from the differently named tapes. You will find that when the time comes to capture all the clips you have logged, Final Cut Pro will go through and ask you for each individual tape by Reel Name as it needs them. If you make all the Reel Names the same, it will be difficult for you to complete the Log and Capture process using more than one tape.

We initialized the Log and Capture window without a tape inserted in the DV device for a reason. This is because we want to get into the habit of giving each separate DV tape a distinct Reel Name, as described above. Although Final Cut Pro will happily allow you to go to the Log and Capture window with a tape already queued up in the deck or camera, doing so causes a default Reel Name of 001 to be inserted in the Reel Name field of the Logging tab. Unless you make a point of changing the Reel Name before you begin Logging Clips, each one will be assigned the Reel Name 001.

This obviously will not work. However, if you get into the habit of inserting tapes only after initializing the Log and Capture window, you will find that each time, Final Cut Pro will greet you with a message, "A new tape has been inserted in the VTR. You may wish to change the Reel." Although Final Cut Pro will not change the Reel Name for you, you will find that you never forget to stop and change it as necessary if you are alerted by this message. The message will reappear anytime you insert a new tape in the deck or camera with the Log and Capture window open.

Look above the Reel box and you will see a broad button labeled Log Bin. Inside the button should be the name "#1 Tape_Logging_Bin," which you assigned as the Logging Bin in Step 24 earlier in the chapter. Although you cannot directly change this Logging Bin assignment to another specific bin, you can open the assigned Log Bin by clicking on the button. You can also create a fresh Logging Bin by clicking on the New Bin button to the right. You can have only one Logging Bin set at a time, no matter how many projects you have open concurrently, so you have to be a little careful about where you are sending your logged clips. Keeping your eye on the ball will keep your clips in order and your project free of clutter and lost work.

The Name and Description fields

Below the Reel field is the Name field. You have two choices here, since for some reason you cannot type directly into the Name field. You can type the name into the Description field, and, when you hit Enter, you will see the information appear as well in the Name field. The other method is to wait until you have selected your Timecode In and Out Points and are ready to actually log the clip to name it.

Next to the Name field is a small checkbox labeled Prompt. If you check this box, on logging a clip you will be greeted by a dialog box that asks you to name the clip. The box will provide fields to describe it with logging notes and mark it as Good or Not Good. Although this method does not give you the option of giving the clip Scene or Shot/Take designations or Markers, you may find it more useful because it forces you to make choices about the naming of your clip, rather than letting Final Cut Pro simply add a number onto the end of the name of the last clip name you logged. Always take responsibility and make choices that are right for your project.

The Markers

Click the sideways triangle to reveal the area labeled Markers (Figure 4-34). As you log each clip, you can also leave special Markers at various points in the clip to help you remember

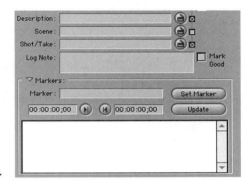

Figure 4-34

where specific sections of the clip are. The Markers are associated only with the timecode of the clip and do not affect your footage at all. But they can make it very easy to divide larger clips into smaller ones, as well as reminding you where important parts of the clip are once you bring the media into Final Cut Pro, as we will see later in the book.

As you log the clip, just hit the Add Marker button when the playhead is parked on a frame you want to mark, making sure that you enter both the first frame and the last frame of the marked sections. If you want to mark only a single frame, make that frame's time-code value both the beginning and end timecode value for the Marker. If you want to mark a range of frames, add a Mark In point, then move to the end of the range of frames you want to mark and add a Mark Out point. Add as many Markers or ranges of marked frames as you want. Name each Marker; you'll be able to go in and change any information in the Marker after the clip is logged and/or captured.

The Capture Buttons

Finally, at the very bottom of the tab you will find four buttons of great importance (Figure 4-35). The three buttons to the right are enclosed in a box labeled Capture. They refer to the different possible methods by which Final Cut Pro can capture media. The three buttons are named Clip, Now, and Batch, and when pushed, they initiate the kind of capture that they refer to. Each one has a different value in the production process. There are instances where Capture Now will be the smartest option, whereas in other situations, Batch Capture will be the most effective. As always, strategy determines which method is best for your project.

Figure 4-35

Capture Clip

The first method, Capture Clip, is probably the least efficient way to gather media. It functions by capturing one clip at a time, based on the present values in the Log and Capture window. Instead of logging many clips and then capturing them en masse, or simply Capturing Now and quickly grabbing a whole section of footage, Capture Clip takes the Timecode In and Out Points that are currently in the window and captures using them. You do not hit the Log Clip button and create an offline clip. You can get only one clip at a time this way, although you must still log your footage to get the timecode values you need to capture with. This option will be most useful if you have paper logs with the timecode right in front of you, but you need only one or two clips at a time.

Capture Now

The second method, Capture Now, is probably the most frequently used method of capturing clips into Final Cut Pro, because to the new user it often appears more intuitive. Push the button and the capture starts. Capture Now begins capturing whatever happens to be streaming down the Firewire tube when you press the button. It will continue capturing from the Firewire tube until you hit the Escape key on your keyboard, or until it reaches the time limit you specified in the Scratch Disk Preferences.

The most important feature of Capture Now is that, unlike Capture Clip and Batch Capture, it does not require timecode data for Capturing Clips. Although you must specify that no timecode is present by disabling Device Control by selecting Non-Controllable Device in the Device Control Presets of the Audio/Video Settings, Capture Now will capture anything coming down the Firewire tube, including a blank blue video screen from a deck with no tape inserted.

Of course, if you are capturing from a DV device that has timecode, Capture Now will capture and use the timecode that is there. But if you are using any of a number of solutions that do not pass timecode data through the Firewire tube, such as DV converter boxes, Digital 8 camcorders that are playing back video recorded on a Hi8 camera, or DV tapes that have extremely spotty timecode data, Capture Now will be the only way that you can capture into Final Cut Pro.

Use of Capture Now prompts two warnings, however. First, capturing clips without timecode data, which is the principal value of Capture Now, is a risky way to work. Once you capture the clip, it will function just like any other clip. But if its media files are somehow lost or deleted, you will never be able to recapture the exact same clip using Capture Now. You can recapture the same area of video footage and painfully reconstruct your edits with the new clip, but you will not be able to use the simple timecode recapture strategies that make Final Cut Pro so resilient and safe to use.

Using timecode on the other hand, would allow you to easily recapture the exact same clip from your timecode DV tape. Capture Now, when used without timecode, starts and stops with about the same degree of accuracy as a consumer VCR, which we described as unsuitable for accurate editing earlier in the book. If you use non-timecode source material such as the examples above, if possible, consider dubbing your source material to a true DV timecode device such as a DV deck or camera. That will provide you with the same material at the same loss of one generation from analog to digital video, but with clean timecode at the cost of an eight-dollar reusable DV tape. You weigh the risks.

The second warning is that, because Capture Now does not require timecode data to Capture clips, it treats those clips somewhat differently when capturing them. You must manually save any Capture Now clips' media files before accessing the clip in a project. As such, there is a specific procedure necessary to get your clips into your project after using Capture Now. If you do not take care to follow the procedure described in the section on capturing, you may lose access to the clip.

Batch Capture

The last button is for Batch Capture. This is the most efficient method of Capture, and takes full advantage of timecode. Using Batch Capture techniques, you will use the Capture window to log as many clips into your Logging Bin as you wish. When you have completed logging them and have saved the project to keep from accidentally losing them, you hit this button and go to a special window where you can capture all of them, or any smaller group of them, en masse. This is a much better way of capturing the clips you need without having to manually go in and get each individual clip. It automates the process so that you can go off and have a sandwich while Final Cut Pro searches the tape and retrieves the exact media that you have specified for your project.

This whole method of logging and Batch Capture is referred to as online/offline capturing. When you enter the timecode data and log the clip, it appears in your Logging Bin as an offline clip, meaning that although a clip exists for the timecode data you've entered, there is not yet any media associated with it. Until you perform a Capture of the offline clip, it will not have any captured media to reference or index to.

However, after you have logged the clips you want and Batch Captured them, they will become online clips, meaning that the clip now refers not only to timecode data, but also to the media you captured using that timecode data. The Batch Capture button initializes this process of capturing offline clips after you've completed the logging process.

Log Clip

The fourth button, named Log Clip, actually creates these offline clips, based on the timecode data you enter in the Log and Capture Window. Once you have given the window a Timecode In and Out Point, instead of simply hitting Capture Clip to capture the clip immediately, you can use Log Clip to create an offline clip in your Logging Bin. Once you have repeated this process enough times to log all the clips you need from your tape, you will proceed to the Batch Capture button described above and capture as many of the Offline clips as you wish. The Log Clip button is the magic switch that converts valid timecode data in the Logging window into Final Cut Pro clips that you can capture and edit with. We'll return to this when we perform the action of Logging and Capturing.

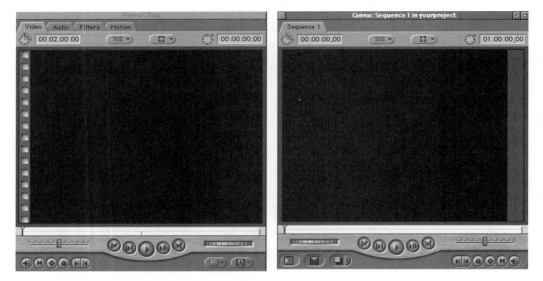

Figure 4-36

Logging and Capturing

Let's log and capture a single clip, using the Log and Capture window to the left of the tabs. We will go through each of the methods for Capturing, first using Capture Clip, next using Batch Capture, and, finally, using Capture Now. All three of these use the same Log and Capture window, although the process differs slightly for each.

Incidentally, if the Log and Capture window is too large and takes up too much of the window, close it and reduce the size of your Canvas and Viewer windows (not the percentage size in the percent drop-down bar but the actual dimensions of the Viewer and Canvas windows). The size of the Log and Capture window is based on the size of those two windows. Keep the Viewer and the Canvas set at 50 percent, and then tighten up their dimensions using the pull tab on the bottom right-hand corner of each window. You want the video window to fit the Viewer and Canvas windows without resulting in scroll bars on either one, so that there is almost no gray area surrounding the video window itself (Figure 4-36). Then reopen the Log and Capture window, and you will see that its size is reduced as well.

The Log and Capture Video Window

The Log and Capture window is dominated by the Log and Capture screen on the left. Unless Final Cut Pro is currently receiving video data through the Firewire tube, the Log and Capture Preview window will simply show color bars.

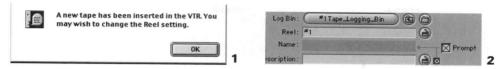

Figure 4-37

STEP 30

Insert a DV Tape. As soon as you do so, you will receive a message recommending that you assign a Reel Name to the new tape that has been inserted (Figure 4-37). Do so, using the Reel Name "#1," which corresponds to your Logging Bin name.

Scratch Disk info

After you name the reel and click through that message, take a look back at the Log and Capture box. At the very top of the window you will see two phrases—Total Free Space and Total Free Time (Figure 4-38). The values for these designations are the amount of free disk space on the current Scratch Disk–assigned drive and the amount of time that free drive space can accommodate at the data rate of the codec you are capturing with. This is merely another warning sign for you regarding your Scratch Disk Preferences. If you see anything unusual here, such as a drive space figure that is too high or low, or a capture limit that is too short, you may have committed an error in assigning the Scratch Disk location or Capture Setting. Final Cut Pro can't fix this for you, but it can help you catch your own errors.

Current frame timecode field

Moving to the top right-hand corner of the window, you will see a small watch dial followed by a field containing a timecode value (Figure 4-39). This field is the timecode for the video frame that the DV Tape is currently parked (i.e., paused) on. If you have a tape in your deck that contains timecode and Final Cut Pro is configured correctly, this window will always report the exact timecode number of the frame the tape head is currently parked on.

You can also use this field to shuttle around the tape. Because the Log and Capture window actually controls the deck as well as receiving video from it, this field works both ways; it tells you which frame is queued up on the tape head, and you can tell it to go to a certain frame by entering the timecode number. This is but one of many ways to navigate to locations on your tape.

Figure 4-38

Figure 4-39

Figure 4-40

The Tape transport controls and J-K-L support

Moving below the video window, you will see the playback buttons, otherwise referred to as transport controls (Figure 4-40). Using these controls you can shuttle back and forth, remotely controlling the deck from inside Final Cut Pro. There are five buttons for controlling playback as well as two knob-wheel–type controllers that mimic physical deck controls.

The most important buttons on the transport controls on any deck are obviously the rewind, pause, play, and fast-forward controls. Final Cut Pro makes using this window—and other windows—easier, by using a keyboard shortcut convention called J-K-L support. What J-K-L means is that the J key initiates rewind, the L key initiates play/fast forward, and the K key simply pauses the playhead where it is. Beyond this, J-K-L has the flexibility of variable-speed playback. If you hit either the J key or the L key repeatedly, you will find that the playback speed increases, while K always stops the playback entirely.

Using J-K-L means that you can leave the mouse alone and forget about looking at the keyboard to find the right combination. Resting your hand on the keyboard with your middle three fingers spread over the J-K-L keys is very natural, which is why it has become very popular. Practice using it to move back and forth to find frames you want, and you'll quickly get addicted to using it.

The other two buttons on the transport are Play Around Current and Play In to Out. Play Around Current simply plays a certain number of frames before and after the frame that is currently on the tape head. Where does this certain number of frames originate? It is determined by the settings in the General tab of the Preferences, where it is referred to as Preview Pre-roll and Preview Post-roll. Whatever values you entered in that Preference correspond to the number of frames that will be played before and after the current frame in the window. Play In to Out is simply a preview of the In point to the Out point based on what is currently entered in the In and Out point fields, which we have not entered just yet.

The Jog and Shuttle wheels

The other two controllers hearken back to the world of physical decks. They are referred to as Jog and Shuttle controls. The control on the left-hand side is the Shuttle controller. If you click on the knob and drag it in either direction, you will find that the deck responds by moving in that direction. The speed of playback corresponds directly to how far in either direction you tug the knob. Shuttling is a great variable-speed method of moving around. The J-K-L controls we just described are referred to as Shuttle controls, because we can determine how fast or slow the playback is by the number of times we strike the same key repeatedly.

The controller on the right-hand side is called a Jog controller. The difference between Jog and Shuttle is that with Jog, one turn of the knob means one frame passes on the tape, no matter how far or with how much force you tug the controller. Unlike the Shuttle controller, which can be pulled only so far in either direction, the Jog controller turns a complete 360 degrees, after which a frame has been advanced. There is also a keyboard frame advance and retreat that corresponds to the Jog controller's frame accuracy. These are the Left and Right Arrow keys, which will advance or retreat one frame at a time.

You will find that Jog and Shuttle have their places and are meant to be used together. Shuttling is useful for moving quickly over large sections of tape, while Jog finds the exact frame you need to edit with. Use Shuttling to get into the neighborhood and then Jog to get the perfect edit frame. But master the J-K-L and Left Arrow–Right Arrow shortcuts rather than dragging the mouse around. Apple will not compensate you for damages caused by carpal tunnel syndrome, so you'd better start protecting your wrist from harm with the keyboard shortcut conventions.

Set the In and Out points for the logged clip

Once you have decided on the In and Out points (i.e., the frames you want to use for the beginning and end of your clip), look below the transport controls to find two timecode data fields, each of which is accompanied by two buttons (Figure 4-41). These are the In and Out Points you will establish for your clip. Prior to setting any In or Out points, these fields will read, "Not Set."

Remember that you have to leave enough room for the deck to Pre-roll into—a number of frames that we specified while setting up the Audio/Video Settings in the Device Control Preset. The recommended Pre-roll setting is five seconds, so you need to make sure that there are five full seconds for the deck to roll back into prior to beginning the capture. Otherwise, you will receive an "unable to lock deck servo" error message when you initiate the actual capture.

STEP 31

When you have selected a suitable frame for an In point, locate the timecode field on the bottom left-hand side of the window. Click on the button to the right of this timecode field and you will see the timecode number appear there (Figure 4-42). This button is the Mark In point.

Figure 4-41

Figure 4-42   **Figure 4-43**

The button to the left of the In point field is the Go To In Point button, which will simply return the tape to the currently established In point should you need to get there quickly.

STEP 32

Now shuttle a little further into your tape, making sure that you do not cross any timecode breaks. When you have found a good location for the end of the clip, look over to the timecode data field on the bottom right-hand side of the window. You will find an Out point button on the left side of this field which corresponds to the In point button on the In point field. Clicking the Out point button inserts a value into the Out point field (Figure 4-43).

As with the In point field, the button on the right of the Out point field is Go To Out point and will quickly roll the tape to that frame if engaged. With the establishment of an In and Out point, you now have a valid, if not yet logged, clip, in the Log and Capture window.

As always, there is a much faster way to insert In and Out points using convenient keyboard shortcuts. The I key and the O key will insert In and Out points into these fields. The convenience is reinforced by the fact that these keys are directly above the J-K-L keys that you use for shuttling. You rarely need look at the computer monitor or keyboard again. Simply keep you hand poised on the J-K-L keys and when you find either an In or Out point, move your fingers a half inch up and strike the key. There's no faster way to log clips, nor one that is less stressful on the wrist. A little practice is all it takes and you'll be able to log clips without thinking about the process at all. Being able to focus on your editing without thinking about which onscreen button you need to access is the key to becoming a good editor.

The Clip Duration Timecode Field

Now that you have entered a valid In and Out point for the clip, look in the top left-hand corner of the window where you will find another timecode data field (Figure 4-44). This is the Clip Duration Field, which calculates the duration of the current clip, based on the In

Figure 4-44

and Out points. You should see a value here now that you have entered an Out point that follows an In point.

Of course, as with the Current Frame field on the upper right-hand corner of the window, this field can actually affect the other fields. If, for example, you already know the In point and you know that you need to log 10 seconds following that In point, there is no need to shuttle forward and find the frame number or even to calculate the Out point in your head. Simply type 1000, which is a shorthand for the timecode value of 00:00:10:00 or 10 seconds, into the Clip Duration field and it will automatically update the Out point based on the In point and the Duration timecode value.

Capture using Capture Clip

Now that you have entered the required information for the clip, you are ready to capture using either of the two logging methods described earlier.

STEP 33

Click the Clip button in the Capture box on the Logging tab side of the window (Figure 4-45), Final Cut Pro will begin the capture process. If you selected the Prompt checkbox on the Logging tab, you will be shown a box asking you to name the clip. Do so, click OK,

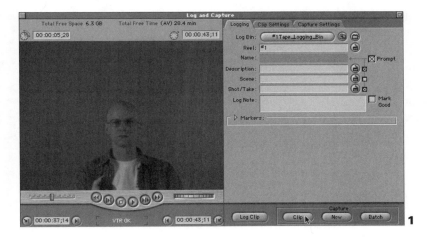

Figure 4-45

Figure 4-46

and Final Cut Pro will wake up the deck and begin searching for the In Point. You will see a black screen with a message at the bottom telling you what Final Cut Pro is doing at the moment. When it finds the In point on the tape, it will begin capturing and continue to do so until your clip is completely captured. It is now available for editing.

If you take a moment and look back into the Log and Capture window, you will see that it has set a new In point that is one frame later than the Out point of the clip you have just finished capturing (Figure 4-46). The new Out point is one frame later than the new In point, because you will need to log a new Out point, no matter what the new In point is. This means that Final Cut Pro is now ready for you to log the next clip you want to capture.

Logging and Capturing as a Batch Capture

Logging offline clips for Batch Capture

Although you have just successfully captured a single clip, it should be pretty obvious that this would be a very slow method if you needed to capture 100 clips, all with widely varying durations. It would be much more efficient if you could log all of your clips one after another, and upon finishing that operation, simply walk away from your station while Final Cut Pro grabbed them all in order.

What you need is the Batch Capture operation. As described earlier in the discussion of Offline/Online capturing, you will log each clip one by one, creating offline clips in your Logging Bin. After you have completed logging all the clips you want from a single tape/reel, you will select Batch from the Capture button selections and let Final Cut Pro do all the work.

Using the Log Clip button

To do a Batch Capture, simply repeat Steps 31 and 32, entering new In and Out point information just as you did with the Capture Clip process. However, instead of clicking Clip in the Capture buttons, select the Log Clip button to the left (Figure 4-47).

STEP 34

After selecting new In and Out points, select the Log Clip button. On clicking Log Clip, you will once again receive a dialog box asking you to name the clip. Do so, and click OK.

After clicking OK, however, Final Cut Pro does not immediately begin the Capture process. If you look into the Logging Bin you established in the Project tab of the Browser

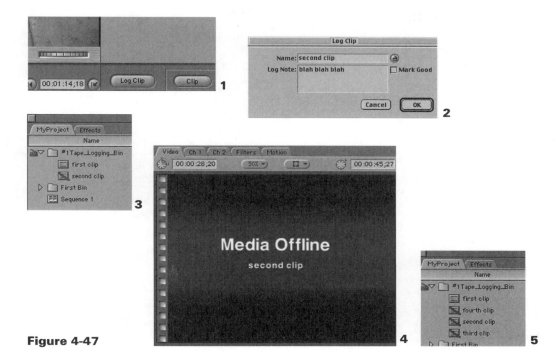

Figure 4-47

window, you will see a new offline clip. Offline clips are identified by a red line slashing through the clip icon. If you double-click the Offline clip so that it loads in the Viewer window, you will see a standard Offline Clip image, indicating that there is no media file yet associated with the clip.

Instead of immediately rushing to capture this offline clip, leverage Batch Capture techniques by logging another few clips using the same logging method. As you proceed, you will see the new offline clips appear in the Logging Bin one after another. After you have logged a few more clips, return to the Log and Capture window and select Batch from the Capture button selections.

The Batch Capture button and dialog box

STEP 35

After logging a number of Offline clips, select the Batch button (Figure 4-48). When you hit Batch, you will be taken to a dialog box that acts as a sort of idiot check for the capture you have initiated. Since the Batch Capture is an automated process that is likely to involve a great deal of unattended action on the part of Final Cut Pro, you are given this opportunity to check all your settings, as well as the number of clips you are about to capture and their expected length.

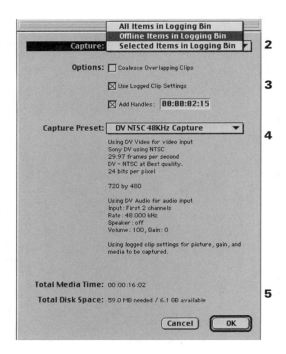

Figure 4-48

Capture Selected or All Clips menu bar

At the top of this box, you will see a small drop-down bar (Figure 4-49). Depending on how you left your Logging Bin, this will allow you to choose to capture either all the offline clips in your Logging bin or just the selected ones. Chances are that if you hit Batch immediately after logging your last clip and without revisiting your Logging Bin, the last clip you logged is the only one actually selected. That is because after you log a clip, Final Cut Pro selects that clip until you log the next one, or otherwise perform another selection.

The Batch Capture dialog box drop-down bar gives you the option of capturing all of the offline clips in the Logging Bin or only the selected ones. Why, you may ask, would you want to capture only selected offline clips in the Logging Bin? Sometimes, it would be counterproductive to try and capture them all. We can log all the offline clips we want, including clips from different tapes. As we saw before, we need to insert a new Reel Name for each different tape we use so that Final Cut Pro can tell that the similar timecode numbers for two clips actually refer to different tapes. If you don't give a clip a reel name, Final

Figure 4-49

Cut Pro will assume the clip should be captured from whatever Reel Name was there from the last time.

If you have logged the clips correctly, including giving the correct Reel Name to each clip that corresponds to the tape it relates to, Final Cut Pro will actually ask you for the correct tape before it begins the act of capturing. Every time it goes to capture a clip, it will check the reel name and make sure that the reel the clip is looking for is the same one as the tape you put in when it asked you for that reel. If a new clip from a different reel is to be captured, Final Cut Pro will temporarily stop the capture process and ask for the new reel the clip is associated with.

If you do all of your logging across many different tapes and Reel Names and you chose All Clips in Logging bin, Final Cut Pro would stop capturing when it reached a clip that was associated with a Reel Name and tape not currently in the deck. As such, you want to select only clips associated with a specific Reel Name and then choose Selected Items in Logging Bin. If you have any clips selected in the Logging Bin, Final Cut Pro will give the option of capturing All Clips in Logging Bin, or Selected. This gives you the ability to let Final Cut Pro do the capturing without interruption for tape changes and, most important, without your attendance. When you come back after the first capture, switch the tape to the next reel name, select the clips associated with that Reel Name, and proceed with the next round of captures.

The next items are the Options checkboxes.

Options checkboxes: Coalesce Overlapping Clips

The first is called Coalesce Overlapping Clips. This refers to how Final Cut Pro will capture the media for any two or more clips that have overlapping timecode values. It's quite easy to accidentally log two clips, each of which use some of the same timecode numbers from the same tape. If you enter an Out point for a clip on one day and then the next day log another clip from the same tape with a beginning timecode value that is before the Out point of the earlier clip, you've got overlapping clips. This isn't the end of the world. You simply need to inform Final Cut Pro of how to deal with it. Since this could easily happen without your knowledge, simply plan ahead and have it set the way that you would want it dealt with if you knew.

To Coalesce the clips means that Final Cut Pro would capture one media file that contained all the data for both clips. The clips in your project would behave the same either way, but coalescing just means that the two clips refer back to a giant media file containing all the video frames for both clips. Disabling Coalesce would create two completely separate clips which each have a copy of the frames of video they share. Disabling is a slight waste of drive space, but there is something to be said for having distinct media files for each clip. Leave Coalesce disabled unless this is a mistake you make consistently, in which case you might want to take a look at your logging methods rather than utilize the Coalesce function.

Options checkboxes: Use Logged Clips Settings

The next Option is Use Logged Clip Settings. For most Firewire DV editors, this checkbox will not make much of a difference. Use Logged Clip Settings simply tells Final Cut Pro to override the Capture preset settings currently in play and to use the ones that were enabled when the clip was logged originally. Since the settings will rarely change for Firewire DV editors, leaving this unchecked will ensure that you capture the clips according to the way you have the system set up as you begin your capture, not the way it was on the airplane as you logged clips on your way back from that business meeting in Zurich.

Options checkboxes: Handles

The next Option is a checkbox that contains a timecode field for entering handles. If you remember, "handles" simply means having the system grab a specified number of video frames during the capture in addition to the number you asked for when you logged In and Out timecode numbers for the clip. If you set a value for handles of 15 frames, Final Cut Pro will always capture an extra 15 frames at the beginning and end of every clip you capture. This will not affect the In and Out points of your captured clip. The frame handles will stay safely out of view until you need to bring them into play. Your In and Out points will always reflect the way that you logged them for the clip until you change them.

You can of course override the option to get handles on your captured clips by disabling the checkbox, but doing so is a pretty bad idea. Once you've captured the clip and you begin editing, you will usually find at least one clip that you need extra frames for, especially in the case of sequences that require transitions like fades and dissolves. It's usually hard to predict exactly how much video you'll need for these purposes, and a good solid 15-frame handle is usually about right for the purpose. In addition, this will protect you from a rare occurrence in which the capture process accidentally duplicates the first frame it captures. Inserting handles will completely eliminate this as a potential threat.

Capture Preset choice bar

The next item is a drop-down bar that allows you at the last minute to switch the Capture Preset you want to use for the Batch Capture. Although it is not likely that you will have to change this, you at least have the option. This would be the right time to make sure that you have selected the proper audio sample rate and to switch it if you have chosen an incorrect one. Below the Capture Preset drop-down bar, you will find a summary of the settings for the current Capture Preset.

Batch Capture idiot check summary

Underneath the Capture Preset bar, you will see your last idiot check summary information. The first item, labeled Total Media Time, refers to the total amount of time to be captured in the entire Batch Capture. It is the sum total of all the clips' durations combined. If

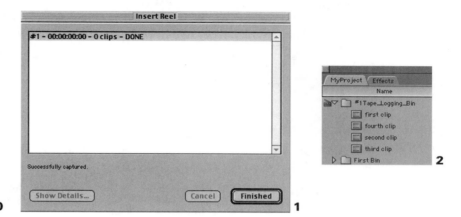

Figure 4-50

you see a particularly odd number here, like 12 hours, you should probably abort the capture and check your clips.

The second item is labeled Total Disk Space and will contain two numbers. The first number is a calculation of the total amount of drive space necessary to capture the amount of Total Media Time based on the data rate of the codec in use. Since Firewire DV editors will be using the DV codec, which we know has a data rate of roughly 3.6 megabytes per second, this number should be easy to mentally calculate. The second number is the amount of drive space currently available on the assigned Scratch Disk Drive. This does not include a Minimum Allowable Free Space on Scratch Disk limit that you may have set in your Scratch Disk Preferences, so don't let a large number here go to your head.

If you see a problem, hit the Cancel button and rectify the issue. If everything is set and checks out, click OK and begin the Batch Capture process. You will be greeted with a box asking you to insert the proper tape. When you do so and hit OK, Final Cut Pro will proceed to capture all the clips. When it has finished capturing, the offline clips in the Logging Bin will have lost their red lines and, having associated media files, will now appear as online clips (Figure 4-50).

Capture Now

Important information about Capture Now

Logging your clips with timecode for immediate capture or Batch Capture strategies is definitely the safest and most efficient method of capturing media. But some Firewire configurations do not have access to timecode. And in some situations, capturing a quick clip will be more productive than logging a clip manually as described in the previous section.

In these instances, Capture Now will be more effective. Final Cut Pro allows for this, although the process is slightly different. Care must be taken, because clips captured using Capture Now are handled differently and it is easy to miss a step in the process, confusing your project and possibly losing the link between the media you capture and the clip you want to work with in your project.

Capture Now and the relationship with DV timecode

First, you need to determine whether you are sending timecode data from the DV device through the Firewire tube to Final Cut Pro. Remember that just because video and audio are getting captured doesn't mean that timecode data is being captured as well.

If you are using a Digital 8 camera, only footage that was shot using the Digital 8 camera has valid DV timecode written into the data of the footage. If you are using a tape that was recorded in a standard Hi8 or 8mm video camera, the footage will not have DV timecode. Although you can still easily capture this footage using Capture Now, you cannot do so using Device Control, and you will not be able to get timecode data for the clip when you capture it. If you are using a DV converter box, there will also be no DV timecode accompanying the DV data stream through Firewire, unless your system has a serial converter of some kind.

Also remember that this refers to any device that offers pass-through operation. Many DV decks and cameras will allow you to connect an analog device to the inputs of the DV device, which will convert the analog signal to DV stream on the fly and pass it directly to Final Cut Pro through the Firewire connection. Such pass-through connections also do not carry DV timecode and act precisely like a DV converter box.

If you are using Capture Now with one of the aforementioned solutions and do not have DV timecode, you may use Capture Now to gather your clips, but you must switch your Device Control Preset to Non-Controllable Device in the Capture Settings tab of the Log and Capture window. This preset will keep Final Cut Pro from looking for the DV timecode that does not exist. If you do not switch this preset, the attempt to Capture Now will result in a long system hang followed by an error message.

On disabling Device Control, you will see that the Clip and Batch Capture buttons have gone gray and become unavailable. This is because they require Device Control and timecode to operate.

Because you have no access to timecode, Final Cut Pro will not be able to control the device. This means that you will have to queue up the footage that you want to capture on the deck by hand. This is a slightly awkward way of working in which you hit play on the device, hit Now in Final Cut Pro, and hope that it begins capturing before the first frame you wanted to work with passes by. Although it may work in a pinch, it becomes frustrating rather quickly in larger jobs.

On the other hand, if your device has DV timecode, you can Capture Now with complete Device Control and with the security that recapturing lost clips using timecode can deliver. In this case, make sure that the Device Control preset is still set correctly for the configuration you want to use and not to Non-Controllable Device.

Special saving conventions

The other major difference between logged clips and Capture Now has nothing to do with the method by which you capture, although that is different as well. The main difference is in how the media is dealt with after the capture is complete. Clips captured through Capture Now are not automatically saved to the Scratch Disk and added to a Logging Bin as are logged clips. Capture Now clips are not associated with a project until you save them and move them to a Project tab in the Browser window. You can even capture clips using Capture Now without a project being open!

This can be very confusing to new users of Final Cut Pro who have not yet mastered the flexibility with which Final Cut Pro can juggle media between and even outside of open projects. To avoid this problem be careful to follow the following steps in using Capture Now.

The Capture Now process

STEP 36

If you have Device Control and DV timecode, make sure that you fill out the Logging tab information to the right, as if you were performing a logged capture. In particular, make sure to enter a Reel Name, since your clip will carry valid timecode data and should be associated with a specific tape in case you need to recapture the exact same clip with timecode later on. Of course if you have no DV timecode, such Logging data will not be stored with the clip anyway.

STEP 37

If you have Device Control, start the tape rolling, using the J-K-L keys as described in the section on Log and Capture. They will perform the same way. If you have no Device Control, simply hit play on the DV Device. You will need to cue up your tape a few seconds ahead of the place you want to start capturing, because it takes a moment for Final Cut Pro to prepare the Scratch Disk for your media. Think of this as a Pre-roll and add at least five seconds so that you are sure you get the In point you wanted, plus a little extra.

STEP 38

As the tape is rolling, hit the Now button in the Capture options. When you do so, the Currently Capturing window will pop up and you will watch what is being captured stream in. When you are ready to stop the capture, simply hit the Escape key. The Escape key is the only way to stop Capture Now, with the exception of reaching the Capture Now limit you set in the Scratch Disk Preference tab in the Preferences.

When the Currently Capturing window disappears, you will see that a new Viewer window has opened containing the clip you just captured (Figure 4-51). The title at the top of this Viewer window will read Untitled 0000, unless you entered a clip name in the Logging tab prior to capture. This is because the clip has not yet been officially saved to your Scratch Disk. Even though the clip is sitting in a Viewer window, it is not yet recognized by a Final Cut Pro project. To be recognized, it must be saved into a project by placing it in one and then saving it to a Scratch Disk folder.

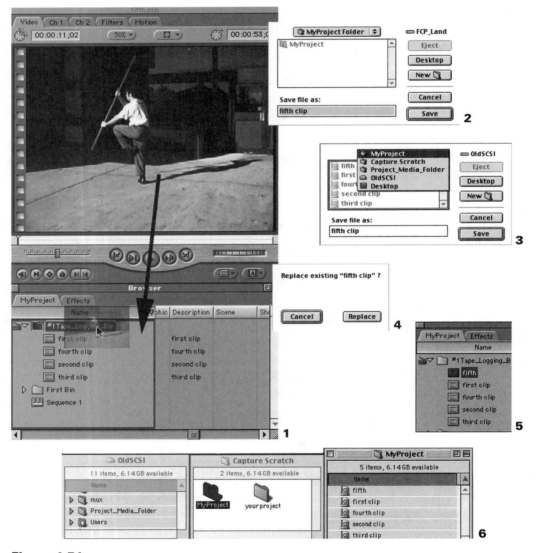

Figure 4-51

STEP 39

Immediately drag the Capture Now clip from the Viewer window down into your Project tab or Logging Bin. When you do so, you will be presented with a dialog box asking you to save the clip. Saving the clip does more than simply keep it from being lost. It establishes a link between the clip in the project and the media file in your Scratch Disk folder, essentially creating a reference for the project to work with.

STEP 40

Navigate the Save dialog box to the folder that you have set as the project's Scratch Disk location. Inside the Scratch Disk Folder, double-click on the Capture Scratch folder. Then double-click on the folder that bears the name of your project. Inside that folder, you will see the name Untitled 0000 (or your clip name if you gave it one in the Logging Tab) grayed out. This file is the temporary capture location that Final Cut Pro will use until you either save the clip, which you are doing now, or decide you don't want to keep it.

At this point, you can name the clip something else. If you were working with DV timecode during this Capture Now session, you could have assigned the clip a specific name in the Logging tab and that name would have stuck through the capture and saving process. But if you were not capturing with DV timecode, you would have been unable to name the clip until this point.

STEP 41

Before saving the clip in this dialog box, change the name to something relevant. If you do not change the name in the Save dialog box, Final Cut Pro will ask if you want to replace the temporary file created during the Capture Now. If you change the name, it will not. Either way is fine; just don't use Untitled anything, and don't skip this process. When you hit OK, the clip will be renamed in the Project tab of the Browser window, and if you navigate back through the Desktop to your Scratch Disk folder, you will see that the Untitled 0000 clip has now been renamed with the name you saved it under.

If you do not save following this method, you will end up with clips saved all over your system. This makes it very difficult, if not impossible, to guarantee that your media will be easy to find in the event you want to get rid of it or that your system works at optimum efficiency with regard to the performance of your media and startup drives. If you hit Cancel during the process and do not actually save the clip, the capture will be lost and the Untitled 0000 clip will vanish from your Capture Folder.

Once the clip has been saved correctly, it is available to your project in the same fashion that the logged and captured clips were.

When was the last time you saved and backed up your project? After you have completed your logging or capturing is the right time to save the project you have created the clips in, because, although the actual media files have been properly saved to the media drive, the clip, which is now associated with the project, has not yet been saved as part of the file. If your system had crashed for any reason, the clip would not be there when you re-opened and you would need to find the media files in your capture scratch folder and manually import them.

It should be obvious that Capture Now is not an efficient system for capturing large numbers of precise clips. Wherever possible, it is far wiser to capture clips by logging their timecode. It is safer, faster, and less likely to give you headaches.

Importing media

Capturing footage from DV tape is not the only way of getting media to work with in Final Cut Pro. Final Cut Pro is quite flexible and can import and export a wide range of media file formats. Final Cut Pro can easily import the following file formats, arranged by file type:

- Graphics:
  BMP
  Flashpix
  GIF
  JPEG/JFIF
  MacPaint (PNTG)
  Photoshop (PSD)
  PICS
  PICT
  PING
  QuickTime Image File (QTIF)
  SGI
  Targa (TGA)
  TIFF
- Video:
  AVI
  QuickTime Movie (MOV)
  Image Sequences
  Flash (video only)
- Audio:
  AIFF
  Audio CD Data (Macintosh)
  System 7 Sound
  Wave (WAV)

Some of these formats may seem obscure, but they all have one thing in common: Final Cut Pro can import them into your project and use them as clips for editing purposes.

Such importing is very common. By using audio tracks from an audio CD or using Photoshop files to create graphics you can easily add professionalism and class to your finished production. Although Final Cut Pro is a really advanced editing solution, there are things that other applications can do much better.

Graphics applications such as Photoshop have tools that are much more suited to developing beautiful text and other visual elements. Audio applications such as Digidesign Protools and Bias Peak are often used to texture, mix and clean up audio tracks that don't sound their best initially. Further, if you deal with clients who provide you with standard graphics or sound elements, you'll have to deal with importing their materials.

You'll be happy to find out that Final Cut Pro could care less what most file types are. It deals with every file as a piece of media just like any clip you capture. The only restrictions on the clip will of course depend on what the file was (audio-only files such as AIFFs, for example, will have no accompanying video); Final Cut Pro will leave this up to your discretion. "Parts is parts," as they say.

The main point is that Final Cut Pro regards each imported file as a clip and a media file linked together after the manner of our captured clips. Because of this, these imported media files need to be saved correctly as were the Capture Now clips. There is no Scratch Disk Preference for importing media files. We must complete that action ourselves. Let's step through the process of correctly importing a couple of common file formats and look at the proper method for doing so.

Importing audio CD tracks

Importing audio tracks from audio CDs is an easy and quick way to add life to your production, but there are a couple of nonintuitive pitfalls that await the editor who engages in such an import. The main two issues are saving the file correctly and making sure the sample rate matches that of your other clips and sequences.

The audio CD format

Audio CDs are not formatted like other media disks.

STEP 42

Pop an audio CD into the Macintosh and it mounts on the Desktop just like any disk. Double-click it and open it up as a window (Figure 4-52). Inside you will find a number of files with track numbers for names. If you double-click on any of them, the Apple Audio CD Player will automatically begin playing the audio track that corresponds to that track number on the CD. It would seem that the file with the track number is the actual audio file on the CD. Unfortunately, this is not true.

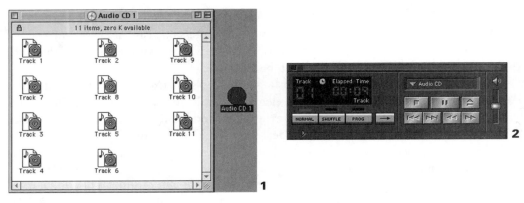

Figure 4-52

STEP 43

If you grab one of these files from the audio CD and drag it to the Desktop, it will copy almost instantly (Figure 4-53). Remove the CD from the Macintosh CD-ROM drive. Then double-click the just-copied file and you will find that the Apple Audio CD Player can no longer find the file the audio track refers to.

The Apple Audio CD Player can no longer find the file the copied audio track refers to because, with the audio CD format, the media file is actually hidden. The audio track file you copied to the Desktop is really just a reference to the media file that is hidden on the CD, rather like our clips and captured media file in Final Cut Pro. The media file is accessed through this file so that you can play the track, but only if the audio CD is present. So much for drag-and-drop music piracy.

Does this mean that you can't use audio CD tracks in your production? Not at all, although you need to be careful of breaking the law by illegally copying and using trademarked material. In order to use audio CD tracks, you need to physically import the track. This involves converting the audio CD file format that is not native to the Final Cut Pro application into an audio file that Final Cut Pro can use. Final Cut Pro does this for you when importing.

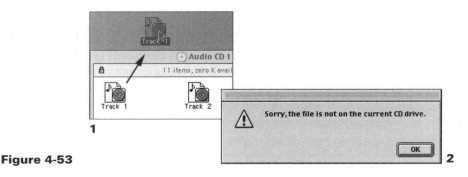

Figure 4-53

CD import part 1: from the CD

STEP 44

Return the audio CD to the Macintosh's CD- or DVD-ROM tray. When the CD has mounted on the Desktop, go to the Final Cut Pro File drop-down menu and select Import, and then File from the submenu (Figure 4-54).

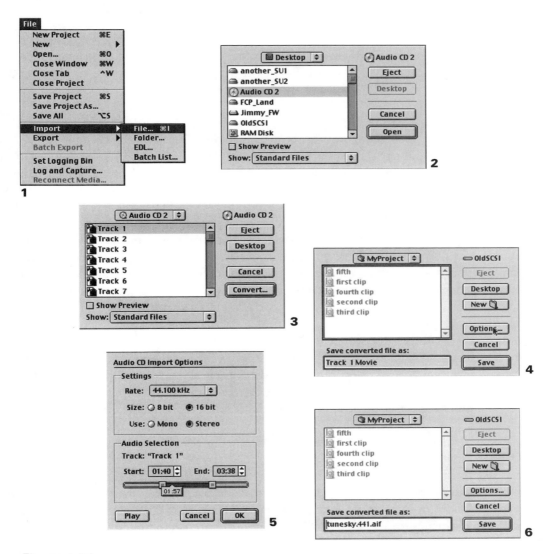

Figure 4-54

STEP 45

You will be presented with a dialog box that asks you to locate the file you wish to import. Navigate the box to the Desktop, locate the audio CD, double-click it, and then select the audio track you want to import.

These track numbers don't tell you much about the audio track you are importing. If you aren't sure you've got the right track, you will be able to preview it right before committing to saving it to your Capture Folder. For now, select the one you think is correct.

STEP 46

Click the Convert button, and you will move to the next dialog box, which contains two important tools. Before saving this track as a file, click the Options button.

STEP 47

You will move to a new box that offers some customizations in connection with the way you import the track. At the top of this box, you will be asked to select a sample rate to convert the Audio CD track to. Audio CD sample rates are always 44.100 KHz. You need to maintain the highest possible sample rate here, so select 44.100 KHz. On importing an audio CD track, you cannot up-sample (i.e., convert the sample rate up from 44.1 KHz to 48 KHz), so maintain the same quality that exists in the original format.

In the bottom area of this box, you can preview the track you are importing by hitting the Play button. Also, you can actually create a range of the track to import by moving the Start and End points, making it easy to trim down the audio track to just the section you want to use. If you realize that this isn't the audio track you wanted, hit Cancel, return to the previous box, and select another track.

STEP 48

Once you have set these Options, click OK to return to the Save dialog box. Now, navigate the dialog box to your project's Capture Folder. Initially, the Save button will be grayed out, because the current Save location is on the audio CD, which is a locked disk. Since this audio file is a relatively large file with a higher data rate, it should be saved with your other video and audio media in the same Capture folder you have designated in your Scratch Disk Preferences.

STEP 49

Once you reach your project's Capture folder on your Scratch Disk, type in a name that corresponds to the file, followed by the suffix .441.aif—for example, "purplehaze.441.aif." When Final Cut Pro converts the Audio CD Track to a format it can use, it will convert to an AIFF file, which is the native audio file format for sound files in QuickTime, the multimedia architecture that Final Cut Pro is based on.

STEP 50

Click OK and the File Format Conversion begins. When it is complete, you will see that a new audio clip, designated by the icon of a speaker, has appeared in your project tab. The speaker icon tells you that the clip is an audio-only clip. If you double-click the clip to load it into the Viewer, you will see that it contains only an audio tab and a filters tab. Now that the clip is imported, you can use it as you would use any clip.

CD import part 2: converting the sample rate

But there is still one complication. Our audio recorded in camera—and therefore the appropriate audio sample rate for our sequence—is either 48 KHz or 32 KHz, not 44.1 KHz. We know that we cannot mix sample rates in our sequences. So how is it possible to get the sample rate of the new 44.1 KHz audio clip up to the 48 KHz or down to the 32 KHz required by our sequence?

The answer is simple; we must export the clip, converting the sample rate as we do so, and then re-import the clip at the new, proper sample rate. It takes only a moment to do so, but not converting the sample rate delivers the risk of audio distortion and possible sync issues. Always work with clips that share the same sample rate with other clips and with the sequence they appear in.

STEP 51

Select the 44.1 KHz clip in the project tab (Figure 4-55). Then go to the File drop-down menu, select Export and then QuickTime from the submenu. A dialog box will appear that requires a couple of changes.

STEP 52

Look to the bottom of the box and locate a drop-down bar entitled Format. Clicking on this menu will offer several different options for Export file formats. Remember that QuickTime is really just an architecture, not a specific file type, so there are separate QuickTime formats for Audio, Video and Audio, MPEG, and the PC variations of AVI and WAV. Since this is an audio file, select AIFF. When you select AIFF from the menu, you will see that Final Cut Pro automatically adds an '.aif' suffix to the file name in the box. You will want to change the name in a moment anyway, so move to the next Option before naming and saving.

STEP 53

Click the Options button and you will enter a new dialog box that asks you for specifics about how to process the file as it exports. This is where you specify the sample rate conversion. Change 44.100 to 48.000, making sure that Compressor stays set to None. Click the OK button to return to the Save dialog box.

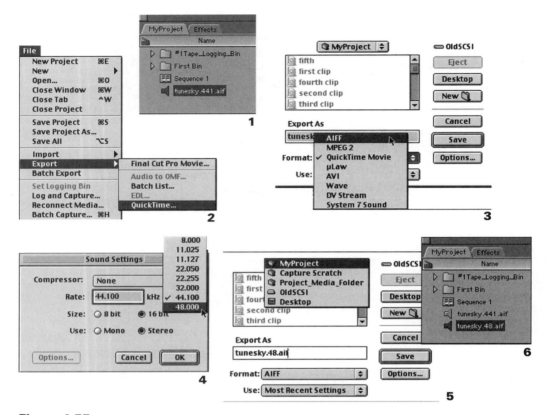

Figure 4-55

STEP 54

Navigate the box once again to the same proper Capture folder in your Scratch Disk folder.
Now type in a name that retains the subject of the clip but specifies that the clip is now 48
KHz and an AIFF file to boot—for example, "purplehaze.48.aif." Click OK, and the con-
version will take place.

Once that is complete, you can import the new 48 KHz AIFF file into your project
and begin working with it. Be careful that you pay attention to the naming. Unless you
throw away one of the two clips you will have a 44.1 KHz and a 48 KHz clip of the same
audio CD track in your project. Some editors simply throw the one with the wrong sample
rate out immediately to avoid confusion. I generally back up everything from my CD track
down to my toenails. Do what is best for your own organization purposes, but remember
to be careful about naming clips and files.

Importing Photoshop image files

Graphics files created in Photoshop are easy to integrate into your Final Cut Pro project. Why settle for the limitations of Final Cut Pro's built-in Title Generator (which we will in fact examine later in the book), when you can come up with well-designed graphics files that incorporate image content as well as stylistic text. Best of all, it's easy to get these into your project.

As with audio files, there are a couple of complications in the import process that must be taken into account. With image files from Photoshop, the two issues are square vs. non-square pixels and accessing Photoshop Layer effects in Final Cut Pro. Failure to take these into account will result in distorted images and/or unexpected features on completion of the import process.

Square vs. non-square pixels

The first issue, that of square vs. non–square pixels, calls for a little explanation about the way different systems display images. A digital image is broken down into individual dots of image detail called pixels, short for picture elements. It is an atomic model in which the pixel is the smallest indivisible section of an image. We measure all digital images using pixels (with the exception of vector-based images, which are addressed in the section that follows), whether that image is a still digital video frame or a scanned image.

Now if all pixels were exactly square, this would make things very simple. And in fact, all computer-driven display systems use square pixels to interpret digital imagery. Most computer applications, like Photoshop, use square pixels. Some video capture cards use square pixels, and video that is to be viewed only on computer screens, such as Web and CD-ROM multimedia codecs, use square pixels as well.

But the problem is that most professional DV editing systems, including Final Cut Pro, use video codecs that require images to be processed using rectangular pixels, which are commonly referred to as non-square pixels. Not all video codecs use non-square pixels, and, as we know, Final Cut Pro can use many different codecs in addition to the DV codec we are using with Firewire editing. The need for square or non-square pixels is determined by the video codec you use. The main thing to consider here is whether or not the image you are creating in Photoshop, or wherever, will be mixed with video compressed by a codec that uses a non-square pixel size. Like audio sample rates, you cannot mix square pixel images with non-square pixel video.

The DV codec that you use with Final Cut Pro uses non-square pixels. If you work with it in a sequence, and your images are not prepared in advance to deal with the difference in the shape of the pixels, the images be stretched vertically and will appear tall and thin. The square pixels of the image that were created in a square pixel graphics application such as Photoshop are interpreted into the non-square pixel shape of the DV codec Final Cut Pro is using.

There is an easy way around this problem. Create your images in your square pixel graphics application using a stretched frame size that Final Cut Pro can scrunch back to the proper pixel shape for mixing with non-square pixel video. The process involves prefiguring the dimensions of the DV frame in the graphics application as if it actually were using square pixels. Then you squeeze the image down in the equal and opposite direction that you know it will be stretched when the square pixels get interpreted as non-square.

Don't let the math or the process bother you. It's a simple series of steps that has been ingrained in every Final Cut Pro editor's head since day one. DV NTSC, which has an actual non-square pixel size of 720x480, would be 720x534 if its pixels were perfectly square. DV PAL, which has an actual non-square pixel frame size of 720x546, would be 768x576 if its pixels were perfectly square.

In Photoshop, you will create graphics images using the pre-figured 720x534 or 768x576 square pixel size of your DV video frame. Then the last step you will take before saving your graphics file will be to use the Image Size command in Photoshop (or the equivalent in whichever application you use) to squeeze the pixel dimensions of the graphics file down (to 720x480 for NTSC or 720x546 for PAL) for import into Final Cut Pro. When you import to Final Cut Pro, the squeezed image will be stretched back out as the square pixels are stretched into non-square pixel shapes. No detail or data is lost or gained. The pixels are just shaped a little differently.

Getting Photoshop layer effects into Final Cut Pro intact

The other thing to understand about importing images is the way that Final Cut Pro deals with Photoshop layers. Photoshop files can be flattened single-layer files or stacked multilayer files. When you create a Photoshop image, you can add imagery to it either on the initial image layer, rather like painting on a flat canvas, or by adding imagery to new layers that exist as discrete images stacked one top of another. This adds a lot of functionality to the Photoshop image, allowing you to edit parts of an image without altering other parts of it and to create one file that contains many different image components, all separated and discrete.

The way this is organized in the Photoshop file is that each layer of a multilayer file is actually a distinct flattened image. Each layer of the Photoshop multilayer file could be viewed as an individual still image, but they are knitted together, one on top of the next so that you see them from the top down as a composite image. This just means that you see the multiple layers of the file from the top down as if it were one image. Anything on the upper layers covers whatever is beneath on the lower layers.

Final Cut Pro is integrated with Photoshop in a special way. It accepts single-layer flattened Photoshop files as you would expect, simply translating the still image into video frames. But it can also import multilayer Photoshop images with the individual layers intact, each layer being regarded as a flattened still image. This is a real bonus, because it allows you to apply Final Cut Pro effects or movement to the individual layers of the graphics you've created rather than having to apply all your effects to an entire image. This can speed up your workflow enormously, especially if you work with a lot of Photoshop-generated text and graphics that must be manipulated and adjusted on a case-by-case basis.

The problem with this is that many of the really excellent effects that can be applied to image layers in Photoshop cannot be directly accessed when a multilayer file is imported to Final Cut Pro. Because many of these effects are Photoshop's proprietary vector-based effects, they are lost when the image is directly imported into Final Cut Pro, which cannot understand them. On the other hand, if you rasterize, or convert the multiple layers to bitmap while still in Photoshop, the layer effects cease to be vector-based and become a part of a flattened bitmap, or raw pixel data. This means that they will still be present when you import the image into Final Cut Pro.

Vector and bitmap graphics

To understand why this is, we need to understand a little about the difference between vector-based and bitmap-based graphics. Vector-based is from Mars, bitmap-based is from Venus, and you are sitting here on Earth. With a few rare interesting examples, applications cannot deal directly with both types of graphics, because they measure and plot space in a distinctly different way. The vector-based method describes the world in a series of geometric shapes, since all complex shapes, like the human body, can be broken down into many smaller less complex shapes. The bitmap method on the other hand is an atomic model, based on the earlier described pixels, in which image matter is composed of an exact number of indivisible individually colored dots.

These two methods of generating an image are at loggerheads. There is no such thing as a single reducible geometric shape. As your high school geometry teacher might have drilled into you, "A point is not a shape." Shapes are possible between points. With a vector-based graphic, you have a point A and a point B. To describe a simple shape like a line, all you have to know is that it is the shortest distance between A and B. To make things more complex, just add another point, or vector, C, and now you've got a triangle. The more vectors you add, the more complex the shapes get until at some point you aren't looking at geometry, you're looking at images.

If you look around at naturally occurring shapes, it's difficult to escape the notion that vector geometry is a natural science, as well as a thoroughly artificial one. Infinitely thin lines and small points aren't physically possible, but nature does generate uncannily geometric patterns. Philosophical pondering aside, it is important to note that vector geometry is an exceptional way of creating computer graphics, one that allows a lot of flexibility and efficiency in adapting shapes in such applications as Photoshop.

Shapes do not exist in the bitmap universe. Neither does empty space. This is the primary difference between the bitmap and the vector. With a bitmap image, each space of the image is occupied by a pixel. The color of that pixel determines what you see in that section of the image. The larger the number of pixels squeezed into an image, the higher the detail (resolution) of the image, since there will be more opportunities for pixels to change color over a distance.

But that is the key; no matter what color the pixels are, they are there, completely filling the image. Vector images, on the other hand, are shapes created from points in an empty space. You see a geometric shape because it has an inside and an outside. A bitmap has no outside, unless you go outside the image file entirely.

Which system is more real or accurate? That is impossible to say, because it depends on the situation. In real life, we tend to see things from a vector-based perspective, because we see through the air around us. When no objects are in a space, we say it is empty, even though we know all about the air molecules in the way. Vector-based images tend to get ridiculously complicated once you get to the level of detail we think looks acceptable for broadcast video. The value of geometric shapes isn't really worth much when you have to express a geometric shape with 6,000,000,000,000,000,000,000,000 different vectors. That could take years just to express one frame of video.

But in the bitmapped view, there is no such thing as empty space, and any number of physicists will come running to join in on the argument. For the bitmap view, outer space, which is almost empty of molecules, would only be a slightly lower resolution of the air-filled empty space on Earth in which we see no objects. The problem with bitmap as a system is that if you don't have enough pixels squeezed into the image—in other words the resolution—you get lesser and lesser detail. It takes a lot of resolution, or pixels per inch, to generate a respectable bitmap image.

The process

The issue for our two applications here is that Photoshop, which lives on Mars, performs some of its image editing effects using vector-based calculation and display methods. At present, video applications, with only a couple of fairly limited exceptions, are all very much living on Venus with a bitmap format. A video frame, as we have said elsewhere, is simply a still image composed of a certain number of pixels of height and width. The series of these still images creates video, And those still images are all bitmaps.

Photoshop has many applications and tools that were not created strictly to address creating images for video applications. For this reason, some of the more interesting visual effects you can create with text and image are vector-based and unavailable for direct use in Final Cut Pro. This does not mean that we have no way of taking advantage of such vector-based effects. It just means that we have to at some point convert the vector-based effects into a bitmapped image data. This is a process called "bitmapping" or "rasterizing." When you bitmap a layer that has vector effects, you take all the geometric shape information and turn it into raw pixel bitmaps that describe the shape that the original vector-based effects looked like.

We will create a Photoshop text graphics file, add some vector-based shading effects, then bitmap it and import it into Final Cut Pro. Let's step through the process of importing a text image from Photoshop.

STEP 55

In Photoshop, create a new image that is 72 pixels per inch (the native resolution for DV video) and 720x534 for DV NTSC projects or 768x576 for DV PAL (Figure 4-56). Since we will want to be able to use this text image over a video clip as a superimposed element, select transparent for background.

STEP 56

Create some text using the Photoshop text tool. Be careful not get your text too close to the edge of the image or it will go out of what is called the Title-Safe Zone.

The Title-Safe Zone and Overscan

The title-safe zone is a rectangle slightly smaller than the actual frame size of the video standard you are using. All video monitors and television have a frame running around the front edge of the screen, cropping what the cathode ray tube displays into a nice tidy 4:3 rectangle. Unfortunately, this crop is not only slightly different on all video monitors, it is also a good deal smaller than what the television tube is actually capable of displaying. The area that is cropped from display is called the overscan.

When you are working in Photoshop, you have to imagine where this title-safe area square should be and get your text inside of it. There is no law about the size of title-safe any more than there is a law about the individual cropping of the television screens by television manufacturers. A safe bet is to use 10 percent of the screen height and width to determine the distance in pixels from each side. For DV NTSC, this would be a centered rectangle in the Photoshop image of 576 pixels wide and 384 pixels high. You can easily set up guides in the Photoshop image window to keep you in the safe area.

You will notice that anytime you create text in Photoshop, the text is created on a new layer, automatically making the Photoshop image a multilayer file. The original layer that

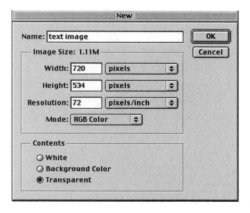

Figure 4-56

was created when you began the Photoshop image is called the background layer (or layer 1). This layer will remain intact as our transparency layer. We will describe the technical part of transparency in the compositing and effects chapter later in the book. For now, just know that it has to be there and that it needs to remain intact and discrete from your text layers.

STEP 57

In Photoshop, select the text layer in the Layer palette, go to the Layer drop-down menu, select the Effects and then in the submenu that follows, select Bevel and Emboss (Figure 4-57). Tweak the settings to your amusement and hit the OK button.

Bevel and Emboss shading effects, being vector-based layer effects, will disappear if we import this Photoshop file as is into Final Cut Pro. Since we want to retain the special layer effects we just added, this will have to be a flattened image file. Since the vector-based effects we want to keep are also layer effects, we need to find a way to convert the layers that have vector-based effects into bitmaps.

The easiest way to change vector-based effects into bitmapped information is to merge two layers into one. When two Photoshop layers are merged, all the separate layer information (e.g., vector-based effects) is bitmapped into the one layer. It becomes a flat group of pixels, rather than a range of vector data.

We could do this by Merging Down or Flattening from the Layer drop-down menu right now. But there's a complication. If we merge or flatten this text layer with the Background layer, we will lose the transparency that the Background layer was going to deliver when we get to Final Cut Pro. The easiest way to get around this problem is to create another empty layer just beneath the text layer and merge down. That way, the text layer gets bitmapped with an empty layer and the Background transparency layer stays intact.

STEP 58

Go to the Layer drop-down menu and select New, and then Layer from the submenu. Click OK in the Layer dialog box that follows. There should be a third and empty layer in your Layer palette above the text layer and the background layer after doing so.

STEP 59

Go to the Layer palette, grab the text layer and drag it up above the new empty layer in the palette. Make sure the text layer is selected.

STEP 60

Go back to the Layer drop-down menu and choose Merge Down. When you choose Merge Down, the text layer will merge with the empty layer directly underneath it, converting the nice shading effects of the text to bitmapped image data. The Background bottom layer will still be intact. Now that we have completed preparing the text for shading and transparency, we need to perform the final square pixel squeeze so that it is displayed correctly in Final Cut Pro when imported there.

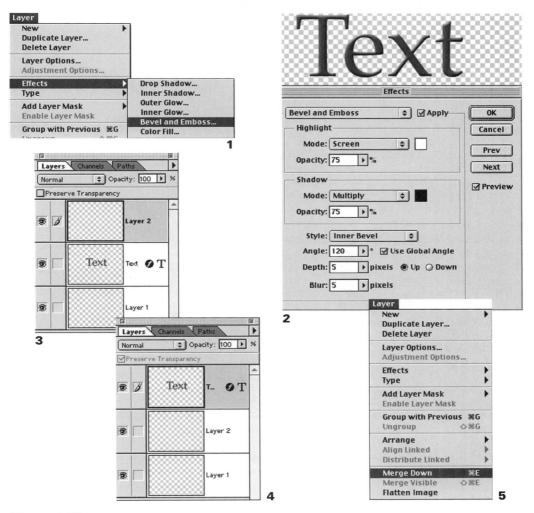

Figure 4-57

STEP 61

Go to the Image drop-down menu and select Image Size (Figure 4-58). Immediately uncheck Constrain Proportions, since we need to squeeze the image vertically more than horizontally.

STEP 62

For DV NTSC projects, leave 720 for the width and type in 480 for the height. For DV PAL projects, enter 720 for the width and 546 for the height. Hit OK. The Image should re-size itself, gently squishing the text slightly out of shape. It is ready to be interpreted by Final Cut Pro as a correctly ratioed non-square pixel image.

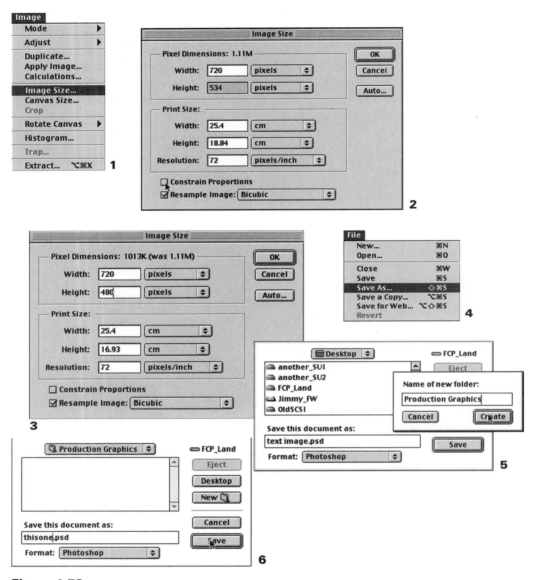

Figure 4-58

STEP 63

Go to the File drop-down menu and select Save As. In the Save As dialog box, navigate to the Desktop and select New Folder. Name the folder "Production Graphics folder," and click Create.

STEP 64

Name the file "thisone.psd," making sure that it is followed by the .psd suffix. Click Save. Quit Photoshop and start up Final Cut Pro.

Importing Photoshop files into Final Cut Pro is simple. You can go through the File menu drop-down process of importing the individual files as with the audio CD, but it is much easier to simply pick up the folder your Photoshop files are in and drag it directly into the Project tab or individual bin of the Browser window (Figure 4-59). The folder you drop becomes a bin, and Final Cut Pro automatically imports all the files it contains.

With no fuss, Final Cut Pro instantly recognizes the files and creates icons for them. The actual Photoshop files do not move from the Production Graphics folder; Final Cut Pro merely creates a link to them from the icons representing them in the Project tab.

Unlike the Audio file we imported and the files we captured, still images need not be saved to a special media drive. This is because the image file will not actually come into play when the clip indexing it is played. Instead, imported still image files must be rendered to play back correctly. The render files that will be created during the render process will go to and be played back from the Scratch Disk location. Thus the original graphics file will not be directly accessed during playback. Just remember not to move the image file's location after importing it to the project or you will break the link between the graphics file and the clip in the project.

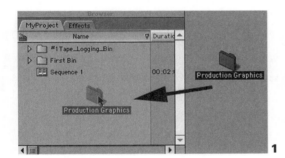

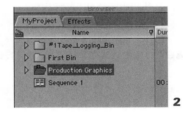

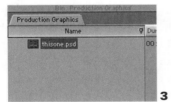

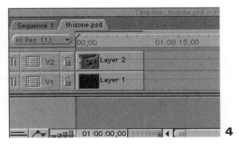

Figure 4-59

STEP 65

Open the new "Production Graphics" bin. Take a look at the icon used for the "this-one.psd" Photoshop image file. Immediately you will notice something interesting. Instead of your Photoshop image file being imported as a normal clip, it is imported as a sequence. The Photoshop file contained two layers when we saved it: the text layer and the transparency layer. When Final Cut Pro encounters a Photoshop image with more than one layer, it interprets it as a nested sequence, a concept we will discuss later in the book. For now, take a look at what is in this sequence.

STEP 66

Double-click the "thisone.psd" sequence so that it opens in the Timeline window. You will see that there are two video layers; one contains the text image, and the other is black (the background transparency layer). This is the best part about importing Photoshop multi-layer images. Because the Photoshop layers are still discrete, you can manipulate them as individual clips. As we will see later in the book, this sequence can be used just like a clip. If you want to see its transparency, you'll need to superimpose it on a video layer above another clip, which is one of the subjects of the upcoming chapters.

As a final note to the introduction of graphic elements to Final Cut Pro projects, imported still graphics must be rendered in order to play back. This is because Final Cut Pro must convert the still image into a video file. An imported graphics file is a single still. For the still to be played back, it must be rendered into a still image for each individual frame in which it appears on a sequence.

Organize your project for media

Much was said in the preceding chapter about ways of organizing the various media in your projects and for good reason. The Project tab is an open slate at the beginning of a project, but it quickly gets confusing as the many different forms of media crowd in.

Simply employing unique naming conventions is not enough. It is important to use the many tools Final Cut Pro offers for keeping your project clean. Make sure to establish bins early, separating not only different types of media, such as imported audio and graphics files, but also the different Reel Names and tapes utilized in your projects.

Also remember that any media you import rather than capture will most likely not have timecode data the way that a captured DV clip does. Although there are exceptions to this rule, any file that does not carry timecode should be backed up to an archive format, such as CD-R or removable disk. Most audio files and production graphics files are relatively small in comparison to video files, and since they cannot be recaptured the way that video can, backing up copies is good insurance against accidental loss. This is especially easy in view of the fact that all new Macintosh models ship with CD burners as the stock CD-ROM drive.

5 What Is Editing?

What is editing?

Now that you have media clips gathered in the project tab of your Browser window, what do you do with them? Edit, of course. But what is editing? Most good editors will tell you that it is telling a story with a series of images and sounds. This is certainly correct. Editing means being able to take isolated pieces of media and arrange them to communicate an idea. That is the gratifying and artistic part of the process.

But there is a technical aspect to editing that has more to do with the tools and less to do with your message. Video Editing is the process of modifying and arranging video and audio resources so that they play back in the linear order you desire. The flexibility and functionality of the tool you use to accomplish this really defines how easy it is to achieve that goal. And there are some fairly standard tools among nonlinear editors for accomplishing this.

Final Cut Pro is a very flexible editing application. It is open-ended enough to allow for many different editing styles. Your own working style will ultimately develop as you get used to the different tools available to you within the Final Cut Pro interface. Unlike the project setup process, which is very rigid and defined by the specific hardware and software configuration you work with, your editing style depends on how you feel most comfortable working and the type of projects you work on.

As we saw earlier, the various windows in Final Cut Pro are linked together based on the media passing through them. We load clips into the Viewer window to manipulate them prior to editing them into a sequence. The Canvas displays what is currently active in the Timeline. And the Timeline is the linear window, displaying the order of clips as they are played back. But this is only one way to view the usefulness of these windows.

The Timeline window is not only useful for arranging the horizontal linear order of clips, it can also be used to stack many vertical layers of video. We can thus control exactly which layer is seen and how much of each layer is seen, much as with the Photoshop layers discussed in the preceding chapter, but over time rather than as a still image. Clips can be loaded into the Viewer window for further manipulation directly from the sequence in the Timeline instead of being loaded from the project tab in the Browser window. You can trim or cut the length of a clip from directly where it sits in the sequence rather than trimming it in the Viewer. The Viewer can do more than simply load and manipulate clips from the project tab and the sequence. You can load up whole sequences as if they were individual clips and manipulate them as well. It will be up to the individual editor to become adept at fully utilizing the many possible methods of completing an edit. Let's step through the process of building a first, simple edit.

Working in the Viewer window

The Viewer window

Editing is a process of arranging pieces of media so that they can be replayed in a particular order. To perform such an action, each media clip must be prepared based on two editing decisions: the clip duration (how short or long each clip will be in the sequence), and the edit type (how the incoming clip will meet the clip occurring before it and after it in the sequence).

Determining the duration of a clip is initially performed in the Viewer by setting In and Out points. When you originally captured the clip, you gave it In and Out points, but they were probably rough decisions. You will want to trim frames away and adjust the clip's duration before you put it into the sequence. The beauty of a nonlinear editor is that such actions are very simple. Since we are dealing with a clip, we can put In and Out points wherever we want, or change them later on a whim if they don't suit us.

STEP 1

Double-click one of your captured clips in the Project tab to load it into the Viewer. You will be happy to find that the playback controls of the Viewer utilize the exact same conventions as the Log and Capture window. The Viewer window, and all other windows in the Final Cut Pro interface, use the same keyboard navigation conventions (J-K-L, I, O, and Right and Left Arrow keys), even though they are playing back clips that exist on a hard drive rather than on an external tape.

As with the Log and Capture window, the top left timecode field displays the duration of the clip loaded into the Viewer. The top right-hand field displays the frame the playhead is currently parked on in the clip. Below the Viewer's video window, you will see that there are no timecode fields for In and Out points, which have been replaced by a few buttons as described next.

Match Frame button

The first button on the left is the Match Frame button (Figure 5-1). If a clip loaded into the Viewer window is already being used somewhere in a sequence, the Match Frame button will call up the sequence in the Canvas and Timeline windows, and it will place the sequence playhead on the same frame of the clip in the sequence that the playhead is parked on in the clip in the Viewer window. It is used for quickly finding the exact same

Figure 5-1   **Figure 5-2**   **Figure 5-3**

frame that is being used elsewhere in a project. This is an invaluable tool if you want to
make sure you are not accidentally reusing footage. You will find that there is also a similar
Match Frame command for the sequence that allows you to call up a clip in the Viewer.

Mark Clip button

The next button, the Mark Clip button, is like a quick reset button for In and Out points
in the clip (Figure 5-2). Clicking it immediately places In and Out points at the beginning
and end of the clip. If you want to quickly clear the In and Out points you have already
placed in the clip, or if you know that you want them to encompass the entire clip, click
this button. You will see two sideways triangles appear in the clip's timeline area defining
the duration of the portion of the clip to be used in an edit.

Of course these points are totally malleable, and setting them is as temporary as you
want it to be. You can grab the points themselves and move them manually, set them using
the In and Out point commands to be described in a moment, or simply Mark Clip them
back to the beginning and end limits of the media file associated with the clip.

Add Keyframe button

Add Keyframe, the next button, allows you to quickly manipulate the clip's physical
attributes that are located in the Motion tab of the Viewer window (Figure 5-3). It's simply
a shortcut for defining a change in the way the clip's video and audio is displayed or heard.
Keyframing will be addressed in more detail in the chapter on Compositing and Special
Effects.

Add Marker button

Add a Marker applies a Marker to the frame on which the Viewer playhead is currently
parked (Figure 5-4). Markers are very useful tools for editors. Although they do not affect
playback at all and are visible only to the editor of the clip in the Viewer window, Markers
allow you to insert detailed information about a frame or entire sections of your clip. Do
not confuse the Markers with the Mark Clip button. Mark Clip addresses In and Out editing points for the clip, while Markers are just accessory information about the clip that is
linked to a frame or range of frames.

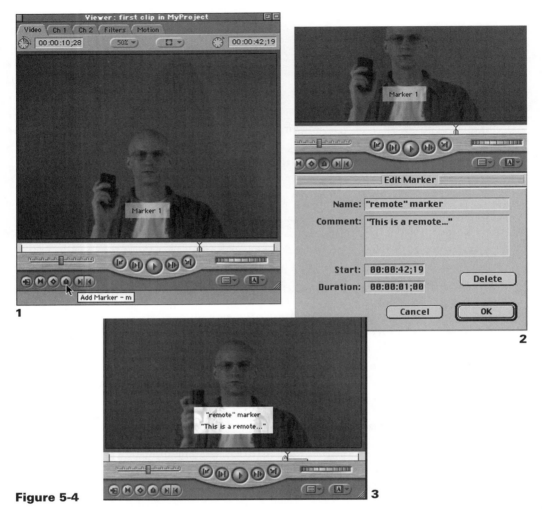

Figure 5-4

STEP 2
Click this button to add a Marker to the clip loaded in the Viewer window. Then, making sure that you see the word Marker 1 in the video window, press the M key (if you do not see the Marker 1 in the Viewer's video window, move the playhead around the Marker point in the Viewer playhead timeline until you see it). This brings up the Edit Marker dialog box.

STEP 3
Here in the Edit Marker dialog box, you can rename the Marker anything you want, type a message about the area covered by the Marker, and give it a duration so that the details of the Marker can be clearly linked to more than one frame. All this information will be dis-

played in the Viewer window whenever the Viewer playhead is in the Markered area of the clip. Best of all, the Marker does not affect your edit at all. It only provides another means of adding information about the clip so that you can be better organized.

Adding Markers to clips is especially useful when two or more editors collaborate on a project. Since they may not be working at the same time, it's often critical to leave notes about specific footage that cannot be missed as the editing process moves forward. Markers, it will be recalled from the last chapter, can also be set in the clip during the Log and Capture process. Those Markers will still exist within the clip after capture, being available to the editor long after the process of logging is completed. This is an excellent way for producers to communicate their desires for a clip's manipulation to the editor, who may start working on a project long after the producer has become unavailable for questioning.

In and Out Point buttons

The remaining two buttons here are the In and Out points for the clips (Figure 5-5). The In and Out points will determine the first and last frame from this original, or Master clip, that will be used when the clip is edited into the sequence. The idea of Master clip and sub-clips is important here. The clip you load into the Viewer window from the Project tab is what is referred to as a Master clip. The Master clip is the original clip brought into the project that contains all the footage you choose your edits from. It likely contains far too much footage and must be whittled down using In and Out points before moving to the sequence.

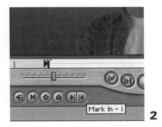
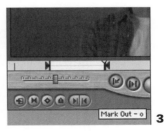

Figure 5-5

A funny thing happens, though, once this Master clip is dragged from the Viewer window to the sequence. Instead of the Master clip itself appearing in the sequence, a subclip, or an edited and shortened version of the Master clip, is created in the sequence. This subclip is exactly like the Master clip in the Viewer it was created from, with one exception. When you load the Master clip into the Viewer window, you see all the frames of the clip, regardless of where the In and Out points are positioned. But the new subclip in the sequence uses the In and Out points you applied in the Viewer to determine the beginning and end of the clip on the sequence.

The Master clips and subclips used here should not be confused with the Final Cut Pro Subclip, which is a special command to be discussed later. The use of Master and subclip here is standard usage for describing the way that any nonlinear editing systems relate an initially captured clip to subsequently edited versions of that clip. The reason for belaboring the point is to make the reader aware that the clip edited into the sequence is not the same clip as the one in the Project tab or the Viewer, although they relate back to the same media file on the media drive. You can create as many subclips from a Master clip as you wish, and indeed you do so each time you drag a Master clip from the Viewer window into a sequence.

Remember that the Log and Capture keyboard conventions apply in the Viewer window as well, so use the I and O keys to set the clip's In and Out points and the J-K-L keys to shuttle around the clip quickly. Hitting an In or Out point clears the previously selected In or Out point, so you can constantly reset them to your heart's desire.

Drag and Drop to the Sequence: Insert and Overwrite

Once you have selected a new In and Out point for the Viewer window clip, it's time to edit it into your sequence. There are a couple of ways to do this, and the best way really depends not only on how you like to edit but what sort of project you are working on. Once you master the two main methods, you'll be able to judge which is best suited to any particular situation.

The first and most basic method is to simply drag the clip from the Viewer video window into the sequence in the Timeline window.

STEP 4

Click in the video window of the Viewer, and drag the clip down to the sequence without releasing the mouse button (Figure 5-6). Keep the mouse button pressed as you move it around the sequence, and watch the mouse pointer change shape as the sequence prepares to receive the clip.

As you drag the clip around the sequence Timeline, a ghost of each of the clip's video and audio tracks appears there, showing you which sequence tracks the clip will be dropped into if you let go. Notice that the ghost clip may also appear to sit outside the visible tracks

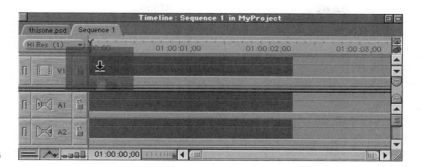

Figure 5-6

in the sequence. This is because Final Cut Pro will allow you to drop the clips where a new video or audio track should be, and then it will automatically create the new track to accommodate it.

STEP 5

Continue to hold the ghost clip so that it appears to sit in Video 1 and its audio tracks sit in Audio 1 and 2. Move the mouse pointer around the base line of the tracks in the sequence, and pay close attention as its shape changes (Figure 5-7).

Notice that depending on where in the sequence track you hold the ghost clip, the mouse pointer sometimes appears as a white arrow pointing either down or to the right and down. This indicates whether the edit you accomplish here will be an Overwrite or an Insert edit. Insert and Overwrite are the two most basic edit actions in a nonlinear system.

An Insert edit, represented by the "right and down arrow" mouse pointer, neatly slips your clip into a sequence. If the sequence already contains clips, Inserting will move all those occurring after the edit In point (the frame where the sequence playhead is parked) over to make room. If the sequence playhead was parked on a frame in the middle of a clip, that clip will be split into two sections, and the section of the clip after the playhead position will be moved to the right to make space for the incoming Inserted clip. No frames or clips are removed from the sequence; clips occurring after the edit In point are merely moved further down the Timeline.

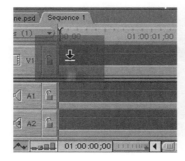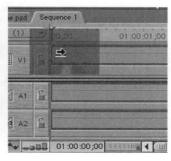

Figure 5-7

An Overwrite edit, represented by the "down arrow" mouse pointer, lays the clip down wherever you drop it in the sequence regardless of what is currently sitting on the Timeline. If any clips occupy an area of the sequence where the incoming clip will sit, it overwrites them. Unlike the Insert edit, which maintained the same clips in the sequence by moving them out of the way, the Overwrite edit deletes any frames of any clips in the sequence that it encounters.

The fundamental difference is that Insert editing changes the duration of the sequence. Since the new clip moves everything over to the right and adds its duration to the sequence's total duration, the result is more footage in the sequence Timeline. Overwrite, on the other hand, keeps the duration of the sequence exactly the same, since it is replacing whatever clip it overwrites with itself. No clips are moved over to compensate, and the duration stays the same unless the overwrite clip extends beyond the clips it is overwriting. These two types of editing require more than one clip to highlight the difference between them.

STEP 6

Using either the Insert or the Overwrite arrow, drop the clip you dragged from the Viewer window onto the first frame at the beginning of the sequence so that there is only one clip in the sequence (Figure 5-8). If there are no clips present in the sequence, the effect of Insert and Overwrite is exactly the same, since there is no clip to overwrite or insert into.

Notice that the clip, once dropped in the sequence Timeline, appears in the Canvas window (Figure 5-9). Wherever the playhead is moved to in the sequence Timeline, the playhead in the Canvas moves there as well; conversely, when you click and drag the playhead in the Canvas, the playhead in the sequence Timeline moves around, mirroring the action in the other window. We say that the Canvas window is a video preview of the sequence in the Timeline window.

For this example, we will edit in the same clip already loaded into the Viewer window, although you could load another clip and set In and Out points for it just as easily. Until you close the Viewer window or load another clip into it, the previous clip you loaded will still be sitting there.

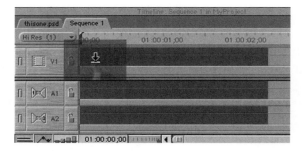

Figure 5-8

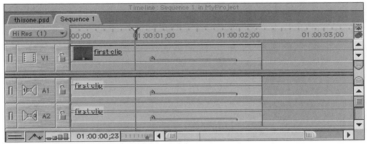

Figure 5-9

STEP 7

Play through the clip in the sequence and park the playhead on a frame somewhere halfway through the clip (Figure 5-10). The position of the playhead will be parked on the target frame we intend to drag the new incoming clip to in the sequence.

Figure 5-10

STEP 8

Return to the Viewer window where your previous clip is still loaded. Once again, click and drag the clip from the video window and move it down to the sequence without releasing the mouse button. When you bring the ghost clip down to the area where you previously parked the playhead in the first clip, the mouse pointer will snap to the playhead in the sequence timeline (if the clip does not snap to the playhead in the sequence window, hit the N key to enable Snapping, which may have been disabled. The Snapping feature will be more thoroughly described further on in the chapter.)

STEP 9

As you move the mouse pointer around this area, you will see the Insert/Overwrite arrows appear (Figure 5-11). When the arrow is pointing to the right and down indicating an Insert edit, let go of the mouse button to perform the edit.

You will see that the first clip was split in two and that the second incoming clip was inserted into the newly created gap between the first and second half of the initial clip. No footage was removed from the sequence; footage after the playhead was moved over to create the space necessary to insert the incoming video clip.

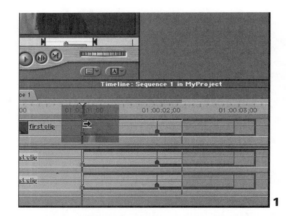

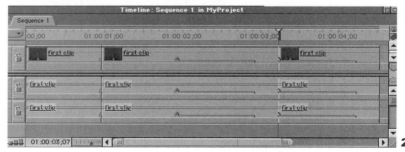

Figure 5-11

Hit Command-Z to undo the Insert edit from Step 9 so that you are once again looking at a single clip on the sequence Timeline. You can always quickly undo the last action performed by hitting the Command-Z shortcut. Once again, play through the clip in the sequence and park the playhead on a frame somewhere halfway through the clip. Now perform the same action again, but this time, perform an Overwrite edit by letting go of the mouse button when the arrow is pointing down.

STEP 10

Return to the Viewer window where the previous clip is still loaded (Figure 5-12). Click and drag the clip from the video window and move it down to the sequence without releasing the mouse button. When you bring the ghost clip down and it snaps to the playhead on the sequence Timeline, you will see the Insert/Overwrite arrows appear again. When the arrow is pointing directly down, let go of the mouse button to perform an Overwrite edit.

Instead of splitting the clip in half, moving the remaining footage over and inserting the incoming clip, Final Cut Pro simply lays down the second clip in the place of whatever was after the playhead position. It overwrites any footage that lies in the path of the incoming video clip.

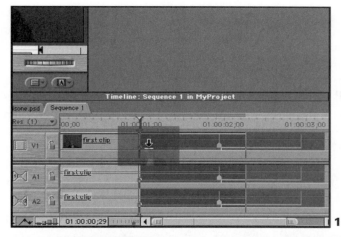

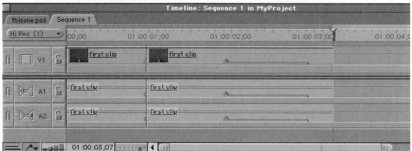

Figure 5-12

Drag and drop to the Canvas: Insert and Overwrite with a Transition

There is another way to perform Insert and Overwrite edits, this time with the option of adding another important element, the Transition. Instead of dragging the clip from the Viewer window directly to the sequence Timeline, we will use the Canvas window. As we perform the edit using the Canvas, we will also address the second important issue in the process of editing: how we will have the clips cut from one clip to the next in the sequence.

We use the venerable old tape-editing terms *straight-cut editing* and *A-B roll editing* to describe the two primarily different ways of getting from one clip to the next in a sequence. Straight-cut editing means that when the playhead reaches the end of the first clip, it immediately cuts to the second clip. There is no transitional area in which both clips are simultaneously in view, and clips do not overlap or fade into one another. One clip stops on one frame and another one starts on the very next frame.

A-B roll editing allows for a different sort of cut between clips called a Transition. A Transition is a juncture between two clips. But unlike the straight-cut juncture, which allows no overlapping, the transition lets us see a portion of both the outgoing and the incoming clip simultaneously. This can be a Dissolve, in which the outgoing clip fades out and the incoming clip materializes. It can be a Wipe, in which a line crosses the screen covering the outgoing clip and revealing the incoming clip. It can be any number of stylistic effects, but the main point is that the two clips are both seen simultaneously for the duration of the transition.

The transition originated in the old film and video tape-based editing systems with a mixing system called A-B Roll editing. With film, an optical printing process was used to mix an outgoing and incoming edit. To perform video A-B Roll editing, you need to have three decks. The first two decks, A and B, contain the outgoing and incoming clip footage respectively that is being edited onto the third deck, the edit (or record) deck. A video mixer lies between A and B decks and the edit deck. The video mixer takes video sources A and B and mixes them together in whichever transition was required as they pass out to the edit deck. A transition from clip A to clip B is performed by mixing out clip A and mixing in Clip B during a short period in the recording. Once the transition is complete, clip A is no longer playing and clip B is being recorded.

The editor decides where the transition should occur and then uses the video mixer to switch from source A to source B, using one or another of the many possible transitions. Which transition is used and how long the transition takes depend on the editor, as does the length of the overlap between the two clips being transitioned. Although the term A-B Roll isn't common any more outside of expensive online tape editing suites, the process of transitioning between clips on the sequence is essentially handled the same way. The A and B decks correspond to the first and second clip that we want a transition to separate in our sequence. As we pass out of clip A, we want to transition into clip B.

There are two things to remember about the A-B Roll editing example. First, there was a video mixer involved in the A-B roll example; second, there was an overlap between the two video clips in order to perform the transition. The mixer from our example will be the transition that we apply to the cut where the two clips abut. And the overlap that is required will be the extra, unused footage that lies beyond the In and Out points of the clips we edited into the sequence. Since the transition requires overlap area between the two clips, you need to make sure there is trimmed footage that the transition can access from beyond the In and Out points of each clip. The extra footage on either side of the In and Out points of the clip is the handles you were just being recommended to capture with your clips.

Once again, we take advantage of Final Cut Pro's drag and drop technique to create this transition.

STEP 11

On the sequence, move the playhead a little further down to an area where there are presently no clips. Now go back to the Viewer window where your clip is still loaded. This time, create new In and Out points, making sure that there are at least 30 frames between the beginning of the Master clip's media and the In point you apply to it. Likewise, make sure that there are at least 30 frames between the end of the Master clip's media and the Out point you apply.

These 30 frame handles on either side of the In and Out points will provide plenty of overlap frames for our transition. The overlap space for each clip is the footage that lies before the In point or after the Out point of the clip.

STEP 12

Click the clip in the Viewer video window and this time drag it into the video window of the Canvas without releasing the mouse button (Figure 5-13). As you hold it over the Canvas, you will see overlay windows appear offering the choice of Insert, Overwrite, Replace, Fit to Fill, and Superimpose. In addition, the Insert and Overwrite areas will have an optional extra window for including a transition.

Because there is currently no video clip in the sequence, we could choose either Insert or Overwrite with equal results. You will perform the exact same edit as you did when dragging directly to the sequence earlier. This time, however, instead of determining the frame to edit into the sequence with the mouse pointer, you will simply edit to the place where the sequence playhead is currently parked.

STEP 13

Hold the clip over either Insert or Overwrite (do not choose the transition option just yet), and you will see the choice highlighted. Release the mouse button, and you will see the clip appear on the sequence exactly where the playhead was parked.

STEP 14

To add the second clip with a transition, first make sure that the playhead is parked at the end of the clip you just inserted into the sequence (Figure 5-14). When you edit a clip into

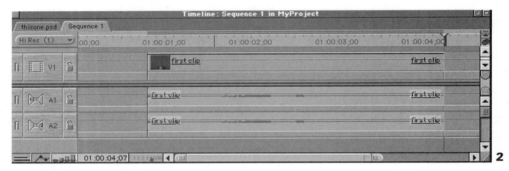

Figure 5-13

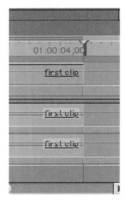

Figure 5-14

a sequence, the playhead immediately moves to the last frame of the incoming clip, since it assumes your next edit will follow that clip. If the playhead has been moved, drag it back over to the end of the clip you just added to the sequence, making sure that it is on, not following, the last frame of the clip.

Transitions, such as the default Cross Dissolve that we are about to use, are difficult to see if the clip being transitioned to is very similar to the clip being transitioned from. Since the Cross Dissolve fades from one clip to the next, you might not even be able to see the transition. For this reason, pick a new clip from your Project tab that looks substantially different from the clip you have just edited into the sequence.

STEP 15
Double-click another clip in the Project tab to load it into the Viewer window. Once again, set In and Out points, making sure that you leave at least 30 frames on either side of the In point and the Out point to provide adequate overlap space for the transition.

STEP 16
Pick up the clip in the Viewer window and drag it into the Canvas window, this time releasing it as it hovers over the Overwrite with Transition area (Figure 5-15). Instantly, the clip appears on the sequence following the first clip, but this time there is a small gray patch connecting the two.

The small gray patch between the two clips is the Transition. It is an area in which footage from both clips is present, although it is actually composed of footage that exists beyond the previously established In or Out points of the clips. In this nonlinear system, Final Cut Pro can reach back into the media file associated with your clip when it needs more footage as long as there is more footage in the clip's media file to access. The benefits of adding frame handles in the capture process of the last chapter become more obvious now that footage beyond our In and Out points becomes valuable as transition overlap footage.

If you receive a message stating "Insufficient Content for Edit," it means that you do not have enough footage beyond the In or Out point of either one or both of the clips for Final Cut Pro to create the transition. Remember that an overlap of footage means that both clips must have at least some footage outside the In and Out points. Although the number of frames outside the In and Out points does not have to be equal, it does have to exist.

By default, this transition is a Cross Dissolve, which will fade one clip out and the other in over the length of the transition. You can change the default transition to something you use more frequently, or choose any other transition available to replace the Cross Dissolve after it has been applied.

After applying the transition, place the sequence playhead directly in the center of the transition (Figure 5-16). You will be able to see both the first and second clips at half their opacity or visual strength, since one is fading out and the other is fading in. If you try to play this short sequence back, you will probably notice that when the playhead crosses the transition, the footage disappears and the word Unrendered appears. Once the playhead is

Figure 5-15

past the transition, the video footage appears again. This is because transitions must be rendered to be played back correctly.

The reason that your normal captured footage can be played back without rendering is because the footage exists on your hard drive as captured media. There is media that the clip is related to. When the playhead encounters the clip, the media is accessed and the video plays back. But the transition is different. There is no media clip yet that corresponds to the dissolve between the two clips. Final Cut Pro knows that the transition footage needs to be created or rendered to be played back correctly, so it warns you that the material is unrendered with this message.

Any footage in a sequence that must be rendered will have a red bar over it in the Timeline window (Figure 5-17). The red bar indicates that render files have not been created for the clip and that it will not be displayed properly upon playback. Anytime you see this red bar, it means that there is an effect of some sort—a Motion tab effect, an Effects filter, or a Transition that must be rendered. The only exception to this is if your Preferences and/or Audio/Video Settings have been incorrectly configured, in which case your

Figure 5-16

captured footage will likely display the red bar as well. In those instances, do not render; go back and fix your settings!

STEP 17

Rendering the transition is simple. Make sure the Timeline window is active, then go to the Sequence drop-down menu and select Render All (Figure 5-18). There are other choices

Figure 5-17

Figure 5-18

available for rendering in this drop-down menu. What choices are available depends on what clips, if any, are selected on the sequence Timeline. If an individual clip or transition is selected, you can choose Render Selection to avoid rendering all clips or transitions you may not be presently concerned with. If no clips or transitions are selected, then Render Selection will render all clips in the sequence.

A render progress bar will come up showing you where the rendering process is. After the render is complete, the red bar will be replaced by a barely perceptible light gray bar, indicating that the material has been rendered. Go back to the sequence and play through the transition. You will now see the dissolve properly played back.

Of course you are not limited to the way this dissolve performs. You are looking at the default settings for the Cross Dissolve transition. You can easily change the parameters of the transition itself, being limited only by the number of overlap frames available past the In or Out points of either the outgoing or the incoming clip.

STEP 18

To change the parameters of the transition, go to the sequence and double-click on the gray transition itself (Figure 5-19). Make sure that you click on the gray transition and not the cut, or edit point, between the two clips (if a Trim Edit double video window appears, close it and try again). When you successfully double-click on the transition, it will load up in the Viewer window, as if the transition were a clip, but with a control tab for the transition itself.

When the transition is loaded into the Viewer window, you can adjust how long the transition lasts, where the middle point of transition occurs, and many other factors that depend on the type of transition you have created. You will see that as you adjust the parameters in the Viewer window the transition in the sequence is updated. You will also see that the red render bar has returned, because the render files you created no longer reflect what the transition does. To view the new transition parameters, you will need to render again.

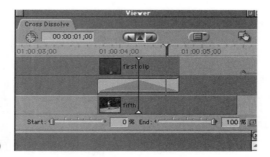

Figure 5-19

More about transitions

Drag-and-drop transitions

The method discussed in the preceding text is not the only way to apply a transition to two abutting clips. There is yet another simple drag-and drop technique for applying transitions.

STEP 19

Clear off the sequence by activating the Timeline window, clicking on the dark gray area of the sequence Timeline and clicking and dragging through the clips with the mouse button pressed (Figure 5-20). This action is called a marquee selection. You will notice that each clip marqueed this way turns brown, indicating that the clip was selected.

You can also quickly select more than one clip by clicking on clips with the Shift key depressed (i.e., Shift-clicking). As you Shift-click more clips, they all stay selected, whereas if you were not Shift-clicking, selecting one clip would deselect the previously selected clip(s). Shift-clicking changes the selection status of a single clip without changing the selection status of other clips. To deselect an individual clip while Shift-clicking clips, simply Shift-click the clip again to remove it from the group of selected clips.

STEP 20

Once all the clips in the sequence have been selected, hit the Delete key to remove them from the sequence. When the sequence is again empty, quickly perform two Overwrite edits into the sequence so that two clips are abutted with no transition but have the previously described overlap frames on either side of their In and Out points.

STEP 21

Go to the Browser window and select the Effects tab. Double-click the bin named Video Transitions (Figure 5-21). Inside the Video Transitions bin that opens up, double-click the Dissolve bin. Inside the Dissolve bin, click the Cross Dissolve icon and drag it over onto

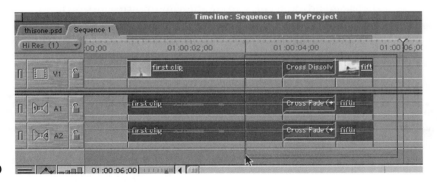

Figure 5-20

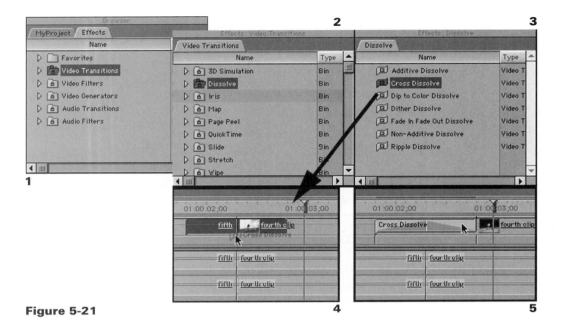

Figure 5-21

the cut between the two clips you just edited into the sequence. When released onto the edit point between the two clips, the transition will pop right in, just as if you had applied it from the Canvas.

STEP 22

Double-click the gray transition you just dropped onto the edit point between the clips so that it loads into the Viewer window. Once again, adjust the various parameters of the transition until you are satisfied with the results. Click on the Timeline window again, hit Command-R (the keyboard shortcut for Render Selection) and after the brief render process, play the transition back.

Using the drag-and-drop method of applying transitions to edit points between clips, you can select any of the available transitions in the Video Transitions bin of the Effects tab.

Creating customized transition favorites

Once you have adjusted the parameters of a transition the way you want them, you can save a new version of the transition that retains the parameters you have changed for quick access in the future. Instead of having to change the parameters for a transition every time you use it, you will make a copy of the transition you have customized and apply that instead of the default one you started with. This is called creating a Favorite.

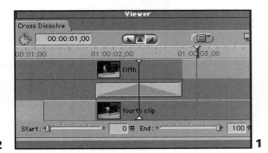
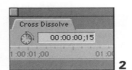

Figure 5-22

As an example, let's say that you prefer for your Cross Dissolve to last for 15 frames instead of the full one second that the default Cross Dissolve transition is set to last.

STEP 23

In the sequence, double-click on the transition so that it loads into the Viewer window (Figure 5-22). In the timecode field to the upper left of the window, change the value from 00:00:01:00, or 1 second (29.97 frames for NTSC and 25 for PAL), to 00:00:00:15, or fifteen frames. The transition will automatically shrink both in the Viewer and in the sequence.

Now that the transition in the sequence has a new customized parameter that differs from the one in the Video Transitions bin, we will create a Favorite of it.

STEP 24

Double-click the Favorites bin in the Effects tab of the Browser window so that it opens up (Figure 5-23). Go back to the sequence, grab the gray transition, and drag it to the opened Favorites bin.

STEP 25

A new Cross Dissolve transition bearing the parameters you changed is in the Favorites bin. Although it is still named Cross Dissolve, it actually has a different length than the Cross Dissolve you originally pulled from the Video Transitions bin. Before you can confuse one for the other, click on the name of the Cross Dissolve transition in the Favorites bin to highlight it. Type in the new name "CD15frames" to distinguish it from the original Cross Dissolve or any further variations of the Cross Dissolve transition you may create as Favorites.

Now whenever you wish to apply a 15-frame Cross Dissolve, you can simply drag this transition from the Favorites bin and apply it. The transition will also be available from the Effects drop-down menu under the heading Favorites (Figure 5-24). You can make this transition the default for use in applying transitions from the Canvas as we did earlier in this lesson by selecting your "CD15frames" transition in the Favorites bin and choosing Set Default in the Effects drop-down menu.

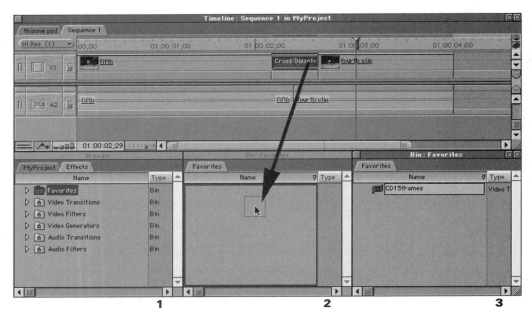

Figure 5-23

You could also have placed the transition on the edit point by moving the playhead in the sequence to the point between the clips and going to the Effects drop-down menu. There, under the Video Transitions submenu, you will find all the Final Cut Pro transitions available to you. The Effects drop-down menu is your access menu for many other effects to be described later in the Compositing and Special Effects chapter.

If it seems arbitrary to have waited until this point to describe the drop-down menu items for performing transitions and effects, there is a reason. Your speed and agility within Final Cut Pro, as well as other production applications, greatly depends on your ability to use the interface of the application directly rather than hunting through menu items one by one to find what you are looking for. The risk of carpal tunnel syndrome is another reason for avoiding the repetitive action of pulling down drop-down menus. Learning the keyboard shortcuts for actions and using the mouse only for things like dragging clips around will save your wrist a lot of work and speed up your process into the bargain.

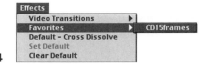

Figure 5-24

Trimming those edits

At this point, you can bring clips into Final Cut Pro, select In and Out points for them, and get them to the sequence in the Timeline window with or without a transition. That is enough to edit with. But it implies that we don't want to change things once they make it to the sequence.

What we have done up to this point isn't really that far removed from the process of tape editing. With tape editing, cuts follow cuts. You lay down an edit onto the edit deck, and then you lay down the next edit, possibly with a transition. You continue doing so until you complete the piece.

But that is linear editing. If you discovered that you've extended a clip a little too long, you have to go back through and re-edit the whole thing from the offending clip on. But we are using a nonlinear editor here, and we should take full advantage of its unique ability to alter clips from within the sequence itself, instead of simply adding edit to edit in a linear fashion.

Of course we've already seen a little nonlinear action with the Insert edit. With the Insert edit, we put a clip exactly where we wanted it to occur without overwriting what was already there. But we can do much more with our nonlinear editor. We can trim our clips with great precision in the sequence itself, using tools that are dedicated to helping you arrive at the most appropriate cut between any two clips that adjoin.

There are a few ways of trimming clips on the Timeline. These include directly grabbing the edge of the clip and moving its In or Out point, using the dedicated Trim Edit window, or using certain dedicated tools that live on the Toolbar. These different methods offer differing conveniences. One is less precise but easily accessible, one is more precise but must be accessed in a separate window, and another is a unique mixture of both.

The key is to master them all so that you can access each one where it becomes valuable. Only experience can instill the quick appropriate use of each one, so try them all until you get the hang of them, and then access them frequently. Trim functions often confuse new users because the relationship between clips and other clips or between clips and their media is unclear.

Trimming is not difficult however, as long as you remember that each clip on the sequence is only a small carved-out section of a larger Master clip that you originally captured or imported into the Project tab. Because it is only a section of the clip, Final Cut can easily change the In and Out points for that section on the fly without the need to revisit the Viewer window's edit tools. As you change the In and Out points for the clip, you are only defining which section of the Master clip the smaller section in the sequence refers to. Trimming a clip that exists in a sequence does not affect the Master clip, which exists in the Project tab.

Mastery of the various Trim functions will turn your Final Cut Pro station from a nice editing package into a real professional editor. It will give you control over every aspect of the sequence and allow you the ability to make quick changes that drastically alter the nature of your production without having to rebuild entire editing structures from scratch. Trim is the secret weapon of all good editors and is an essential tool to master.

The easiest trimming technique

The simplest trim function you can perform involves trimming the In or Out point of a clip directly on the edges of the clip itself in the sequence. This trim is so basic it does not require more than one clip. You are simply changing the In or Out point for the clip based on the amount of media associated with the Master clip.

STEP 26

Start with a fresh clean sequence in the Timeline by selecting and deleting all the clips there. Insert edit a clip onto the sequence. Move the mouse pointer near either the beginning or end of the clip until you see it change from the generic Selection arrow to a vertical line with an arrow pointing in either direction (Figure 5-25). When the mouse pointer is shaped like this, it has assumed a trim function.

STEP 27

Click on the end of the clip and drag in either direction. You will see that the length of the clip adjusts following this tool. When you let go of the mouse button, the new In or Out

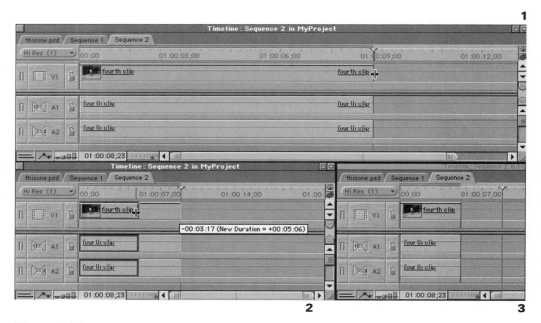

Figure 5-25

point for the clip will stick and you will have changed the duration of the clip. Also notice that as you adjust the trim, a small text box appears next to the clip indicating the number of frames you have added or subtracted to the clip in the trim operation and the new total length of the clip measured in duration.

STEP 28

Click the clip edge again and perform the same drag motion, dragging as far to the left or the right as the tool will allow. At some point, you will reach the limit of the media file that the Master clip refers back to. Obviously, you cannot extend a clip beyond the limits of the media file or the Master clip it refers to, so this is the limit of the trim for the clip.

This tool can be very limiting for a couple of reasons. When you trim a clip, you are usually doing so to refine an edit point, something that you want to approach with precision that may require removal or addition of only a single frame. Unfortunately, this direct trimming action can be very unwieldy for precise editing. Much of the time, you will be looking at your Timeline scaled out so that you can see several seconds or even minutes at a time. In such situations, trimming a single frame using the direct method is practically impossible, like using your entire fist to dial a number on a touch-tone telephone. Although it can be done, the edit make take several hair-raising and frustrating moments, during which all your creative juices leak away.

Another problem with using the direct trimming tool is that you can trim only one end of one clip at a time. This may not seem limiting at first glance. But what if you have two clips abutted as a straight cut in a sequence and you want to take one frame away from the first and add one frame to the second? Using the direct trim method described above would require clicking on both clips, probably more than one time. And if you weren't sure that's what you wanted, you'd have to go back through the same motions to change it back.

More trimming precision: the Trim Edit window

Final Cut Pro has a solution for such issues. It's called the Trim Edit window, and it is a dedicated window for allowing extreme precision in selecting the exact point at which two adjoining clips meet each other. Using the dedicated Trim Edit window, you will be able to quickly add or remove frames from the Out point of the first clip and the In point of the second clip. You will also be able to easily preview the trimmed edit from inside the window so that you can adjust it again before committing to it. And best of all, you can choose to adjust the first, the second, or both clips simultaneously. It is very flexible.

Calling up the Trim Edit Window

To use the Trim Edit window effectively, you need to have two abutting clips.

STEP 29

Clear off the sequence of clips using Select and Delete so that you start with a fresh sequence. Perform two edits so that you have two clips abutting on the sequence in a straight cut.

STEP 30

Go to the sequence and double-click the edit point between the two clips (Figure 5-26). Doing so will immediately open a large double video window called the Trim Edit window.

    The two video windows of the Trim Edit window each correspond to one of the video clips; the left window is for the first or outgoing clip, and the right is for the second or incoming clip. Inside this window are all the tools you need to prepare a frame-specific edit between any two clips.

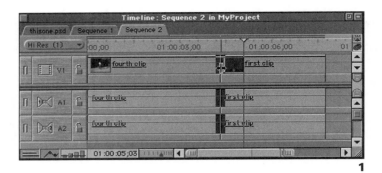

1

2

Figure 5-26

Some of the navigation tools will appear familiar, but make no mistake; this window functions according to its own rules. What you do here updates the sequence it was opened from, as one would expect. Until you are familiar with the functioning of the Trim Edit window, make sure to keep the Timeline window in view so that you can monitor the effect the Trim Edit window has on the sequence it originated from.

On the top left-hand and top right-hand corners of the Trim Edit window, you will find the familiar clip duration fields. These give the present duration of their respective clips, based on the current In and Out points of the clips in the sequence. This number does not reflect the duration of the Master clips from the Project tab that the clips from the sequence reference. The duration will change as you trim the clips using the tools in the window. The name of each clip is included for reference as well.

The Track drop-down bar

In the center of the top of the window is a Track drop-down bar that lets you select which video or audio tracks you are basing your trim adjustments on (Figure 5-27). You can actually apply the same trim action to more than one track at the same time, which can be a big help in keeping two or more clips equal in duration or alignment.

There are many reasons for trimming multiple edit points at once. Chances are that you will encounter situations in which you need to extend or retract the edit points for more than one clip to keep a group of clips aligned with each other. By Command-clicking the additional edit points between two abutting clips in the sequence, those edit points are added to the choices in the Track drop-down bar at the top of the Trim Edit window.

The only two restrictions on this activity are that there can be only one edit point selected per track and that the trim applied in the Trim Edit window is applied equally to all selected edit points, regardless of which edit point is selected in the Track drop-down bar. This lets you base the trim effect of many different edit points across all the video and audio tracks on how you trim one critical edit point in any one of the video or audio tracks.

STEP 31

Leave the abutting video clips in the Video 1 track. Next, edit in a new clip from the Viewer window, but this time, drag it from the Viewer and drop it into the sequence in the space above the Video 1 track so that it creates a new Video 2 track (Figure 5-28). Make sure that this new clip's edge is not vertically lined up with the edit point of the two clips you have abutted on the Video 1 track.

Figure 5-27

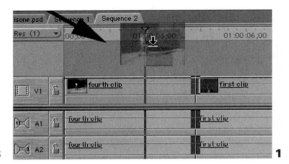

Figure 5-28

STEP 32

Single-click an edge of the clip in the Video 2 track, then Command-click the original edit point division between the other two clips in Video 1 (Figure 5-29). You will see that each edit point becomes brown as it is selected.

STEP 33

Double-click one of the two selected edit points to open up the Trim Edit window.

STEP 34

Go to the Trim Edit window that opens and click on the Track drop-down bar at the top and center of the Trim Edit window. You will see that you now have an option between using the Video 2, the Video 1, or any of the four accompanying audio tracks as the basis for the trim.

Any trimming that takes place will be applied to all the clips that are selected. And since any trim is applied to all these edit points, this also means that any clip with media file limits will limit the amount of trimming that can be applied to any of the clips. If you had more tracks and you wanted to add additional edit points to the group already selected and present in the Track drop-down bar, just Command-click to add them to the selection. But remember that you can have only one edit point selected per track.

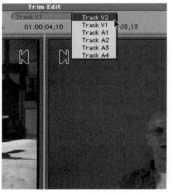

Figure 5-29

Figure 5-30

You will also notice that as you switch between the choices in the Track drop-down bar, the timecode value just below it changes and the playhead moves from edit point to edit point in the sequence. This center value always reflects the sequence frame number that the edit point currently selected in the Track drop-down bar is positioned on. As you trim the edits, this number will likely change as you move the In and Out points back and forth.

STEP 35
Make sure that the Track drop-down bar is set on Video 1, so that there is both an outgoing (left) clip and an incoming (right) clip.

Look directly below the two video windows in the Trim Edit window to find a little timeline track containing a couple of items. The playhead of the track is a little vertical line topped with a yellow triangle. Accompanying this in the track will be an In or Out point marker, depending on the video window in question. There will be an Out point marker on the clip in the left, outgoing window and an In point marker on the right, incoming window.

STEP 36
Move the playhead around in this little timeline and you will see that it is previewing the media available for use by the clip, giving you the opportunity to quickly scan through it for the optimal new edit point. Make sure that as you move the playhead around, the mouse pointer is shaped like an arrow; otherwise, you are probably adjusting the trim. Do not touch the In or Out points just yet. Use Command-Z or undo to quickly negate an accidental trim action.

The range of frames you can preview for the new edit point is equal to the limits of the Master clip and the media file associated with the clip. The lighter area of the little timeline is the range of frames that are already in the clip in the sequence. The darker area is the range of frames that exist in the Master clip and the media file associated with the clip but are not yet included in the clip in the sequence.

Navigating the clips in the Trim Edit window
Underneath this track bar are the familiar tools for navigating around a clip (Figure 5-31). There is a Jog and a Shuttle for moving around. Predictably enough, there is J-K-L support for each separate window, although the response will not be nearly as rapid as in the Viewer or the Canvas. The Trim Edit window is best suited for precise, hair-splitting, one-frame

Figure 5-31   **Figure 5-32**

cuts, so it is not optimized for normal playback out to video. There is also a Play button
and an advance and retreat one frame button.

Beneath each of these video windows is an Out Shift or an In Shift button for the outgoing or incoming video windows, respectively (Figure 5-32). This button will move the
Out or In point to where the playhead is parked in the Trim Edit's little timeline. It is
accompanied by a timecode value that displays how many frames the new Out or In point
is from the original edit point position when the Trim Edit window was first open. Of
course, this In Shift and Out Shift uses the I and O keys as keyboard shortcuts, moving the
In or Out point to wherever the playhead is parked in the little timeline of the Trim Edit
window, just as with the In and Out functions in the Viewer and the Log and Capture windows.

Trimming edits using this window can be done in two ways. As is usual for Final Cut
Pro, there is a direct click-and-drag method and a more precise numerical entry. Care must
be taken when performing the trim, though, because you have to tell Final Cut Pro which
of the Incoming or Outgoing clips you want to trim, or if you want to simultaneously trim
both.

It is pretty easy to identify which clip is to be affected by the trim action you initiate.
Above each video window in the Trim Edit window is a long green bar. If the bar is present
above a video window, trimming actions will affect the clip in that window. You can choose
to have one, the other, or both clips enabled. To enable the green bar and trim actions for a
clip, simply click in its video window. You will notice that clicking in one video window
enables it while disabling the other. To enable both video windows so that trim actions are
applied to both clips, either Shift-click the second video window or click in the thin space
between the two video windows.

When you move the mouse pointer over the video windows, you'll notice that its shape
changes again. This time it appears in one of three shapes, dependent on where in the Trim
Edit window you have it (Figure 5-33). If it lies in the left outgoing clip window, it appears
as a little filmstrip roll with the tongue hanging out to the left. This shape, referred to as a
Ripple Edit, is a metaphor for a film take-up reel. It's called a Ripple Edit because as you
trim using it, the change in duration ripples back through the rest of the sequence the trim

Figure 5-33

applies to, shortening the length and eliminating the possibility of creating a gap between any of the clips. When a clip is ripple trimmed, as it is shortened or lengthened, all other clips in its track are moved over either to give it more room or to cover any gaps the trim might have created.

If the mouse pointer is lying in the right video window, predictably, the mouse pointer becomes a Ripple Edit take-up reel for the incoming video clip. It performs exactly the same function as the other Ripple Edit tool, except that it applies to the incoming clip. Once again, the tongue is hanging in the direction of the clip that will be trimmed by the tool.

The third possible shape is actually just a combination of the two and is called a roll edit. If you place the pointer between the two video windows, you will see a combination of the two Ripple Edit spools and tongues, symbolizing that the trim function will now be applied to both outgoing and incoming clips. Do not be confused though, because the roll edit actually is just moving the edit point relatively for each clip. Rather than taking away a frame from or adding a frame to each separate clip, the roll edit takes away from one clip and adds to the other. The net effect of this action is that the new edit point occurs earlier or later in both clips. The clips themselves are not moved, and the total duration is not changed, only the edit point at which one of them stops and the other starts.

The gist of this is as follows. If you want to trim frames only from or to the outgoing left clip, you click on the left video window to select the outgoing Ripple Edit. If you want to trim frames from or to the incoming right clip, click on the right video window to select the incoming Ripple Edit. If you want to trim from one clip and add to the other so that you move the edit point between the clips without moving the clips themselves, click in between the video windows and select the roll edit. You can always easily tell which window is enabled for trimming by seeing which window has the green bar on top.

There are two ways to perform the trim, one of which is imprecise, but fast, the other of which is frame-specific but much slower. After you have chosen the roll edit or either of the Ripple Edits, you will notice that clicking in the video window has no effect other than enabling that window for trimming.

STEP 37

Click on the left video window to enable the Outgoing Ripple edit (Figure 5-34). Position the mouse pointer over the Out point in the little timeline track of the left video window. Clicking on the Out point and dragging it will trim the incoming clip. Continue to drag the Ripple tool in either direction and watch the sequence Timeline update as you adjust the trim.

STEP 38

Click on the right video window to enable the Incoming Ripple edit. Position the mouse pointer over the In point in the little timeline track of the right video window. Clicking on the In point and dragging it will trim the outgoing clip. Continue to drag the Ripple tool in either direction and watch the sequence Timeline update as you adjust the trim.

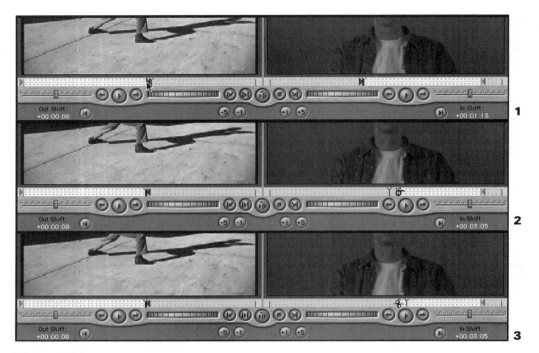

Figure 5-34

STEP 39

Click in between the two video windows to select the Roll Edit. Position the mouse pointer over the In or the Out point in either of the two little timeline tracks. Clicking on either point and dragging it will trim both clips. Continue to drag the Roll tool in either direction and watch the sequence Timeline update as you adjust the trim.

As you trim using these tools, watch the Out Shift and In Shift numbers in the bottom left- and right-hand corners respectively to see how far you have adjusted in either direction as you drag. As you will notice, this sort of click-and-drag trimming is very imprecise. Once you've moved the edit point for either clip, it's difficult to get the Out Shift and In Shift numbers back to zero such that the points are in the exact same position as they were when the Trim Edit window was opened. If you made a change to the edit points and then decided you didn't like it, it could take you all day to get the original edit points back.

For this purpose, Final Cut Pro has a much more precise method. At the very bottom and center of the Trim Edit window, you will find four buttons with the values −5, −1, +1 and +5 (Figure 5-35). These buttons simply perform the same trimming action described

Figure 5-35

previously but for the value of the button. The –1 button, for example, will move the edit point one frame to the left for either or both clips, depending on whether a Ripple or a Roll is enabled. The +1 button will move the edit point one frame in the opposite direction. The –5 and +5 buttons move the edit point even more quickly. You still need to select which side of the Trim Edit window you want to affect using these buttons if you're not performing a Roll edit—that is, applying the trim to both sides equally.

You will in all likelihood find that using the precise buttons is easier and faster than the click-and-drag technique. Since it's pretty easy to count as you mouse-click, you can always quickly click in corrections without having to struggle and wear your wrist out. To make things even easier, there is a keyboard shortcut for these precise trim buttons. The [and] keys correspond to the –1 and +1 buttons. Shift-[and Shift-] correspond to the –5 and +5 buttons. You can actually customize the number of frames that appear in the –5 and +5 buttons by setting a value higher or lower in the Multi-Frame Trim Size field in the General tab in the Final Cut Pro Preferences.

Previewing the trim before leaving the Trim Edit window

Finally, there are a few buttons for previewing your edit before leaving the Trim Edit window. Remember that when you close this window, the Out Shift and In Shift values will reset. Once the In or Out Shift value is gone, it can be difficult to get your edit point back to where it was before you changed it. If you catch a mistake before leaving the window, it's easy to make the adjustment based on the number of frames that the new edit points have shifted from what they were when the window opened. Previewing the edit to make sure it is what you want is a good way to avoid having to go back and piece together your previous edit points.

There is a tool specifically for this purpose. Just below and between the two video windows are a few play buttons (Figure 5-36). The one in the center is called Play Around Edit Loop. When you press this button, Final Cut Pro previews the edit by playing from a few

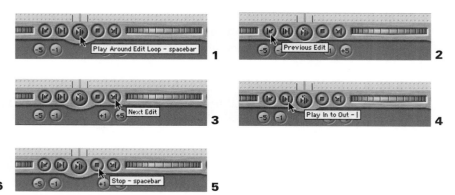

Figure 5-36

seconds before the edit to a few seconds after it. You actually customize the amount of Play Around either before or after the edit by changing the values for Preview Pre-Roll and Preview Post-Roll in the General tab of the Final Cut Pro Preferences.

The furthest left of the five buttons, Previous Edit, is for moving the current edit point loaded into the Trim Edit window to the previous edit. Be careful, though, because it refers to the previous edit point that occupies the same track as the current edit point. If you were looking for the previous edit point on another track, you might end up quite far from where you expected to be. The furthest right button of the five, Next Edit, performs the same function, except that it moves forward in the track rather than going in reverse.

The button just to the left of the Play Around Edit Loop is the Play In to Out. This will preview from the beginning of the outgoing left clip to the end of the incoming right clip, regardless of how your Preview Pre- and Post-Roll Preferences have been set.

The preview playback is a loop, so to stop the loop, there is a button to the right of the Play Around Edit Loop. You can also hit the Spacebar to start or stop it.

Trimming is fast and easy using the Trim Edit window. Just double-click on the edit point you want to trim in the sequence. When the Trim Edit window pops up, select which clip you want to Ripple edit, or you may choose to perform a Roll edit of both clips. Click the plus or minus frame buttons, or the [and the] to define the points the way you want them. Preview the cut to make sure it performs the way you want it to. Then close the window, and you're done.

The Timeline and the Toolbar, a deadly duo

A lot of the material that has been discussed previously has been about how Final Cut Pro allows you to edit in the Timeline from other areas of the application. We created In and Out points in the Viewer window to edit clips into the sequence Timeline. We saw Insert and Overwrite editing from the Canvas window. We learned to trim our sequence's edit points from inside the Trim Edit window. We'll even see later on how to load up a sequence in the Viewer as if it were just another clip.

But don't get me wrong. The sequence and the Timeline window it lives in are powerful tools. Many of the preceding trimming actions can be accomplished directly on clips within the sequence itself. It's necessary to include the Toolbar in this discussion as well, because, with only a few exceptions, the Toolbar tools work directly within the Timeline window. Although editing on the Timeline can sometimes feel a little sloppy and imprecise, there are times when quickly grabbing a clip and nudging it into place will do the trick. It's the editor's job to figure out the fastest, most efficient way of working so that more thought process goes into the "why" of editing and less into the "how."

Let's quickly go through the Timeline window and take a closer look at its features and controls. From the top left-hand corner of the window, the tabs for the open sequences are displayed. If you have more than one open sequence, you can switch between them by clicking on their tabs. You will notice that when you switch tabs, the tab in the Canvas window switches to match. Every sequence in the Timeline has a sister tab in the Canvas window. The Canvas will always reflect what is taking place in the sequence, and until we address how Compositing and Effects work, it should be regarded as a video preview monitor for the sequence.

The Timeline tracks

Moving down into the Timeline itself, we find tracks (Figure 5-37). The default number of tracks when you create a new sequence was set back in the Timeline Options tab of the Preferences. Unless you changed that Preference, the default for a newly created sequence will still be one video and two audio tracks.

The video track area is separated from the audio track area by a thick gray double line. You can adjust how much of the window is dedicated to the video tracks versus the audio tracks by clicking this line and dragging it. If you have only a few tracks currently in a sequence, you won't see much difference from moving it around. But when you have 10 tracks of video and audio, you'll see that being able to prioritize the window for video or audio tracks will make a big difference.

As we saw in the previous section on trimming, it's easy to create a new video or audio track in the sequence by simply dropping a clip into the space where the next video or audio track should be. When you let go of the clip, the track will be automatically created. You can also create new tracks or remove tracks you don't want by selecting Insert Tracks or Delete Tracks respectively from the Sequence drop-down menu. You can have up to 99 tracks of video and 99 tracks of audio in any sequence.

The only serious limitation to the number of possible tracks in your project is the number of audio tracks that Final Cut Pro and your Macintosh hardware can play at once, or what is referred to as real-time audio mixing. We know, for example, that when video must be displayed that does not exist yet (e.g., a dissolve or a wipe), it must be rendered. The render is a mixdown of all the video elements into one composite video clip that can be played out to Firewire.

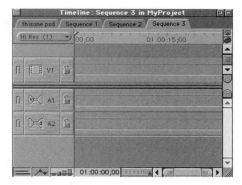

Figure 5-37

Video has a much higher data rate than audio, and the processing to show a real time playback of such unrendered material requires extra hardware and processing power from a real-time video card, such as the MatroxRTMac or the Promax RTMAX. Audio, on the other hand, can be mixed and output to Firewire without rendering, up to a certain point. The rule of thumb is that eight tracks of audio that have no effects applied to them can be mixed together down to two channels and sent out to Firewire along with video. This is referred to as real-time audio mixing.

So what do you do if you need more than eight tracks? You can simply perform a quick Mixdown Audio from the Sequence drop-down menu for the sequence. The ensuing render is quite fast; it's much faster than video rendering. At the end of the render process, the audio tracks in the sequence will still be discrete and editable, but the audio tracks in the sequence will be linked to a render file so that everything above eight tracks can be mixed together and played correctly. If you have too many audio tracks to play back without Mixdown Audio, you will hear a pulsing beep alarm as you play back.

Sometimes, your system may not have the wherewithal to mix all the eight tracks of audio that it is generally capable of doing. This can yield Dropped Frame errors and can be the result of disk fragmentation, sluggish system performance, or any of various other reasons. In these instances, a quick Mixdown Audio will usually take the stress off of your processor or drives and allow your system to do what you want. Remember that the number of tracks you assigned in the Real Time Audio Mixing field of the General tab in the Preferences is the number of tracks you are limiting Final Cut Pro to attempting to mix real time in the sequence. Setting this number higher than eight will not make your system perform better; it will just try to real time mix more tracks than most Macs are generally able to. Keep this preference set to 8 tracks or less.

The track portion of the sequence is divided into two parts: the Target Track Selection Area to the left and the actual Timeline area to the right. The target track selection area is of crucial importance for editing, because of the flexibility it allows you when editing into the sequence. There's more than one way to skin a cat, but some ways provide quicker skinning than others. Properly controlling the target track selection area can save you lots of time.

The target track selection area

To the far left of the target track selection area, there is a green light for each track (Figure 5-38). This is called the Track Visibility switch. Clicking off this green light grays out the track and makes it unavailable to the sequence. It's basically a way to hide a track's contents temporarily. Any audio in the track would not be heard, and any video in the track would not be seen. Some care must be taken with this; turning tracks off irrevocably breaks the link between a clip and its render files. If you have rendered a clip, you want to avoid turning its track off.

The middle item in the target track selection area is the Target Track assignment. For video tracks, this is displayed as a small film frame next to the track's name. For audio

Figure 5-38

tracks, this is displayed as two little speakers pointed in opposite directions. In either case, the track of which the icon is displayed in yellow is a current target track. A target track is the track into which clips will be accepted when they are brought into a sequence. When you perform an Insert or Overwrite edit from the Canvas, the clip is automatically routed into whichever tracks are set as the target tracks.

In our examples of Insert and Overwrite earlier, there were only an initial video and two audio tracks in the sequence. This was sufficient for the requirements of our standard, DV-captured clips that are composed of one video and two audio tracks. But in some instances, you will need to create more tracks than this. To make sure that your edits get sent to the proper tracks, you must make sure that they are targeted.

Clicking on a track icon turns it yellow and sets it as the target track. For audio tracks, there can be two target tracks, since the Firewire connection with our DV device supports two channels out. The target track allows for a designation of a left and right, or stereo pair, in the target track box, since most clips will have two channels of audio and many will also have hard panning, or extreme left-right differentiation between the two tracks. Only one track of video can be targeted at a time, since an individual clip will have only one associated layer of video (although through Nesting, which is covered further on, it's easy to include many layers within that one layer).

Setting the target track is critical for editing when using more than one video and more than two audio tracks. Target track assignments do not change until you change them. If you are using the Drag to Canvas method of editing, the target tracks are the only way Final Cut Pro knows where to send the incoming clips. As you speed up your editing process by accessing keyboard shortcuts and automated features such as Drag to Canvas editing, you can easily screw up by not defining things like target tracks.

Finally, there is a Track Lock feature to the right of the target track selection area, symbolized by a padlock. Normally, this is in the unlocked position, given that you want to be able to work in the tracks, which locking them would disallow. But Track Locking can be very useful. No matter how experienced you are with Final Cut Pro, it can sometimes be tricky to see all the consequences of a single action. One false move could undo hours of precise cutting through an accidental series of edits. Locking tracks in these instances is a good way of making sure that nothing happens to a track or tracks when you are sure they are complete. It's also a good way to lock two tracks that have been manually synced so that they cannot be accidentally shifted by subsequent edits.

Timeline customization controls

Below the tracks are various controls for adjusting the Timeline so that you can customize what you see and how much you see of the clips in your sequences. There are tools for changing the time scale of the Timeline so that you are looking at larger or smaller increments of time, a timecode field for indicating where the playhead is at the moment, and, finally, buttons that can provide additional control and display of the tracks themselves.

The Time Scale of the sequence Timeline is the variable increment your sequence contents are displayed with. Because your clips may range in length from several seconds to several minutes, it can sometimes be difficult to view all the contents of the sequence at once. Similarly, you may need to edit your clips on a very large scale so that you can distinguish between frames in a clip. You need to be able to adjust the time scale of the Timeline so that you can alternate between looking at all the contents of your sequence and focusing on a smaller section of it.

The time scale bar

At the bottom right-hand corner of the Timeline window, there are two controls that adjust the time scale of the Timeline: The Zoom Slider and the Zoom Control Meter. The Zoom Slider is a manual scrolling bar control with a handle on either end (Figure 5-39). The Zoom Scroll bar only works when there is a clip on the Timelime to change the scale of. If you grab either one of the handles and drag it, you will see that the increments above the tracks become larger or smaller. Drag the bar out longer, and you will be looking at seconds—even minutes, depending on the length of the clips in the sequence. Drag either of the handles in to make the bar shorter, and you will be looking at smaller increments, down to single frames of video. This can be especially helpful when you need to make very specific adjustments to positioning of clips or if you need to select many different clips at once. Adjusting the time scale allows you to put as little or as much of your sequence contents into the viewing area as you need.

This bar can actually move the viewing area of the Timeline as well. Grabbing the bar anywhere between the handles and dragging moves the whole viewing area forward and back along a sequence's length, allowing you to move to areas quickly without having to get there by moving the playhead. Remember that this movement does not move the position

Figure 5-39

Figure 5-40

of the playhead itself and that you will need to move it up to the window by clicking within the view area.

Navigating on the Timeline efficiently is a skill in itself. There is no correct timescale. The appropriate timescale depends as much on the relative lengths of your clips as it does on the type of editing you need to do at the moment. Your ability to quickly change the scale and move to the next area of the sequence takes time and practice, but to work efficiently you really need to master it.

The Zoom Control meter

Because Timeline navigation is so important, Final Cut Pro provides a couple of other tools for quick and precise Timeline redefinition (Figure 5-40). The next tool to the left of the navigation bar is the Zoom Control. It is shaped like a meter and redefines the time scale based on where the marker is in the Zoom Control box. When the marker is further to the right, the time increments will be larger, meaning that you will be looking at whole minutes and can see more of the sequence at a time. Moving the marker to the left makes the increments smaller and more precise. Move the marker to the extreme left and you will see each individual frame; move it to the extreme right and you will be looking at a much larger scale that allows you to see every clip that sits in the sequence. In between these two positions are jumps in scale that you can access based on your needs.

Playhead Timecode Field

To the left of this meter is the sequence Playhead Timecode Field (Figure 5-41). This field has two major functions. The first is to display the frame that the playhead is currently parked on. Since this is a timecode value for the sequence and not for the clips within it,

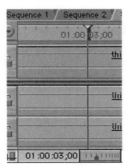

Figure 5-41

Figure 5-42   **1**   **2**   **3**   **4**

the first frame of the sequence will be the Starting Timecode value you entered in the Timeline Options tab in the Preferences. If you did not change the default value there when you set the Preferences, the Starting Timecode value will be 01:00:00:00. You can of course change this value to any number that you want, given your system of organizing timecode values for your project.

The second function of the field is to manually position the playhead at a specific timecode location (Figure 5-42). Navigating to a specific frame to make an edit can be crucial. Entering values into this box is simple; you don't even have to click on the box to do so. Because this box is the only place you can enter numeric information into the Timeline window, simply typing numbers when the Timeline window is active will enter them there.

Final Cut Pro makes things even easier. Although timecode values require colons or semicolons to delineate hours, minutes, seconds, and frames, Final Cut Pro can insert these for you. All you need to do is type in the numbers and make sure that you enter zeros where applicable—for example, type 01000300, and Final Cut Pro will convert this to 01:00:03:00, or the third second of the Timeline. When you hit the enter key, Final Cut Pro will insert the colons and move the playhead to the new location. It can even intuit that the earlier values stay the same if you enter only a few. For example, typing in the above value lands the playhead on 01:00:03:00. Then typing in 15 will move it up to 01:00:03:15. Typing in 215 will then move it to 01:00:02:15.

To make things even more convenient, you can type in a certain number of frames to retreat or advance. Instead of having to do the math to figure which frame to move to, simply type in + or − followed by a number, and the playhead advances or retreats that many frames, seconds, minutes, or even hours. For example, if you are at 01:00:02:00 and you want to move up to 01:00:03:00, simply type in +100, or plus one second and zero frames.

Using the numeric keypad for navigation is the easiest way to work. Rather than wear your wrist out using the mouse to drag a playhead around, move it exactly the distance you want. But what if you want to advance or retreat only one frame? The keyboard shortcut for advance/retreat one frame is the Left and Right Arrow key. Holding the Shift key and hitting the Left and Right Arrow key changes this to advance/retreat one second.

Using each of these three tools, it is possible to quickly jump around your sequence and hone in on the areas you need to work with at the time scale you need to work with. While general and imprecise, the Zoom Bar is a quick-and-dirty method of getting your sequence in roughly the area and scale you want. The Zoom Control meter provides a more precise method of choosing the time scale. Using the timecode field is the most precise way of moving your playhead around to new locations. Mastering all three techniques of Timeline navigation will enormously speed up your operations.

Moving to the left of the time scale and playhead location controls, there are three buttons that allow further customization.

Figure 5-43

Figure 5-44

Track Height switch

The first is called Track Height (Figure 5-43). Track Height simply gives you four options for the size of each track within a sequence. If you are working with 12 layers of video, then you'll probably want to keep the track height pretty low so that you can see most of them simultaneously. If, on the other hand, you are working with straight cut editing on a single video track and you've got a lot of screen real estate to burn, expanding the track height may make things a little easier on the eyes. This setting can be switched on the fly and has no effect on your sequence or clips other than making them more or less visible in the Timeline.

The Clip Overlays button

The next button on the lower left-hand side of the Timeline window is incredibly useful. It's called the Clip Overlays button in Final Cut Pro parlance, but it's popularly known as a *rubberband tool* (Figure 5-44). We first saw this mentioned in the section on Preferences. The Clip Overlays button overrides that Preference whenever it is clicked and enabled or disabled. Rubberbanding allows us to alter audio levels (loudness) and opacity (the amount of transparency) within a clip in the Timeline. In addition, when used with another tool to be discussed with the Toolbar functions, rubberbanding allows us to *keyframe*, or change these audio and opacity levels over time. This is a quick creative way to customize video fades and to even out audio tracks that exhibit too much variation in audio levels.

When the Clip Overlays button is engaged, all clips within a sequence display a thin, colored line—black for video tracks and red for audio. These lines describe the level for audio or video opacity in a video clip.

STEP 40

Open a sequence on the Timeline, edit a clip into it, and enable the Clip Overlays button so that lines appear in the clip. Move the mouse pointer over the Clip Overlays lines in the audio tracks. The mouse pointer will change into two parallel horizontal lines with arrows pointing up and down. While the mouse pointer is shaped as shown in Figure 5-45, click the line and drag it up and down without releasing the mouse button.

Figure 5-45

STEP 41

As you drag the line up and down, you will see a small box appear with a dB value that changes as you move the mouse position. Release the mouse button when the dB box reads –12 dB. Now perform the same action for the other audio track in the same clip (if the audio tracks for a clip are stereo, changing the level of one track will change the other as well). If you go back and play this clip from the beginning, you will notice dramatically lower audio levels.

When you adjusted the levels, you probably noticed that the entire line for the clip adjusted evenly across the clip. This means that the level adjustment evenly affects the entire clip (i.e., the same level of boost or cut is being applied for the entire clip). When we cover the Toolbar, we will discover a way to vary the change in level within the clip itself through keyframing.

Video opacity can be affected the same way that audio levels are adjusted. This is easier to see when the adjusted video clip is *superimposed*, or occupies the video track directly above another video clip. Using the Canvas window Drag and Drop edit method, we will perform a Superimpose Edit of a new clip into a Video 2 track (make sure you don't use the same or similar clips for both video tracks or you will fail to see much of an opacity change after rubberbanding).

STEP 42

Position the sequence playhead at the beginning of a clip in the Video 1 track. Set the Target Track indicator for the Video 1 track.

STEP 43

Load a new clip into the Viewer window, then drag it to the Canvas window and drop it into the overlay box named Superimpose (Figure 5-46).

The Superimpose edit sends the video portion of the clip being edited into the track above the currently selected Target Track. The newly edited clip should be in the Video 2 track, so that it exactly covers the original clip in the Video 1 track. The incoming superimposed video clip is also automatically trimmed down to the exact length of the clip it is being superimposed above. If there was no Video 2 track to superimpose the clip into, Final Cut Pro automatically creates the track and then performs the edit.

STEP 44

Move the mouse pointer over the line in the video portion of the superimposed clip in the Video 2 track. When the mouse pointer changes to the familiar parallel line/arrow shape, click the mouse button and drag the line down. The same box will appear next to the mouse pointer, this time giving a percentage value for the opacity of the clip (Figure 5-47).

A value of 100 means completely opaque, or visible. A value of zero means completely transparent, such that anything on a layer underneath the video track is completely visible.

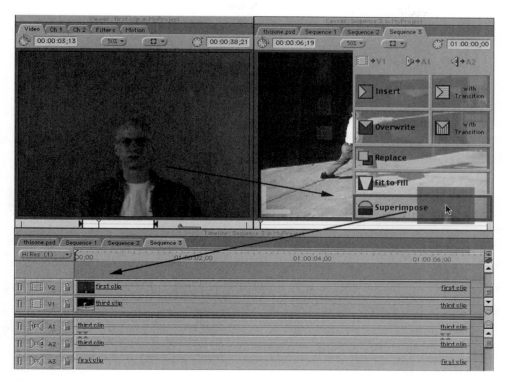

Figure 5-46

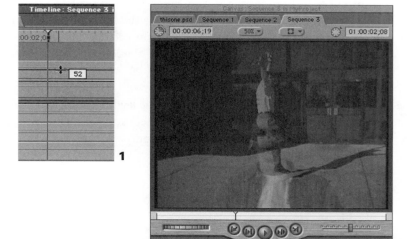

Figure 5-47

STEP 45

Drag the line to around the 50 percent point and release it. A glance back at the Canvas reveals that the clip in Video 2 track is half transparent, revealing the layer underneath. Readjusting the position of this opacity line gives varying degrees of opacity.

You will notice that the red line has appeared over the clip that you applied the opacity change to. This is because once again there is no video file to associate with the visual mixture of the clips in Video tracks 1 and 2. The clip must be rendered to play back correctly. The audio level change does not require any rendering or mixdown because Final Cut Pro can process the level change and real-time mix the audio out to Firewire.

Clip Keyframes

The button on the extreme bottom left of the Timeline window is called Clip Keyframes (Figure 5-48). In addition to, and perhaps more important than displaying any keyframes, it displays any opacity or filter adjustments that have been made to the clip. When a clip's physical attributes change, such as audio level or opacity, a line will appear underneath the affected clip in the sequence, blue for video clips and green for audio. Likewise, if any Special Effects filters like those to be explored in the next chapter are applied to the clip, a green line will appear underneath the clip.

Many times, subtle changes in color correction or brightness and contrast that you apply to a clip are easy to forget about. Not all effects you use to manipulate a clip are obvious; some are corrective in nature. As such, sometimes after altering a clip and rendering it, you might actually forget that you did anything to it.

Since some editing actions can break the link between your clip and its render files, it is always a good idea to have an indicator of clips that have been rendered. This indicator can also alert you to possible settings problems. If you see the red render bar above your clips, yet when you enable the Clip Keyframes button, you can see that no effects have been applied and no attribute changes have been made, that's a pretty good sign that your sequence or capture settings were erroneous.

Once we begin to utilize keyframes, this indicator will take on even more significance. The blue and green lines will reveal the positions of keyframes in the attributes and/or effects applied to a clip, thus giving you hints as to where actions begin, vary, and end without your having to open up the clip in the Viewer window to find them.

Figure 5-48

Toggle Snapping button

In the top right-hand corner of the window is a button with a small green triangle. This is the Toggle Snapping button (Figure 5-49). Snapping is a behavior that enables quick editing on the Timeline by automatically joining manually dragged clips with other clips or the sequence playhead. It's chief usefulness is as a tool for avoiding gaps between clips and accidental Overwrite and Insert edits from imprecise drag-and-drop operations. Snapping is disabled by default. Clicking the Snapping button or hitting the N key toggles it on and off.

When Snapping is disabled and clips are manually dragged to the sequence, they are placed exactly where the mouse pointer is when the mouse button is released. This is fine for Insert edits in which a large degree of exactitude is necessary. As you move the incoming clip around in the area of the sequence you want to put it in, the Canvas window will show the frames it is currently over, allowing you to find the exact point you want to drop the new clip in.

STEP 46

Open a sequence that already contains at least one clip. Go to the Snapping button and make sure it is disabled and light gray in color. Grab a clip from the Viewer and drag it into the Timeline and without releasing the mouse button, ease it through another clip already in the sequence. Watch the Canvas window update as you move through the clip. When you locate a frame that would make an acceptable edit, confirm that the mouse pointer will perform an Overwrite edit and drop the clip. The clip will drop right where you released the mouse button, as you would expect it to.

But there are some instances where such exactitude can be a hindrance. If you are moving a clip around on the sequence and you simply want to place it immediately following a preceding clip, dragging and dropping it against the last frame of the previous clip can be tricky. It's easy to accidentally drop the clip a few frames over the end and overwrite or insert into the first clip rather than at the exact end of it.

Snapping automates this process for you. If Snapping is enabled, when the playhead or the In or Out point of a clip comes within a certain distance of any other In or Out point or the playhead, small triangles appear at that point on the sequence, alerting you that if you drop the clip it will snap to that position (Figure 5-50).

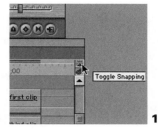

Figure 5-49   1   2

Figure 5-50

STEP 47

Now utilize Snapping by clicking on the Snapping button and turning it green. Grab a new clip from the Viewer and drag it to the sequence where the two clips are. Move the clip around the sequence. As you do so, you will see the incoming clip's edges jump to the ends of the clips already there, as well as the playhead.

STEP 48

As the incoming clip snaps to an end or to the playhead, triangles appear above and below to indicate that the clip has been snapped to another clip. Pick one of the Snap-enabled In or Out points and drop the clip, performing an Overwrite edit. The clip will edit in, snapping tightly to the chosen edit point.

Snapping is a feature that should be used based on your editing situation. It is chiefly useful when your Timeline scale is very large. At such times, it is difficult to spot clip edges and remove gaps between clips in a track. In addition, Snapping functions between clips on different tracks. This allows you to make sure that In and Out points of clips join up together all across the sequence, rather than simply within a single track as other edit functions allow.

Toggle Linking button

The button underneath the Toggle Snapping button is called the Toggle Linking button (Figure 5-51). Linking involves the way a single clip behaves in its relationship with other clips that are connected (linked) to it. When clips are linked, any trimming edits applied to one part of the clip will be applied to the other linked clips. When linking is enabled, the two chain links on the Toggle Linking button are unbroken and green. When Linking is disabled, the symbol becomes a single white broken chain link.

What is linking? When we captured clips, we did so by capturing video and audio at the same time. Final Cut Pro treats the two as a linked clip from the inception, and unless you change that relationship, it will always do so. But the video and audio are really not one integrated clip. The clip really contains one video clip and two audio clips that are linked by Final Cut Pro at capture.

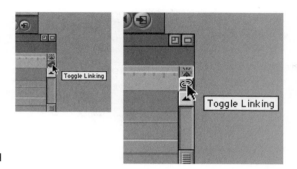

Figure 5-51

Although Final Cut Pro does this at capture, audio and video clips do not have to stay linked to each other, and they can be in turn linked to other clips. The only restriction to this is that a single video clip can be linked only to either one or two audio clips. Two video clips cannot be linked, and more than two audio clips cannot be linked (although they can easily be grouped by the process of nesting as we will see later).

Linking is a completely artificial state. You can enable or disable it for clips in a sequence at any time by selecting the linked clips and toggling the Link command in the Modify drop-down menu. Toggling this command this way creates or breaks links between clips based on their current link status; linked clips will become unlinked, and unlinked clips will become linked if they fit the restrictions (you can break multiple clip links at once, but you can only create links based on the 1 video clip–2 audio clips limitation).

Using clip linking, it's easy to link up a video track with other audio sources. This is useful if you are using audio recorded by another device such as a minidisk recorder instead of the camera mikes, a technique called double-system sound. It is useful also if you want to sync up music tracks from an audio CD and make sure they stay locked to the video clip.

Let's quickly perform a bit of sophisticated linking and unlinking. We will take the video portion of one linked clip and link it to the audio portion of another linked clip. We will break all the links between two captured clips, and then we will delete the video clip for one and the audio clips for the other. Finally, we will link the video clip from the one and the audio clips from the other together to form an artificially linked clip that contains the audio and video we want, regardless of what those clips were captured with.

STEP 49

First, clear off the sequence by selecting all clips and deleting them. Next, edit two different captured clips into the sequence. Select both clips using the marquee selection technique or by shift-clicking them (Figure 5-52). When they are both selected, go to the Modify drop-down menu, and look for the Link command.

The Link command in the Modify drop-down menu is actually a Link toggle switch as well. If the Link command is checked in the menu, it means that the currently selected clip in the sequence is linked with other clips. If it is not checked, the currently selected clip or clips are not artificially linked.

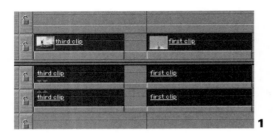

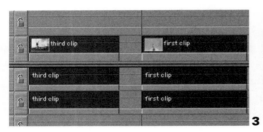

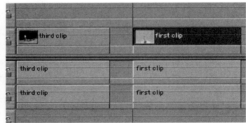

Figure 5-52

STEP 50

Toggle off the links between the selected clips by selecting the Link toggle switch in the Modify drop-down menu. This will break all links between any currently selected clips.

Notice that the names of the newly unlinked clips in the sequence are no longer underlined. When a clip is linked to another clip, its displayed name in the sequence is underlined to signify its linked status. Although all the clips are still brown, because they are still selected, they are no longer linked to each other.

STEP 51

Click anywhere on the Timeline to deselect all the clips. Then, click on any individual video or audio clip. Notice that only the clip you click on is selected, not its sister clips that were captured with it.

When you click on and select a video or an audio clip, it no longer automatically selects the other video or audio portions that were originally captured with it. This is because the initial artificial link between these three individual clips of video and audio has been manually broken. Although we get used to thinking of captured video and audio as a single clip, they are actually separate clips that can be broken up as necessary.

STEP 52

Now that the clips are completely unlinked, shift-click to select the two audio tracks of one of the clips and then return to the Modify drop-down menu to toggle Linking back on for

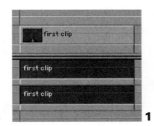 

Figure 5-53   **1**   **2**

the selected audio tracks (Figure 5-53). If the tracks were originally part of a stereo pair as well, you can also select Stereo Pair from the Modify drop-down menu to reconnect them in that manner as well, allowing you to create panning effects between them.

Note that the underline returns to the name of the audio tracks when you relink them. If you reenable the Stereo Tracks toggle for them, little triangles appear in the audio clips, indicating the special stereo relationship between the two.

STEP 53

Click anywhere on the Timeline to deselect the freshly relinked audio clips, then click on the video track above them that they were originally captured with and linked to. Hit the Delete key to remove that video track from the sequence Timeline (Figure 5-54).

STEP 54

Marquee-select or Shift-click–select both the audio tracks for the other clip that was also just unlinked, and delete them as well, leaving the video portion still sitting in the sequence (Figure 5-55).

 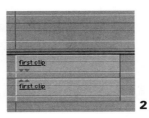

Figure 5-54   **1**   **2**

 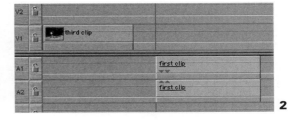

Figure 5-55   **1**   **2**

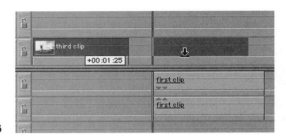

Figure 5-56

STEP 55

Click on the remaining video clip in the sequence, then pick it up and drag it above the freshly relinked audio tracks (Figure 5-56). Use snapping to make sure that the beginning of the video clip you are moving lines up with the beginning of the two linked audio clips you are moving it over.

STEP 56

When the video and audio portions are lined up, marquee-select or Shift-click select to select the video clip and the two newly relinked audio clips. When all three are selected, return to the Modify drop-down menu and once again toggle the Link switch (Figure 5-57).

STEP 57

If you deselect them all and then click on any one of them, the others are automatically selected along with the one clicked on. Notice that the underline has also returned to the name of the video clip, signifying that it is now linked with audio track(s).

This exercise should demonstrate the malleability of Linking. Although editors many times take video and audio captured and linked clips for granted, they really are individual pieces that can be broken up and reassembled on an as-needed basis.

Why then offer a Toggle Linking button in the Timeline? Because sometimes it will be necessary to quickly unlink a clip or clips to make a minor adjustment. Rather than physically unlinking many clips and then having to go in and relink them individually after adjusting them in the long process just described, Final Cut Pro allows you to suspend the links for all clips in a sequence for the moment. When you've completed your adjustment, you turn the Linking back on, and everything functions as before. The link between the clips was never lost; it was only disregarded for the moment until you choose to reinstate it.

Figure 5-57

 1    2

This ability to quickly unlink and relink clips becomes very useful when performing what is referred to as a Split Edit or an L Cut. This popular sort of edit involves audio clips beginning before their linked video clips in an edit. Often an editor will need the next clip's audio to precede the edit by a short duration. Because we can unlink our clip, extend the In points for the audio clips as much as we want, then relink them to the video clip, it is very easy to deliver this effect. Audio and video tracks do not have to line up to be linked; in fact, they can be minutes apart from each other on the sequence Timeline. The only limit for linking is that one video can only be linked with a maximum of two audio clips. Final Cut Pro offers a way to do Split Edits from within the Viewer window, but doing so on the sequence is just as functional.

Sync issues with linking and unlinking clips

You may be a little alarmed about the ease with which Final Cut Pro allows you to link and unlink clips. If one didn't know better, this would be an invitation to audio and video sync disaster. If you unlink a clip, change the position of the video or the audio slightly, and relink, then your sync will be off forever. As long as the two are linked, you cannot move the video or audio without moving the other portion of the clip equally. In truth, the main reason that clips are linked at capture is to avoid such situations, maintaining a rock-solid link between video and audio from capture to mastering to tape.

Fortunately, Final Cut Pro throws in another feature as a guarantee against loss of sync, even in the event that your unlinked clips are adjusted and relinked. Whenever audio and video clips are captured together, their media in your scratch disk is integrated in special ways. Final Cut Pro can sense this, so anytime that two clips were captured together and are placed in a sequence but are forced out of sync, you will see a small red box at the head of each clip stating the number of frames by which the two are out of sync.

STEP 58

Clear off the sequence by performing a marquee-select or shift-select of all clips and hitting Delete. Edit a captured clip into the sequence (not a relinked clip, this clip must have been captured).

STEP 59

Visit the Toggle Linking button in the top right-hand corner of the Timeline window under the Snapping button and make sure that it indicates Linking enabled with two green chain links. Now go to the clip and select any part of the clip's individual audio or video clips. Selecting one selects all the individual clip items because they are linked and because linking is toggled on.

STEP 60

Now go to the Toggle Link button and disable it (Figure 5-58). When you return to the clip and click on any of the individual audio or video clips, you will notice that only the

Figure 5-58

individual video or audio clip is selected, just as if you had unlinked them using the Modify drop-down menu Link switch.

STEP 61

Pick up either a video or an audio portion of the clip, and move it over slightly so that its edges still overlap the items that it was previously linked to (Figure 5-59). You will see a small red box appear in each of the items that were previously linked (if an item is selected, its color switches to brown, and the red box will also change, appearing blue).

The number in the red box reveals how many frames and seconds out of sync the items that were originally captured together are. A positive number of frames indicates that the clip is ahead of its companion pieces; a negative number indicates that the clip is behind them. Final Cut Pro offers you a quick tool to move the clips back into sync.

STEP 62

Hold down the Control key and click on the red "out of sync" box for a moment (Figure 5-60). A contextual menu will appear offering to Move or to Slip the item back into sync. For the moment, select Move Into Sync. We will discuss Slipping techniques in the next section, which focuses on the Toolbar.

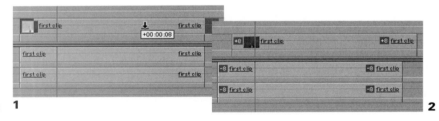

Figure 5-59   **1**   **2**

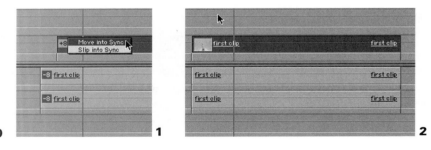

Figure 5-60   **1**   **2**

Some care must be taken here for a couple of reasons. The "out of sync" box appears only when such previously connected items actually overlap each other. If the items are isolated from each other over time, Final Cut Pro will not assume that they are out of sync, but are simply recurrent uses of the same clip. Further, if you are using the same clip twice in the same stretch of frames in a sequence, Final Cut Pro can become understandably confused as to which clip items are supposed to be resynced to which, yielding mixed results.

The Toolbar

The small strip referred to as the Toolbar is full of little tools that will make your life more interesting in Final Cut Pro. Many of the tools are operational within the Timeline window, although some also have functionality across the application. Some functions cannot be performed without switching to the proper Toolbar tool.

Does this mean that you should constantly be clicking back to the Toolbar window to switch tools? Absolutely not. Using the mouse to access that thin tiny little window will do a number on your wrists and isn't a particularly fast way to work anyway. The answer is the keyboard shortcut. For every tool in the Toolbar, there is a simple keystroke that will change the mouse pointer to the tool you want to access. There's no faster way to work, and you'll find that most experienced nonlinear editors rarely use the mouse for anything other than positioning of onscreen items, preferring to fully utilize the keyboard shortcuts that really speed up the workflow.

If you are just beginning your editing life, memorizing all those shortcuts may seem daunting. There are a couple of options that make that process much easier. Each copy of Final Cut Pro comes with a set of keyboard overlay stickers that you can use to learn the shortcuts. Each overlay sticker shows the primary tool that the keystroke will enable. Because the stickers are translucent, you can still see the letters underneath the stickers, allowing you to use the keyboard normally.

With all due respect to Apple, using the stickers is a pretty bad idea, as they tend to stick to everything except for the keys they are intended to fasten to. Invariably they end up in a big gummy pile in front of the keyboard.

What other options are there? There are free keyboard layout charts available that show where the keystrokes are, although that involves looking away from the keyboard, which can be very distracting and obstructive to the learning process. Still, you can't beat the asking price, and it does help you to learn the keyboard shortcut system.

If you are willing to spend a few dollars, some companies manufacture special Final Cut Pro–dedicated keyboards. These keyboards not only contain the name of the tool that the keystroke enables but are also color-coded based on the type of function that the keystroke performs (e.g., an edit tool or an effects tool). Since they are basically just normal Macintosh keyboards with specially colored and labeled keys, they usually perform without extra software or hardware. Beginning editors can quickly pick up the functionality of the keyboard in a fraction of the time it would take to learn from accessing the manual. Either way, learning the shortcut keystrokes is very important and will save lots of time searching for tools in the menus or the Toolbar.

The General Selection tool

The first tool on the Toolbar is the General Selection tool (Figure 5-61). This is the generic mouse pointer arrow that is enabled when you start Final Cut Pro. The Selection tool is simply used for selecting items and moving them around. This tool, as contrasted with the following set of selection tools, is used for selecting individual items. When you want to pick up, move or load up a clip in the Viewer, for instance, this is the tool to engage.

Hidden toolsets

Although there is only one position for each toolset on the Toolbar, Final Cut Pro uses a click-and-hold method of choosing the particular tool from any toolset.

STEP 63

Click on the button below the General Selection tool, and hold the mouse button down briefly (Figure 5-62). Two more tools will appear next to the Toolbar with the currently selected button.

The Selection toolset

The toolset button below the General Selection tool on the Toolbar engages a different set of selection tools that are more specific to particular tasks. Click and briefly hold the button. You have a choice of three specialized selection tools here.

STEP 64

Edit Selection tool
Click on the first button of the toolset directly beneath the General Selection tool (Figure 5-63). If you move the mouse pointer over the Timeline window, you will notice that the mouse pointer has switched from the arrow to a small crosshair. This button activates the Edit Selection tool, which performs the same action as double-clicking on an edit point

Figure 5-61

Figure 5-62
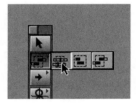

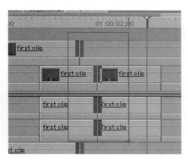

Figure 5-63

between two clips on a sequence. When you click on such an edit point, it automatically opens up the Trim Edit window containing the edit point you clicked on.

Why use this tool instead of just double-clicking with the General Selection tool as we did earlier? As we saw when discussing the Trim Edit window, we can have more than one edit point in the Track drop-down bar within the Trim Edit window if more than one edit point was selected. With the General Selection tool, this meant command-clicking each additional edit point to add them to the Track drop-down bar to be used in the Trim Edit window.

The Edit Selection tool makes this task easier. Clicking anywhere around the edit points you want to use and marquee-selecting them selects any edit points and then adds them in the Trim Edit window. Instead of command-clicking around your tracks and then double-clicking to open the Trim Edit window, simply choose this tool and marquee-select around the edit points you want to work with. If you forgot a couple of edit points and want to add them to the current tracks loaded into the Trim Edit window, simply return to the sequence, Command-click and marquee-click the other edit points, and they will be added to the Track drop-down bar in the Trim Edit window.

The Group Selection tool

The middle tool in the selection toolset is the Group Selection tool (Figure 5-64). Once again, the mouse pointer will assume the shape of a crosshair. This tool behaves very much

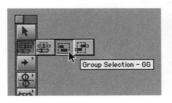

Figure 5-64

like the Generic Selection tool. When you click and drag a marquee with this tool, it selects everything it touches, as well as any clips or items linked with whatever it touches. If the marquee you drag with the crosshairs touches any part of a clip, that clip and its linked clips become part of the active selection.

Range Selection tool

The final tool to the right on this toolset is the Range Selection tool (Figure 5-65). This unique little tool lets you select a section of one or more clips in a sequence. But unlike the Group or Generic Selection tools, the Range Selection tool does not force you to select an entire clip.

Once again, as with the Edit and Group Selection tools, the mouse pointer displays as a crosshair. To make a selection using this tool, click and drag a marquee over an area of one or more clips within a track. Unlike the other selection tools, though, the Range Selection tool does not see clip edges and allows you to select a section of a clip and/or a range of clips. This means that you can select a smaller part of a clip rather than the whole thing. You could, for example, select the last half of one clip and the first half of the next clip.

It does, however see track boundaries, so the range of frames you select with it all need to live within the same track. Because it is a selection tool, it selects linked clips with respect to the way you have the Linked Selection button set; Linked Selection enabled will result in material in other tracks being selected along with the material you select in the track, while Linked Selection disabled will result in only the one track's material being selected.

The tool is referred to as a range tool because it effectively allows you to discriminate footage over a range of frames rather than based on a clip's In and Out points. This can be very helpful when exporting clips as QuickTime movies or in other situations where you need to isolate a specific section of your sequence, regardless of which clips are included. It is literally a selection of frames in a track, rather than a selection of clips in a sequence.

This tool can be used to load up just a section of a clip from the sequence into the Viewer window for manipulation, instead of loading up the entire clip. When you select only a portion of a clip, you can selectively apply effects filters to just the section of the clip, rather than applying them to the entire clip.

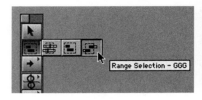

Figure 5-65

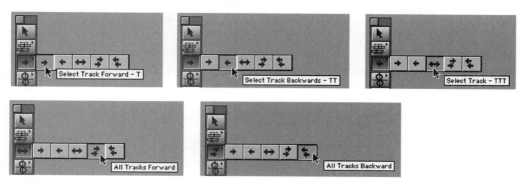

Figure 5-66

The Track Selection toolset

The next button below the special selection toolset is another toolset dedicated to selection, but this time the selection criteria is for clips ahead of or behind the sequence playhead within a track. Clicking and holding the button reveals five possible track selection types here.

The arrow pointing to the right will select all items in a track occurring to the right of the frame the playhead is currently parked on. The next performs the same action, but to the left of the frame the playhead is currently parked on. The middle choice selects all clips in the track. The fourth and fifth choices on this toolset, symbolized by the double arrows on the button, allow you to select all clips on all tracks to the right or to the left of the selection point respectively.

This tool does not override the Linked Selection button in the sequence, so if all your video clips are linked with audio clips, as they will after initially being captured, selecting items in the video track using a track selection tool will also select their linked audio clips in the audio tracks.

Ripple and Roll Edit toolset

The next toolset below the track selection tools contains the Ripple Edit and Roll Edit tools (Figure 5-67). We have already encountered these functions in the Trim Edit window.

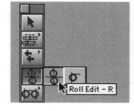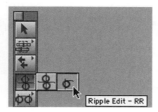

Figure 5-67

The difference is that with these tools selected from the Toolbar, their trim functionality can be utilized directly on clips in the Timeline window rather than from within the Trim Edit window. By clicking on an edit point with either of these two tools, you can trim edit points and dynamically view the results before settling for an edit.

The Roll Edit tool

The Roll Edit tool moves the In and Out frames for the two clips of the edit point chosen in the sequence. Although you do not have access to the large video preview windows and the In and Out timecode tools that are available in the Trim Edit window, when you use the Roll Edit tool, the Canvas window itself switches to a mini–trim window, showing the In and Out point frames as you adjust them. The Timeline is also dynamically updated as you drag. Should you wish to use the Trim Edit window for greater precision at any point, double-clicking the edit point with the Roll Edit tool will open it up just as the Generic Selection tool did.

STEP 65

Edit two abutting clips into the same track on a sequence so that they have an adjoining edit point. Disable Snapping in the Timeline temporarily (hit the N key to toggle it off) so that you can engage the Roll Edit tool from the toolset without imprecise jumping from edit point to edit point.

STEP 66

Click the mouse pointer on the edit point and drag in either direction (Figure 5-68). As you drag, keep an eye on the Canvas window, watching the Out and In points change as you move the pointer. Also note that in the Timeline window, as you move the edit point around the clip there is an indicator line demonstrating where the current Out and In points for the two clips are. Release the mouse pointer, and observe that the clips now meet at different Out and In points respectively.

As with the Roll Edit in the Trim Edit window, this trimming action shortens the length of only one clip and adds equally to the length of the other. The duration of the two clips combined remains the same. The Roll Edit cannot roll either clip past the media limits of the media file it refers to, thus limiting how far either clip may be roll edited.

The Ripple Edit tool

The Ripple Edit tool offers the same trimming functions as it does within the Trim Edit window. As with the trimming action there, the Ripple Edit tool only extends or shortens the duration of either the outgoing or the incoming clip, whichever the tool is being applied to. Since the tool is Rippling the edit point in only one direction, it affects the duration only of the clip being trimmed; if it affected two clips' durations, it would be a Roll Edit Tool! It does, however, affect the overall length of the sequence, because it is adding to or subtracting from the length of one of the clips in the sequence. As the Rippled

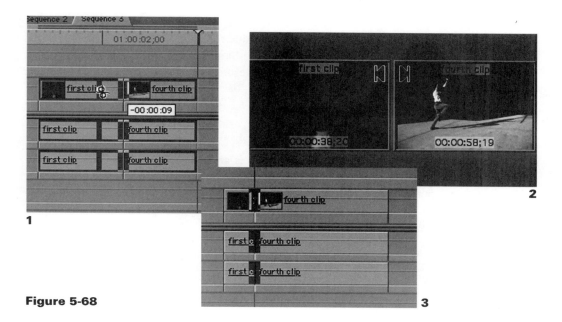

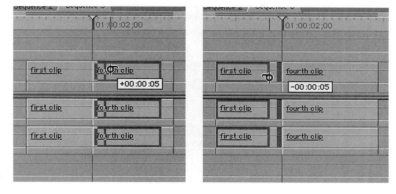

Figure 5-68

clip is trimmed, the rest of the clips in the sequence move over to avoid creating a gap or being overwritten by the duration change of the Rippling clip.

In the Trim Edit window, the clip being Rippled is determined by which video window the green indicator was lit over. With the Ripple tool in the Timeline window, an edit point will only show either the In or the Out point selected. Clicking near the edit point of either the Outgoing or the Incoming clip will switch the edit point that is selected to be trimmed. Then moving the mouse pointer over either the left, outgoing clip or the right, incoming clip will switch directions of the Ripple Edit tool's "tongue," indicating which clip of the edit point the trimming action will be applied to. Select the edit point you want to affect, move the mouse pointer over the clip you want to shorten or lengthen, click and drag, and your clip will be trimmed (Figure 5-69).

Figure 5-69

As with the Ripple Edit action, the Canvas window will show a continuous update of the new Out or In points the trim action is producing. Not surprisingly, the keyboard shortcuts that applied in the Trim Edit window, the [and] or Shift-[and Shift-] keys, will work on the Timeline as well, provided the Ripple or Roll tool is activated and an edit point is selected.

The Slip and Slide toolset

The next toolset below the Roll and Ripple tools is another combination of convenient editing tools, the Slip and Slide tools (Figure 5-70). These two tools can make quick, one-stop adjustments that would otherwise take several steps and a couple of windows to perform. The Slip and Slide tools allow you to adjust either the length or the In and Out points of one clip within a group of clips without affecting the overall duration of the group of clips. They allow you to quickly shuffle the length or edit points of a clip or clips so that you see the range of frames you want without changing what can often be a sequence duration predetermined by other factors, such as a client's wishes or the length or rhythm of a musical score.

The Slip tool

The Slip tool focuses on one clip surrounded by other clips in a sequence. If you edit a clip to a sequence and then realize that you wanted to use an earlier part of the Master clip than the later part you set In and Out points for, instead of reloading the Master clip in the Viewer and editing in a new version of the clip, you can simply "slip" the contents of the clip you already have on the sequence. Thus a clip on the sequence that was edited from the later part of its Master clip when it was created in the Viewer window could be "slipped" so that the earlier part of the Master clip is now used in the clip.

The Slip tool's effectiveness as a fast-and-dirty editing tool is best displayed with a group of four adjoining clips. To line up four clips quickly in a sequence, we will use a technique referred to as storyboard rough cutting.

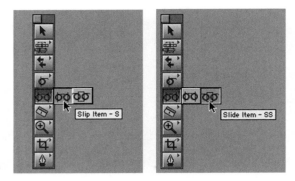

Figure 5-70

Storyboarding rough cut technique

Storyboarding is an old film term that describes a comic strip–style panel drawing of different scenes to be shot (Figure 5-71). Because film and video production is expensive, directors usually "storyboard" the progression of their planned shots in comic-book panels to think out the order and rough look or angle of each shot. These storyboards are then used as a shooting template for everyone in the production. The cinematographer knows how to construct a shot, the art director knows how to arrange the set, and all involved can quickly and conveniently get a sense of what the director wants them to do to produce shots.

Storyboards are more than useful in productions. In addition to keeping everyone on roughly the same page, they help the director actually think out the story line. The difference between productions that use storyboards and those that don't is that the few that do not storyboard usually produce unconvincing, disorganized, and weak results. A short storyboard session prior to a shoot answers most of the on-set questions and eliminates unnecessary surprises.

The usefulness of storyboarding does not end with the shoot. Storyboards also help later to construct the edit, since they also generate a linear progression of the storyline through the series of shots being edited. Once you've captured all the clips from a shoot and they are gathered in a bin in your project, you can simply order them based on their occurrence in the storyboard and create a quick rough cut of the entire project. This technique is an efficient way to generate a rough cut of your project. For this example, you will need at least four separate captured clips.

STEP 65

Log and capture four new video clips, each of which is at least 15 seconds in duration, making sure to include frame handles of at least 5 full seconds. In normal practice such long handles are not necessary, but shorter frame handles will make it more difficult to see the slip action this exercise is investigating. Don't forget to figure this long handle time in with the necessary pre-roll time when you log the clips to avoid "unable to lock deck servo" messages. Also don't forget to use a Logging Bin for these clips so that they get stored in a special bin of their own.

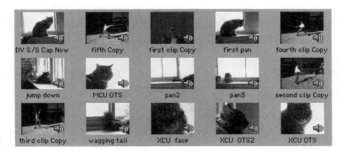

Figure 5-71

To use the Storyboard Rough Cut technique, we will use the Large Icon view of the Logging Bin to create a sort of comic strip panel–style storyboard of our clips. We will arrange them in the order that they should occur, left to right and top to bottom, just like the Sunday newspaper funnies. Then we will take the whole group of clips and deposit them in order on the sequence Timeline in a rough cut. All that will be further required is the trimming actions of the editing toolsets of the Toolbar.

STEP 67

After the capture process is complete, open the Logging Bin and look at the four captured clips. With the Logging Bin window active, go to the View drop-down menu, select Browser Items and, in the submenu that appears, select As Large Icons (Figure 5-72).

Normally, the Browser window is most effective when viewed as a list so that all the detailed information about your materials is available. But when you choose As Large Icons, you are given a thumbnail of each item. This thumbnail by default shows the first frame of a video clip, although you can change which frame it displays by scrubbing the clip and setting the Poster Frame, which is discussed subsequently.

STEP 68

Now that your clips are all displayed in large icons in the Browser window, arrange them left to right, top to bottom, as if you were reading them like newspaper comics. With only four clips, you'll only be able to create two rows of two columns, so if you really want to see this as a functional storyboard rough cut, feel free to add more clips to the Logging Bin. Four clips is really just the minimum required to demonstrate the technique.

STEP 69

Once all your thumbnailed clips are in the proper comic strip–panel order, create a new sequence. Go to the File drop-down menu, select New, and, in the submenu that appears, select Sequence.

Figure 5-72

STEP 70

Click in the Logging Bin window and drag a marquee selection so that all the thumbnail clips are selected (Figure 5-73). When they are all selected, pick up any one of the clips and drag it to the sequence. When you do this, you will see that all the clips are carried over simultaneously and laid down on the sequence with no gaps between them. You could also simply pick up the bin itself from the Project tab in the Browser window and drop it in the sequence for the same effect.

Although we have not done any precise trimming yet, the clips in the sequence are now in a linear order that follows the left-to-right and top-to-bottom order that they maintained in the Logging Bin. All that remains is to go through and individually trim the clips and their edit points to meet our needs.

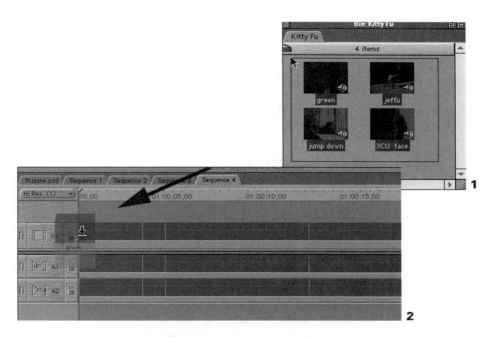

Figure 5-73

The Trim Edit window or the Ripple and Roll tools are perfect for editing these clips when you need to add or remove frames of video from the beginning or end of a clip, specifically to shorten or lengthen it. But in some circumstances, the fastest way to correct the edit points would be to simply adjust the contents of the clip instead of moving its edit points back and forth.

STEP 71

Go to the Toolbar and make sure that your Slip tool is enabled. Then go to the sequence where your rough-cut sequence of clips is sitting. Click on any clip that is surrounded by two other clips, and click and drag to the left and right while holding the mouse button (Figure 5-74).

Overlaying the clips on the sequence Timeline, you will see a ghost image of the Master clip that the "slipping" clip originated from moving around following the mouse pointer. The ghost Master clip shape is there to let you visually know how far you can slip the clip within the current In and Out points of the subclip on the sequence Timeline.

STEP 72

Move the ghost clip around somewhat and then release the mouse button, you will see that the In and Out points, and therefore the duration, of the clip have not changed. However, the portion of the Master clip that these In and Out points relate to has changed. Literally, Final Cut Pro is slipping the subclip in the sequence inside the Master clip so that you can hone in on the footage you want the subclip to use. This is much faster than reloading the Master clip into the Viewer window, setting new In and Out points for the clip and then replacing the old clip in the sequence.

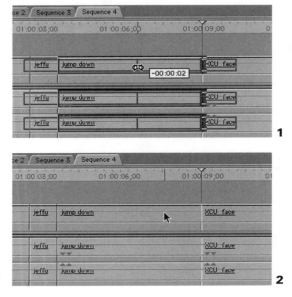

Figure 5-74

The Slip tool is primarily useful for what is referred to as B-Roll footage. Remember from our discussion of A-B Roll editing that B Roll is footage that is incoming footage following an outgoing cut, many times being accompanied by a transition of some sort. B-Roll, as a production term, follows this concept in that B-Roll is extra, relevant but nonspecific footage you might use to break up long static cuts.

For instance, if you are editing a talking-head interview with a school administrator, cutaway shots of kids playing in a school playground and school buses would make excellent B-Roll footage, because they reference the interview but can just be picked up at random rather than scripted out. Interesting and well-shot B-Roll footage is something you can never have too much of, and it can really help smooth out the editing rhythm in a sequence.

With a quick clip adjustment using the Slip tool, all you need initially are your subject footage, long B-Roll clips, and a sense for when to cut away from and back to your main content. The duration that the B-Roll clips should last will largely be determined by your content. If you capture lots of different B-Roll shots in a single Master clip, all you need to do is use the same B-Roll clip frequently in your sequence. When you need a specific shot to appear in the B-Roll cutaway, just slip the B-Roll clip around until you get to the specific section of the clip you need.

Slip into Sync

As a final note, it should be remembered that the Slip function was also just encountered earlier in the chapter in the "Move into Sync/Slip into Sync" box when we Control clicked the "out of sync" box in the Linking example. We chose "Move into Sync" there to move the Control clicked clip back into sync with its related clips.

The "Slip into Sync" choice would have performed an interesting variation on this. Instead of moving the out of sync clip to a new position in the sequence to resync it, the "Slip into Sync" would perform a slip action to the content of the out of sync clip such that it is correctly synced although it does not squarely sit over its companion clips (Figure 5-75). Although the resynced clip does not change its position on the Timeline, its frames are slipped so that the frames that do overlap the other associated clips are in sync. This can be tricky to work with and can easily make sync matters worse if used improperly, so be careful to select the Control-click option that is appropriate to your situation.

Figure 5-75   1   2

The Slide tool

The Slide tool is a very convenient tool that also does not change the duration of a group of clips when it is applied to one of their number. But where it differs from the Slip tool is that it actually moves the clip around between the clips surrounding it, lengthening the clip on the one side and shortening the clip on the other so that the edit points of the clip's edges remain intact and no gaps are created between clips.

The In and Out points of the clip you click and drag with the Slide tool stay exactly as they are in relation to the Master clip they refer to. In fact, the clip you click and drag is not changed at all, except with regard to where it sits in the sequence Timeline. But in order for the clip to occur earlier or later in the group of clips of which it is a part, the clip preceding and following must have their outgoing and incoming edit points simultaneously trimmed to accommodate the "sliding" clip.

STEP 73

Select the Slide tool on the Toolbar and click on a clip that is surrounded by two adjoining clips. Drag it back and forth, and you will again see a ghost image of the clip you are moving, but this time it will maintain its sequence subclip length and instead show its new changed position within the surrounding clips (Figure 5-76).

STEP 74

When you release the mouse button, the clip will jump to the new position you have moved it to, but the edit points for the surrounding clips will adjust so that no gap is introduced. The Out edit point for the clip preceding the "sliding" clip and the In edit point for

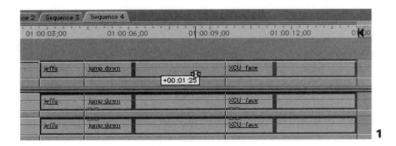

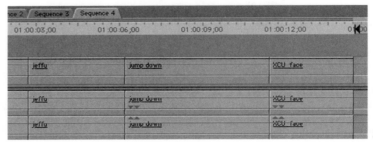

Figure 5-76

the clip following the "sliding" clip will have shifted, being shortened or lengthened as necessary so that no gap is introduced on either side of the "sliding" clip.

This sort of tool will also be very useful with B-Roll footage, but this time for maintaining the exact length of the B-Roll clip. If you wanted to make sure that a 3-second and 15-frame B-Roll shot of a school bus is used in its entirety and that we cut to it as the school administrator says the words, "School bus," this would be the perfect tool. Just "slip" the B-Roll shot up to the footage of the bus and "slide" it right into position at the frame in which the administrator begins the words. Nothing could be simpler or faster.

Monitoring the Slip and Slide tools in the Canvas Window

When you use the Slip or Slide tools, the Canvas displays the familiar double video window to show you the frames of the clip that you are adjusting (Figure 5-77). With the Slip tool, the first video window shows the new frame used for the In point of the clip, and the second window shows the new Out point frame, since the Slip tool will change both In and Out frames of the "slipped" clip. With the Slide tool, the first window shows you the new video frame used for the Out point for the preceding clip, and the second shows you the new frame being used for the following clip's new In point, because the changing frames here will be the preceding and following clips, not the "sliding" clip itself.

There are limitations to what you can do with this tool of course. You can't Slip or Slide past the media limits of the clips you are working with, which is why we made sure that there were adequate handles on each clip in our example. Since the Slip tool will move the In and Out points of the clip you are moving and the Slide tool will change the outgoing and incoming edit points of the surrounding clips, you need to make sure that these clips have enough media to move around in. Final Cut Pro will not tell you you've reached the media limits; it will simply not let you move past them. As usual for the trim tools, a small timecode box opens up as you adjust, specifying how many frames you have adjusted for precision work.

1

2

Figure 5-77

The Razorblade toolset

The next toolset on the Toolbar is referred to as the Razorblade tool. The Razorblade tool is a very intuitive tool that can be picked up very quickly. The Razorblade simply cuts a clip into two separate clips in the sequence at the point where the tool is clicked. This creates two clips from the original: one clip preceding the Razorblade point and one clip following it. The last frame of the first clip occurs right before the first frame of the second clip.

This toolset consists of the tool itself and only one variation on it—the single Razorblade, which performs the cut just described for the single track that is clicked (this includes any linked items if Linking is enabled), and the Razorblade All, which performs the same sort of cut across all clips in all tracks.

Using the Razorblade tool

To illustrate the Razorblade tool, we will cut a clip. Then, using a shortcut called the Ripple Delete, we will delete the part of the razorbladed clip we don't want and snap the clip following it up to the razorbladed edge, eliminating any gap in the sequence.

STEP 75

Select all clips in the sequence and delete them. Next, perform the storyboard rough-cut technique again so that there are contiguous clips in the sequence.

STEP 76

In the sequence, move the mouse pointer over a clip. Using the J-K-L shuttling technique, find a frame in the clip where you want introduce a cut separating the clip into two parts (Figure 5-78).

STEP 77

Select the Razorblade tool on the Toolbar, and then return to the sequence. Make sure that Snapping is enabled in the Timeline window, and then move the mouse pointer along the track of the clip you want to cut.

Figure 5-78   1   2

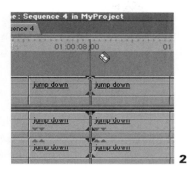

Figure 5-79

As you move the Razorblade mouse pointer along the clip in the track, you will see a small vertical line indicating which frame the tool will cut if it is clicked. Notice that the line appears in both the video and the two audio clips you drag it over. This is because the Linked Selection tool forces the Razorblade to perform the Razorblade action on all linked items.

STEP 78

When the Razorblade is near the playhead position in the clip, it snaps to it, enabling you to cut precisely on the frame of the clip. Click on that frame, and the Razorblade will cut the clip in two (Figure 5-79).

Now that the clip has been divided into two separate clips, we want to get rid of the second clip that contains footage we don't want. At the same time, we want to have any clips occurring later in the track to snap up to the newly cut clip end to eliminate a gap in the track. If we just selected the second clip and deleted it using the Delete key, there would be a long empty gap between the first razorbladed clip and any subsequently occurring clips in the track. Instead we will use the Ripple Delete function.

STEP 79

Move the mouse pointer over the second clip in the track that we want to delete. Press the Control key and click on this clip (Figure 5-80). You will be presented with a contextual

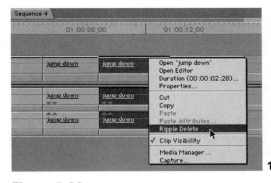

Figure 5-80

menu with many options. Choose the option Ripple Delete. When you do so, the second clip will disappear and the subsequent clips in the track will shuffle back to fill in the gap created by the deleted clip. The keyboard shortcut for this is Shift-Delete.

The Zoom toolset

The final toolset we will be looking at in this chapter is the Zoom toolset. This toolset is a pretty common feature in most image-editing applications. If you click on the toolset button in the Toolbar, you will see four options: Zoom In, Zoom Out, Hand, and Scrub Video. These tools allow you to customize and navigate the windows of Final Cut Pro in ways that do not affect the clips and sequences of your project. They help you refine the way the various windows appear and how much of them you can see at a given time. Rather than using the mouse or menus to change the shape of Final Cut Pro, learn to take advantage of this toolset.

The Zoom In and Zoom Out tools

The primary Zoom tool, shaped like a magnifying glass with a plus sign, simply increases the scale of whatever object it is applied to (Figure 5-81). If you click on a sequence with this tool enabled, the scale increases as if you had adjusted the scale controls of the Timeline window itself. The Zoom Out button next to the Zoom In button provides exactly the opposite effect, shrinking the scale of the window, so that you can quickly adjust the exact scale of any window you are working with. Holding the Option key down while the Zoom In tool is enabled converts it to the Zoom Out tool, making it a quick and precise customization tool. Just hit the Z key to enable the Zoom In tool, then the option key to pull back out if you Zoom In too far.

The Zoom In and Zoom Out tools have uses outside the Timeline window as well. The Viewer and Canvas windows can be boosted or shrunk in scale based on a percentage of the actual size of the frame. This number is located in the top center of each window. Generally speaking, this number should be left at 50 percent for optimized playback. Having the window at 100 percent can cause playback problems in machines with less video processing horsepower, and 100 percent is sometimes a bit extravagant when screen real estate is so precious. 50 percent gives a large enough image to look at without getting a headache; it doesn't overtax the processor on playback, and it leaves plenty of room for all the other windows of Final Cut Pro.

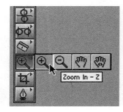 

Figure 5-81

The Hand tool

Sometimes it's important to get a closer look at an image you are working with, however, as we will be doing in the following chapter on compositing. So it's possible to Zoom In and Out much further than you would normally think necessary. But once you zoom in to 800 percent in the Canvas or Viewer, how do you navigate around the image to get to the section you wanted to look at? There are scroll bars on the edge of the window, but that will wear out your wrist and frustrate you as you try to get to a place you can't see using only up-and-down or left-to-right scroll bars.

Predictably enough, Final Cut Pro has a tool to simplify this task. The Hand tool simply moves the window around so that you can go directly to the part of the window you want without fooling with scroll bars. It does not make any adjustments to clips or sequences; it functions only to move what is visible around the Final Cut Pro interface windows.

STEP 80

Load a clip into the Viewer window. Then select the Zoom In tool from the Toolbar. Click inside the Viewer window several times, each time watching the number at the top and center of the window jump in percentage points. Keep clicking till you see the number reach 400 percent (Figure 5-82).

By this time you should have noticed that the image is far larger than the window can accommodate. There are scroll bars to the bottom and side, but instead, use the Hand tool from the Zoom toolset of the Toolbar and grab the image, moving it around by clicking and dragging. You may have also noticed that after 100 percent, the image appears to disin-

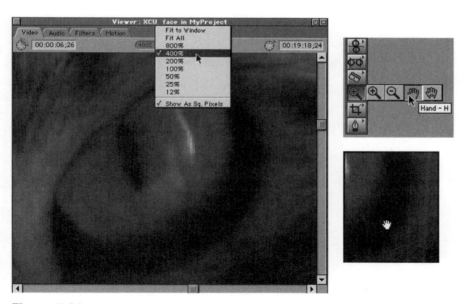

Figure 5-82

tegrate slightly, taking on a blocky appearance that is anything but the beautiful DV image you started out with.

Relax, there's nothing wrong with your television. You are seeing *pixellation*. Each frame of DV is made up of pixels (i.e., picture elements), which are the smallest indivisible sections of a digital image. A single frame of DV NTSC video contains 720 pixels across and 480 pixels high, or 345,600 pixels. When you zoom in to over 100 percent, you begin to see these individual pixels as the rectangular colored blocks that they really are, similar to the dots you would see in a newspaper picture if you looked at it under a magnifying glass. Once you zoom back out, you'll see the pixels bleed back together into the lush imagery you were expecting. Just hit the Option key and Zoom Out to return to 50 percent.

The Scrub Video tool: Scrubbing a clip's thumbnail

The last tool of this toolset is a really handy tool called the Scrub Video tool (Figure 5-83). Although its use is very specific, it can be a lifesaver if you are doing a lot of clip duplication from one Master clip. The Scrub Video tool offers a fast efficient way to set Poster Frames for your clips in the Browser window. A Poster Frame is the small image in the clip you see when your Browser Items are set to display as Large Icons (see View drop-down menu>Browser Items>Large Icons). The default Poster Frame for any given clip's thumbnail will be the first frame of video in the clip.

This can be very confusing if a lot of your footage looks very similar, especially near the beginning of the clip. If you have five clips that all start with roughly the same shot, the Poster Frames in the thumbnails will all look exactly the same (Figure 5-84). Changing the image of the Poster Frame to some other more distinct frame from the clip can make it much easier to identify clips at a glance.

There are a couple of different ways to set the Poster Frame for a clip. If a clip is loaded into the Viewer window, you can set it by finding the frame you want to use with the play-

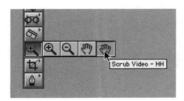

Figure 5-83

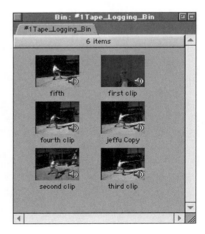

Figure 5-84

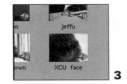

Figure 5-85

head, then going to the Mark drop-down menu and selecting Set Poster Frame, or Control-P as the keyboard shortcut (Figure 5-85).

But doing this requires that the clip be loaded up into the Viewer window. Using the Scrub Video tool on the other hand allows you not only to look through the footage of the clip without loading it into the Viewer, it allows you to set the Poster Frame there as well.

STEP 81

Open the bin with your storyboard-style thumbnailed clips. If you have reset the bin to display as a list, or with no thumbnails, set it to show the bin as Large Icons by going to the View drop-down menu and selecting Browser Items followed by Large Icons. Your clips need to be visible as large thumbnails for this example (Figure 5-86).

Figure 5-86

STEP 82

Go to the Toolbar and select the Scrub Video tool from the Zoom toolset. Now return to the Browser window and pick a clip that has a fairly wide range of imagery. Click in the center of the thumbnail, and drag to the left and right. As you do so, you should see the thumbnail scrub through the video frames of the clip.

This is a good way to quickly preview a clip before working with it, but it isn't radically faster than loading the clip into the Viewer window, where you could get a much better view of the contents of the clip. The main value of the Scrub tool is not in previewing the footage in the clip but in selecting the Poster Frame. But when you scrub with the tool and release the mouse button, the Poster Frame snaps back to the first frame of the clip.

Setting the Poster Frame with the Scrub Video tool

To actually change the Poster Frame, you need to perform a little intricate Control-clicking.

STEP 83

First, click and drag on the thumbnail as you just did previously, scrolling through the frames of video in the clip. But as you drag, press and hold down the Control key.

STEP 84

When you find the image you want to use as the Poster Frame, first release the mouse button while still holding the Control key. Then, release the Control key and your Poster Frame will have changed to the frame you want to use (Figure 5-87).

All this may seem like gratuitous effort just to get a little unique thumbnail to look at in the bin window. And with Large Icon view, you lose access to all the great detail information in the columns that you get in List view. Furthermore, you will find that it is far easier to mentally organize Master clips and subclips in the infinitely customizable List view rather than the loose arrangement of the Icon views.

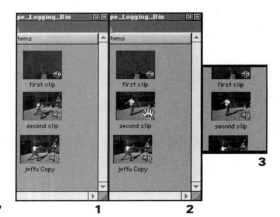

Figure 5-87   1   2

Master clips and subclips

What are Master clips and subclips? A Master clip can be thought of as a clip that relates directly to the entire media it references. When you capture video footage, Final Cut Pro creates a QuickTime Movie media file and then a clip in your project that references that media file. This clip has access to the entire media file that you captured. We call this a Master clip. But what if you wanted to divide up that Master clip into smaller bite-sized portions? Perhaps the large captured clip contains several different takes, or maybe you just want to use selected bits of the same take repeatedly.

When you load the Master clip into the Viewer and give it an In and Out point to edit with, the Master clip retains these In and Out points. If you later on wanted to use a different section of the clip, you would need to change these and make new In and Out points. This is the way that clips work. They retain edit points until we change them.

But our minds generally do not work this way. Its usually easier to organize a project if you have isolated the various parts of your project in such a fashion that you can look at them en masse. How could you possibly storyboard rough-cut edit, for example, if all your material was contained in one Master clip? Clearly such a method will work only in very small, uncomplicated projects.

A much better way to organize a complicated project with Master clips is to create specific subclips and treat them as clips of their own, even though they ultimately reference the same media files as their fellow subclips and the Master clip from which they originate. Final Cut Pro certainly doesn't care; media files are media files, clips are clips, "parts is parts." The key to successful editing is both project organization and the ability to put the pieces of media together in a pleasing way. You can't have one without the other.

The Final Cut Pro Subclip

There are a couple of different ways to create subclips from initial Master clips, each of which carries its own implications. The first way is through the Final Cut Pro Subclip command. Final Cut Pro allows you to do exactly as we are suggesting. When you load a Master clip into the Viewer window and set In and Out points, rather than immediately edit with this In and Out pointed–master clip, we can make a subclip based on the In and Out points you added. A new clip appears in the Browser that is limited to just the In and Out points you specified. And you are free to reload the master clip in the Viewer and create other subclips until you've completely divided and organized the master Clip into workable clips.

STEP 85

After switching the bin's view mode back to List in the View drop-down menu's Browser Items, double-click a long clip and load it into the Viewer window. Play through this Mas-

Figure 5-88

ter clip and set an In and Out point for what you would consider to be a separate clip within the larger clip (Figure 5-88).

Set your In and Out points wide of the actual frames you want to edit; the Subclip command does not kid around and really considers its subclips as having no access to the rest of the media files they share with the Master clip. You will not be able to Trim Edit outside of the In and Out points of the subclip you are creating.

STEP 86

After setting the In and Out points in the Viewer window, go to the Modify drop-down menu and select Make Subclip at the top of the list. Instantly, a Subclip icon will appear in the bin with the Master clip. A Subclip icon can be distinguished by the shaggy right- and left-hand edges of its icon when viewed in List view mode.

STEP 87

Before moving forward, immediately click on the temporary name "Clipname Subclip" and type in a new name for the subclip, identifying which Master clip it came from and what it contains (Figure 5-89). Also remember that there are Comment and Description columns in the Browser window for adding details. There's never a good excuse for forgetting what a subclip is or where it came from.

Figure 5-89

STEP 88

Double-click the new subclip in the bin to load it into the Viewer. You will see that the length of the clip is exactly the length of the In to Out points you set when you created it from the Master clip. If you edit it into a sequence, you will further see that the subclip can not be trimmed out beyond the frames you established as the In and Out point. Final Cut Pro regards anything created with the Make Subclip command as if it were a captured clip rather than a clip carved from a larger Master clip.

Of course, Final Cut Pro offers an override for this limit to the subclip's ability to access the rest of the Master clip it was created from. At any time, you can select the subclip, return to the Modify menu, and choose Remove Subclip Limits (Figure 5-90). This restores the original access to all of the media to which the Master clip had access. The subclip is still a subclip, but it now references all the media that the Master clip does. Of course when you do this, you lose the range of In to Out that you set up the subclip for in the first place.

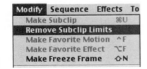

Figure 5-90

So, using this process, you can divide your Master clips up however you like. The process can be a little clumsy, though, if you are still in the rough-cut stage, because many times you don't know yet how long a subclip should be. If you create subclips that are too short, there's no way to add a little extra footage to the subclip short of restoring the entire media limits to the clip. That limitation can defeat the purpose of creating the subclip in the first place.

Another subclip technique

A second way of creating subclips from Master clips is really much simpler and more functional as long as you keep your eye on the ball and watch your naming conventions for your clips. Instead of creating a Final Cut Pro Subclip using the command from the Modify drop-down menu, we will simply duplicate the Master clip, and give the duplicate a new name along with the In and Out points we want to associate with it as a subclip. For the sake of avoiding confusion, Subclip capitalized will refer to Subclips created using the Final Cut Pro Make Subclip command. The word subclip with the lower case S will refer to this alternative method of creating subclips.

STEP 89

Create a new bin in your Project tab, and name it "Master clip>subclips." Double-click this bin so that its window is open and available as you proceed through the next few steps (Figure 5-91).

STEP 90

Select the Master clip in your Logging bin, and load it into the Viewer window. Click the Mark Clip button in the Viewer, or go to the Mark drop-down menu and select Clear In

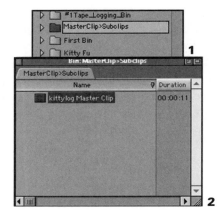

Figure 5-91

and Out. This is to clear off any previous In or Out points; you want to start your subclips off as blank slates with no In or Out points.

STEP 91

Once your Master clip is loaded in the Viewer and cleared of irrelevant edit points, set an In and Out point in the Master clip just as if you were going to create a Subclip via the Make Subclip command. Isolate the content you want in the clip, and surround it with In and Out points.

STEP 92

When you have In and Out points selected, grab the video window of the Viewer and drag it to the "Master Clip>subclips" bin window (Figure 5-92). When you drop it there, you will have an exact copy of the Master clip but one that contains In and Out points specific to what you want to use in the subclip.

Figure 5-92

STEP 93

Rename this new subclip in the "Master Clip>subclips" bin with a name appropriate to the section of the Master clip that you want to isolate in it. Technically, the name you give the subclip doesn't matter as much as the fact that it be named uniquely. As you bring more subclips down into the bin this way, each one will carry the name of the Master clip, since you are really only creating a clone of it in the bin, rather than creating a totally new clip.

STEP 94

Return to the Viewer window, where the Master clip is still loaded. Repeat Steps 91 through Step 94 until you have created as many subclips as it takes to divvy up all the different content in the Master clip.

STEP 95

After you've completed the creation of all the subclips you want in your special subclip bin, double-click one of the subclips and load it into the Viewer. Observe that all the frames of the Master clip are still available as handles outside the In and Out points of the subclip, but that each subclip retains the special In and Out points that were set right before you cloned the Master clip into the subclip bin. Load up other subclips to further demonstrate this fact.

Creating thumbnails in the List view

You may find yourself asking, "Why can't I have a clip thumbnail in the List View?" All the foregoing processes occurred in the List view mode, where by default there is no clip thumbnail. And although thumbnails are just icing on the cake for the really well organized, there's no reason why we shouldn't help ourselves to as much icing as we can get.

There is a way to enjoy the convenience of the Large Icon view's thumbnails in the List view mode. Although using List view will not allow you to arrange your clips in storyboard fashion as is possible in Large Icon view, most editors find that the List view combination of thumbnails and informational columns along with the ability to arrange clips by any criteria you can think of is far more important than assembling storyboard panels. If you need to look at things from a storyboard perspective temporarily, there's no reason you can't switch View modes briefly, then switch back.

STEP 96

With your "Master Clip>subclips" bin set in the View drop-down menu>Browser Items to List, Control-click on the first column header of the List view, which by default is Duration (Figure 5-93). As you hold the Control key down, a small menu list appears, at the bottom of which is an option for Show Thumbnail.

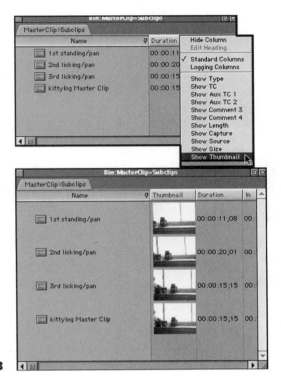

Figure 5-93

STEP 97

Select the Show Thumbnail option and release to add a Thumbnail column to the lists. The Duration column simply moves one space to the left to make room. You can now set the Poster Frames for the thumbnails of your new subclips. The Scrub Video tool from the Toolbar will function precisely the same for this thumbnail as it does for the Large Icon view. The only complication is that the subclip you created still has access to the entire Master clip's frames, meaning that when you scrub for the Poster Frame, you will have to carefully scrub to choose a frame that matches the In to Out range of the subclip. If that is too difficult to do, you can always use the more mechanical method of setting the Poster Frame, involving the Mark drop-down menu.

STEP 98

Double-click on a subclip in the "Master Clip>subclips" bin to load it into the Viewer window. Shuttle through the range of frames between the In and Out points until you find a frame that really distinguishes that subclip from others. With the playhead parked on that frame, go to the Mark drop-down menu and select Set Poster Frame (or Control-P on your keyboard). The subclip's thumbnail will change immediately to that frame (Figure 5-94).

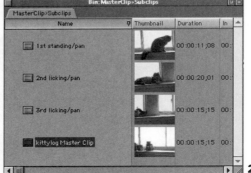

Figure 5-94

The Crop/Distort and Pen toolsets

The two remaining toolsets in the Toolbar, Crop/Distort and the Pen toolset, are best covered in the next chapter, since they relate almost entirely to image manipulation rather than the cutting and trimming functions that define editing.

6 Compositing and Special Effects

What exactly is compositing?

Periodically, we have mentioned the word *compositing* in the course of this book. What does this term mean? Compositing, in its most literal sense, means taking a stack of items and making them into one item. When we discussed analog video standards, we referred to one such standard as Composite. Composite video, if you recall, is a video signal in which all the different luma and chroma electrical signals are integrated into one electrical waveform. They become one composite signal.

Compositing in Final Cut Pro and other applications involves the process of taking more than one image and integrating the various images into one image. As one Final Cut Pro editor puts it, "Compositing is everything above Video Layer 1." As we discovered when we worked with the sequence Timeline, Final Cut Pro allows up to 99 separate video tracks. The good news is that in this manner Final Cut Pro gives you fantastic flexibility in customizing the way those separate layers work together when they are composited into the flat image you see on the video screen.

To be successful at compositing, you really have to rework the way you think about making a visual image. Compositing is the foremost of the tasks in which there is never any one right way to accomplish a goal. Part of that lies in the fact that compositing, more than any other element, requires a mental vision of the results. Your uniqueness as a human and an artist cannot help but shine through.

No two compositors generate the same results, just as no two painters use the same brushstroke. Why must you undo your prior knowledge of image making? Because sometimes, adding red to an image is more about taking away green than it is about painting with red. Final Cut Pro gives you the ability to obscure or reveal imagery using not only the various features of light, such as opacity (visible to invisible), hue (the shade of the color) and saturation (density of the color), but also the way that obscuration or revelation occurs over time, also known as *keyframing*.

We'll start with an exploration of layering, track opacity, and the Composite Modes. In the course of setting up and using these effects tools, we will also learn to use Wireframe, Keyframe Interpolation, the Text Generator, Matte and Masking tools, and the Alpha Channel.

Compositing layers

From our definition of compositing, we already know that we need more than one layer in a sequence in order to create a composite. Layers in a sequence are stacked, one on top of the other. What you see in the Canvas window is actually what lies in the sequence as seen from the top layer down. Imagine that your sequence of video layers is a messy stack of different types of paper. Some sheets cover others; some are more transparent than others, allowing you to see through them to the next layer underneath. When you edit video clips into a new layer on the sequence, you are inserting a new sheet into the stack.

Setting up a sequence for compositing layers

STEP 1

Create a new sequence in your project. Edit an initial video clip into the sequence in the Video 1 track.

STEP 2

Choose another video clip, drag it to the sequence, position it in the Video 2 track, and drop it (Figure 6-1). If no Video 2 track yet exists, simply place it where the track should be, and Final Cut Pro will automatically create one.

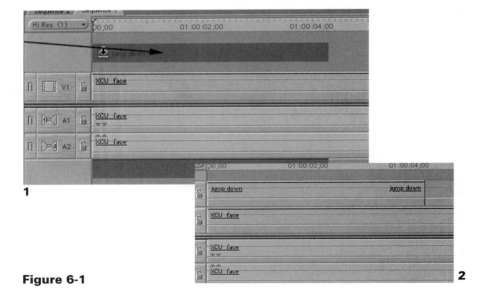

Figure 6-1

STEP 3

Make sure that the sequence playhead is positioned somewhere within the clips. A glance at the Canvas window will reveal that the video clip in the Video 2 layer is covering the clip in the Video 1 layer. This is because, by default, all video clips are at 100 percent opacity. We have already explored this opacity feature in our discussion of the Clip Overlay button. When we adjusted the opacity in that chapter, there were no other video layers involved. Since the default background color for a Final Cut Pro sequence is black, changing the opacity of the clip simply resulted in a fade to black. But now that we have more than one video layer, changing the opacity of the top layer will reveal the layer below.

STEP 4

First, make sure that the Clip Overlay button is enabled in the Timeline window. Remember that the Clip Overlay button, though a part of the Timeline window, is actually a function of sequences within the Timeline window. Just because it is enabled within one sequence does not mean that it is enabled in others. If you do not see the rubberband line going through the clips in a sequence, rubberbanding is not enabled.

STEP 5

Choose the General Selection tool on the Toolbar (shortcut Command-A) and adjust the rubberband line of the clip in Video 2 layer (Figure 6-2). As you pull the line down, the red render bar will appear as expected, but looking at the Canvas window will reveal that the clip in the Video 1 layer is beginning to appear through the Video 2 layer. If you drag the line all the way down to zero, layer 2 will be transparent, totally revealing the clip beneath it.

If we render the clip, playing back will reveal that the entire clip carries the opacity level we set the rubberband line at. But how useful could this be in everyday life? Setting a static, unchanging opacity level in a clip might be useful in some situations, but in most, we need to be able to control the level of opacity over time within the clip. To do so, we need to introduce the concept of *keyframing*.

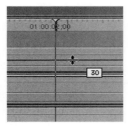

Figure 6-2

Keyframing in the sequence timeline

Keyframes come into use with almost all visual effects in Final Cut Pro, as well as other popular editing and compositing applications. Keyframing means that for a certain feature in a clip, such as opacity, a certain value exists at a certain time. For instance we could set a keyframe at the tenth frame of the clip for an opacity value of 25 percent. Any other frame of the same clip could have the same or a different percentage value of opacity. On any frame where we establish a specific, definite value, we create a keyframe.

In the digital universe, everything has a value. The example we are working with, opacity, has a value at any given time of between 0 to 100 percent. At 0 percent, the clip is transparent; at 50 percent, it is half-transparent; and at 100 percent, the clip is opaque. No keyframes are necessary for a clip to have an initial value; all clips do. But establishing a keyframe and giving a specific frame a value gives us the ability to change a value in a single clip over time.

If we assign the first keyframe for a feature in a clip, the value of the feature for the duration of the clip still does not change. Even though we have a keyframe, the value of this initial keyframe doesn't change to a new value. The lesson is that one keyframe in a clip expresses only one value. To express a change in values in a clip, you must have two or more keyframes, each of which has a different value. We set a keyframe for the value we want the feature to have initially, say 100 percent opacity, then we set additional keyframes for the change in values we wish to see over the duration of the clip, say 50 percent, then 25 percent, and then 0 percent. If we do not want a value to change over time, there is no reason to insert a keyframe at all.

To insert keyframes into a clip's opacity in a sequence, we need to have two features enabled. First, we need Clip Overlay to be enabled, since this is where our sequence keyframes will be inserted. Next, we need to go to the Toolbar and enable one of the last two toolsets, the Pen tool. If you click on the Pen tool and hold the mouse button briefly, you will see three options: the Pen tool, the Pen Delete tool, and the Smooth Pen tool (Figure 6-3). We will discuss the Smooth Pen tool subsequently, but for now we are concerned with the Pen and Pen Delete tools.

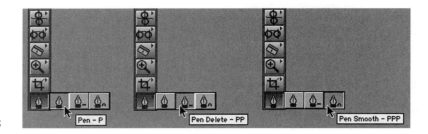

Figure 6-3

STEP 6

Select the Pen tool and return to the sequence. Look to the clip in the Video 2 track that you adjusted the opacity of previously. Move the mouse pointer to the beginning of the clip and place it near the rubberband line (Figure 6-4).

You may want to go into your Preferences and change Thumbnail Display from Name plus Thumbnail to Name. This will remove the thumbnail from the clip in the sequence, making it easier for you to work with the rubberband line at the beginning of a clip.

When you move to the beginning of the clip, the mouse pointer will change from the familiar arrow into the shape of the Pen tool. This indicates that the Pen tool is able to assign a keyframe for the frame you are presently hovering over.

STEP 7

Click on that rubberband line, hold the mouse button down and drag the keyframe up and down. As you do, the value box appears next to the mouse pointer informing you of the current value of the keyframe you have just created by clicking on the rubberband line.

STEP 8

Now, without releasing the keyframe, drag it from left to right and back. A timecode value box appears instead, informing you how far you have moved the keyframe from its original clicked position.

As long as you have the keyframe under the thumb of your mouse button, you have total control of the value it represents, the time it occurs within the clip, and the sequence the clip is included in. If you let go of the mouse pointer, but you want to change the keyframe's value or position, simply grab it and drag again. After you have established a keyframe, the Pen tool displays as a crosshair when suspended over it.

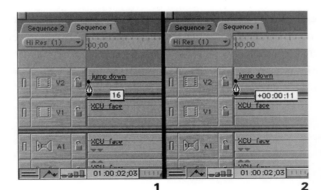

Figure 6-4   1   2

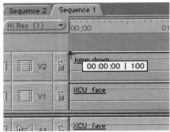

Figure 6-5   **Figure 6-6**

STEP 9

Drag the keyframe all the way to the left, so that it occurs on the first frame of the clip. Then before releasing it, drag it all the way up to the top of the clip, so that its value is 100 percent (Figure 6-5).

Now we have one keyframe. But as we mentioned earlier, one keyframe is really no different from having none. To establish a change in value, we need to add a second keyframe that occurs either before or after the initial keyframe. Since our first keyframe is at the beginning of the clip, we will move up several seconds to add the second one.

STEP 10

Send the sequence playhead back to the first frame by hitting the Home key on your keyboard. With the Timeline window active, type in "+300" which will move the playhead up 3 seconds from its present position (Figure 6-6). Move the Pen tool up to the sequence playhead, and watch for the Pen tool to bulls-eye the yellow triangle above the playhead indicator. When it does, click and drag the newly created keyframe down until it hits 0 percent (Figure 6-7).

Now you will have two keyframes, each of which contains a different value. The first keyframe is at the beginning of the clip at 100 percent, and the second is three seconds later at 0 percent. More interesting than this is the fact that between the two keyframes is a long slanting line denoting the change in value of opacity for all the frames in between the two keyframes. This is referred to as interpolation.

Figure 6-7

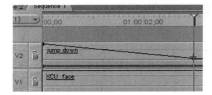

Keyframe interpolation

Interpolation is a process by which the computer comes up with all the values between two known values, saving you a lot of calculation. If you know that the opacity value for the clip at the first frame was 100 percent and at the end of three seconds it was zero, you'd have to do some quick math to figure what percentage of change occurred during each frame of video in between the first and the last. You could do it, but like most people, you probably hate long division and fractions, and frankly, setting all those intermediary frames wouldn't be worth it (except for rotoscoping animation artists).

Luckily, we are dealing with a computer here, a machine designed to crunch numbers. We give Final Cut Pro the values we want the clip to have at the beginning and end keyframes in the clip, and it figures out the values for each frame in between in less time than it would take you to remember where you left your calculator. If you move the playhead back through the area of the clip that is between the two keyframes, you will see that the Video 2 layer clip slowly and evenly goes transparent as the opacity keyframe values go from 100 percent to 0 percent over 3 seconds. Each frame between the two keyframes is displayed with a progressively smaller percentage of opacity. And all you need to have for this is the two opacity keyframes that change.

Composite modes

The next sort of layer compositing we need to look at involves the way in which the various luma and chroma values of the images on a layer interact with the values of those on the layer or layers underneath. This topic is one of the most poorly understood features of Final Cut Pro and image editing applications in general. But Composite Modes should not be avoided because they provide a tool of incredible flexibility when generating layer effects.

Part of the confusion about using the Composite Modes is in understanding what they do and what needs to be in place in order for them to show results. The first requirement is obviously that the clip must be in the sequence. Since a Composite Mode is an effect that relies on a clip's relationship with other clips, changing its Composite Mode outside of a sequence will not display its true appearance. Although you can change a clip's Mode when it is loaded in the Viewer by itself, you won't see the ultimate effects until it is layered above the intended clip in the sequence.

Next, Composite Modes applied to a clip always interact with the clip or clips on the layer directly beneath it. This means that when you choose a Composite Mode for a clip, you need to make sure that there is a clip underneath it that will change the clip in the way you want. If no clip exists underneath the clip whose Composite Mode you change, there may be no apparent change in the clip. Because the background of a sequence or a clip is always either Black or White or Checkerboard, based on how you have it set, you may not see much change. Composite Modes alter a clip by using the variations, or lack thereof, of the underlying clip to influence the appearance of the clip the Composite Mode has been applied to. Thus using straight Black or White backgrounds for a clip's Composite Mode complement could yield unpredictable results or no change at all.

When you first engage a clip, its Composite Mode is set by default to Normal.

STEP 11

Clear off your sequence from the preceding example and add two stacked clips, one into the Video 1 track and one into the Video 2 track. Single click the clip in the Video 2 track to select it (Figure 6-8).

STEP 12

Go to the Modify drop-down menu and scan to the bottom to find Composite Modes. When you move to this submenu, you will find a list of many different Composite Modes (Figure 6-9). Each name describes how the luma and chroma values of the underlying clip will influence the appearance of the clip you apply the Mode to.

The problem for most users is that the names of the Modes do not accurately describe what the effect on your clip will be. Unlike Effects filters such as Gaussian Blur and Fish-Eye, which give some indication of their effect, the Composite Modes only describe the method by which Final Cut Pro will take the luma and chroma values of the underlying clip and use them to alter the appearance of the clip you apply the Mode to. This means that the effect resulting from the application of a Composite Mode depends on the contents of both clips and the Composite Mode chosen.

That said, there are some Modes that come into frequent use. Perhaps you want to lighten or reverse the colors in one area of your clip. You may want to enhance the effect of an Effects Filter by drawing out the luma and chroma values of a clip before applying the Filter. Some people are masters at using the Composite Modes as a type of unique paint-

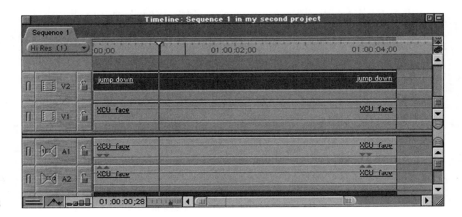

Figure 6-8

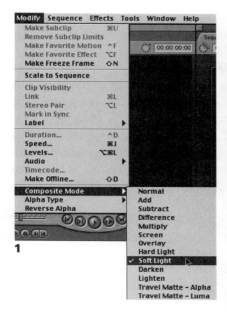

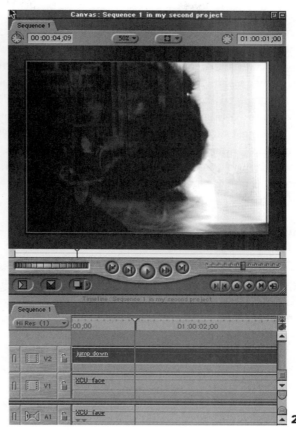

Figure 6-9

brush, one that is able to make complex and beautiful patterns of color and light from the mixture of math and image.

The default setting for a clip's Composite Mode is normal, which means that the clip is not influenced by the clip below it. If you select any of the Composite Modes other than Travel Matte Alpha or Travel Matte Luma, you will begin to see interesting results.

STEP 13

In the Composite Modes submenu of the Modify drop-down menu, select a Composite Mode for the selected clip in the Video 2 track. Refer back to the Canvas window and look at the effect that the Mode creates based on the clip beneath it.

Some of the Modes show an even mixture of the two images, while others show only the darker sections of the one or the lighter sections of the other. Some reverse areas of similar or different detail, yielding odd color combinations. Each Mode gives different results based on the two clips that you have juxtaposed.

STEP 14

Continue choosing Modes from the list for the selected clip, noting how each Mode yields both different and sometimes similar results, depending on the two clips and the Mode.

Feel free to change the Composite Mode for a clip as often as you'd like. Of course, any clip that has its Mode changed will require rendering, because the video to be played will have changed. To turn off the Composite Mode for a clip, you must return the Composite Mode to Normal. Composite Modes are not keyframeable.

Almost alone among Final Cut Pro's features, Composite Modes have no keyboard shortcuts or palettes, and are only available from either the Composite Modes submenu in the Modify drop-down menu or the Item Properties dialog box in the Edit drop-down menu. Hopefully, in future versions Apple will make this feature a little more accessible.

Mattes, masks, and stencils

Two of the most valuable Composite Modes were skipped in the preceding text because of their unique function and value to Final Cut Pro editors. These are Travel Matte-Alpha and Travel Matte-Luma. These tools offer the ability to frame the images of one video layer inside a shape that appears above the background layer. This is great for more advanced picture-in-picture effects, such as the ability to show one video clip through the letters of a word that uses another clip as a background.

The effect is rather like using a stencil to paint letters. If you lay the stencil on a background you want to paint the letter onto, you can paint with impunity above the stencil itself, knowing that the paint will reach only the background in the spaces that the stencil allows it to get through. Thus, three layers are involved in the process: the background, the stencil, and the layer of paint that is being used for the letters applied through the stencil.

A stencil is part of a group of tools that all perform roughly this same task. Whether they go by the name of Stencil, Mask, or Matte, these tools are part of a family referred to as *mattes*. Mattes allow you to obscure or reveal areas of an image based on the criteria that defines the matte. In the stencil example above, our criterion is very simple. The stencil blocks the path of the paint, restricting it to the shape of the letters the stencil is constructed to paint. Another example of a matte would be a mask that covers a light source to create shadow shapes. This tool, called a *cucaloris* (*cookie* for short) is used on a theatrical, film, video, or photography set for simulating shadows like window panes and the like.

With most real-world mattes, the intention is to block some physical element, like light beams or paint, so that a visual effect is achieved. Our mattes in the digital realm are no different, although we have more flexibility there because we are not limited to physical objects. Our mattes in the digital world can be generated by various means and can have just as many possible effects on the objects being masked.

Travel Matte-Luma

Using the Text Generator as a matte

In order to use a Travel Matte, we need to create a matte. Our matte will be a text clip generated by Final Cut Pro's built in Text Generator. We will put the text clip onto a Video layer between the background clip and the foreground clip. This text layer will perform as a stencil, only allowing us to see the parts of the foreground clip that occur in areas where there is text on the matte layer.

STEP 15

To make sure that your Modes are set correctly, start with a clean slate. Select and delete all clips in the sequence. Then edit what will be our background clip onto the Video 1 track. For purposes of demonstration, choose a clip with darker imagery. Once you understand the process, you can experiment with any sort of imagery you want, but a greater contrast between your clips will be better for revealing what the matte does in this process.

Next we will generate a text clip using Final Cut Pro's built-in text generator, although you could just as easily use an image created in Photoshop or another image editing application for this purpose.

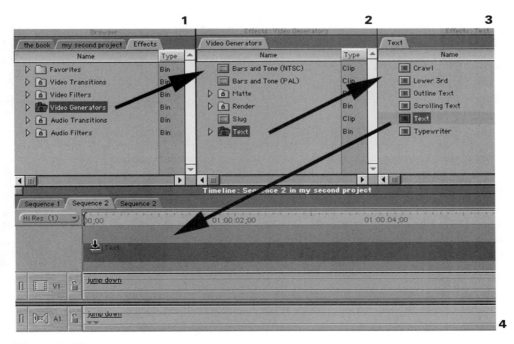

Figure 6-10

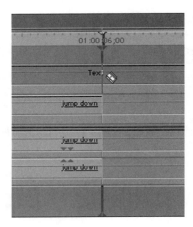

Figure 6-11

STEP 16

To create the text clip using Final Cut Pro, go to the Effects tab of the Browser window. Double-click and open the bin named Video Generators. Locate the bin labeled Text and double-click to open it (Figure 6-10).

STEP 17

In the Text bin that appears, grab the Text clip and drag it over to your Video 2 track in the sequence so that it covers the clip you edited into the Video 1 track. The length of the text clip will probably not match the length of your clip in the Video 1 track, but you can extend or trim edit it to whatever length you need. Trim the Text clip so that its length matches the background video clip in the Video 1 track (Figure 6-11).

Arranging and refining the clip using the wireframe and the Control tab

A glance into the Canvas window will reveal that a line of text has appeared directly in the center of the window. Since this is probably not the placement you want, we will move it using the Wireframe tools in the Canvas window. Wireframe refers to the bounding box around an object in Final Cut Pro. This bounding box does not exhibit any of the bitmap, or pixel, data of the image. It only displays the physical dimensions. It's rather like looking at the chicken wire and structural beams of a building rather than the concrete and smoked glass windows.

Normally, this Wireframe view is disabled, because we want to see the video we are working with, not its edges. But any direct physical adjustment we make to the image, such as its size, position, and rotation must be made using the Wireframe edges. Final Cut Pro gives us the flexibility to view our clips using either Image, which is the default view; Image+Wireframe, which shows us both; and just Wireframe, which, though uncomfort-

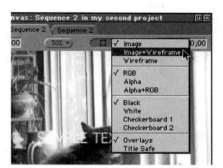

Figure 6-12

able, can be useful when the image is so complex and distracting that it makes it difficult to see and use its Wireframe edges.

STEP 18

To view the Wireframes, go to the Canvas window next to the Window percent Scale menu and look for the drop-down bar that contains a small square (Figure 6-12). Click on this bar and the menu will open, revealing a long list of options. At the very top are three choices: Image, Image+Wireframe, and Wireframe. Select Image+Wireframe.

When you do so, you may or may not see a bounding box appear in the Canvas Video window. This is because Wireframes are displayed only when an object is selected.

STEP 19

Go back to the sequence Timeline, click on an empty area to deselect all clips, then single-click the text clip (Figure 6-13). A bounding box will appear in the Canvas window that covers the entire viewing area. You could also have clicked in the area of the Canvas window itself that is occupied by the clip you want to select. However, this can sometimes be difficult when dealing with text clips and other objects that are largely transparent.

STEP 20

Next, make sure that the General Selection tool is chosen in the Toolbar and go back to the Canvas window. Click on the window and drag, and you will see that the text layer moves along with the mouse pointer (Figure 6-14). The bounding box follows this movement, because you are actually moving the entire object, of which the text is only a part. Position the text in some portion of the window that feels more comfortable, such as the lower portion of the window (in the broadcast industry, titles in the lower portion of the television screen are referred to as "lower thirds"; there is another text generator in the Effects bin named Lower Thirds specifically for this purpose).

Experiment with moving the layers around using the Wireframes. Notice that you can cause the clip in the Video 1 track to contain the bounding bars by clicking it in the sequence instead. Also notice that if there are no clips selected and you click on an area in the Canvas window, one of the clips becomes selected. This is because the clip that is high-

Figure 6-13

Figure 6-14

est in the sequence stack of layers becomes selected once you click on any area of it in the Canvas. Notice that if both clips are selected in the sequence, both Wireframes show in the Canvas and adjustments made to one clip apply to the other. Obviously if two clips are selected simultaneously and they directly overlap each other, as our two clips would initially have done, you will only see one Wireframe. Although they are both there, one clip's Wireframe obscures the other clip.

Notice particularly that if a layer is selected but you click on an area outside of its Wireframe, the lower clip becomes selected instead. This is a major reason that Final Cut Pro requires that Wireframes be enabled before making adjustments in the Canvas or Viewer windows: without seeing them, you could never be sure what was selected as you clicked around.

STEP 21

After you've played around with Wireframes briefly, make sure that the clip in the Video 1 track is lined up correctly with the frame window in the Canvas. Leave the text clip in the position you want it to occupy.

Now let's make some adjustments to the text itself. Obviously we need to change the text in the layer, but we can also customize the size, color, font, leading, and kerning.

STEP 22

Double-click the text clip in the sequence to load it up into the Viewer window (Figure 6-15).

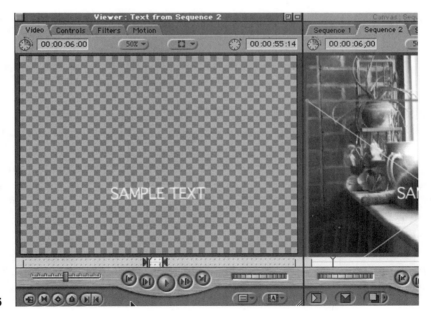

Figure 6-15

Make sure that you load the text clip by double-clicking the clip in the sequence and not the one in the Effects tab from the Browser. Changes you make to clips loaded into the Viewer directly from the Browser window do not apply to the clip in the sequence. Once you place a clip into a sequence, it becomes its own clip. Although it may have been originally generated by the Text Generator in the Effects tab, it is now a separate clip and adjustments and effects must be directly applied to it. Adjustments applied to text clips loaded into the Viewer window from the Effects tab will not have any effect on clips edited into a sequence.

When you double-click the text clip and load it into the Viewer window, the Viewer window will update, revealing a new tab and missing some that you would have expected to see. The first tab, Video, works just as it did with our captured video clips. The only difference is that here we see only our text and a checkerboard background. This is because Final Cut Pro assumes we want our text to appear as an overlay of the layer underneath. As such, it includes transparency information in the clip, a topic we will cover in detail later in the chapter.

Working with your text can be difficult if the background is distracting, so Final Cut Pro offers you the ability to change the color of the background while you work.

STEP 23

In the Viewer window, look for the same drop-down bar you adjusted the Wireframe view with earlier in the Canvas window. Click it and look for a list of background choices that read Black, White, Checkerboard 1, and Checkerboard 2 (Figure 6-16). Selecting any of these will change the background color to that selection. The correct option will depend on what color text you are working with. Since our text is currently White, select Black as the background color.

A glance at the Canvas window will reveal that this has no effect on the composite image in the sequence as displayed in the Canvas. The Background color you use in the Viewer window is just a convenience of that window, not a part of the clip itself.

STEP 24

Just for fun, return to the drop-down bar in the Viewer window and switch the Wireframe mode from Image to Image+Wireframe (Figure 6-17). You will have noticed that when you opened the text clip up in the Viewer window, the text was positioned in the same place as it is in the Canvas. This is because when we adjusted the clip's position in the Canvas, it updated this information in the clip in the sequence. When we open up the clip from the sequence in the Viewer window, it reveals the position of the image based on the changes we made in the Canvas window.

Likewise, any changes we make to the position in the Viewer window now with the clip loaded from the sequence will be displayed in the Canvas window as the clip in the sequence is updated. This is an example of *global adjustment*.

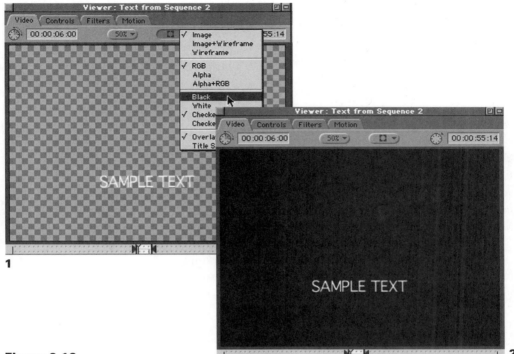

Figure 6-16

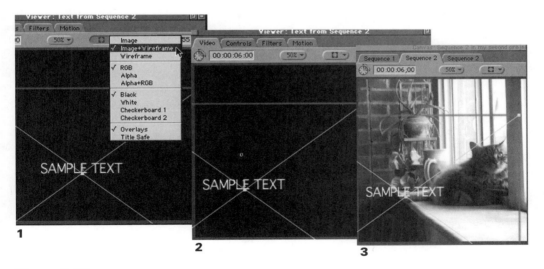

Figure 6-17

STEP 25

With the Viewer window's drop-down bar set to Image+Wireframe, click on the Wireframe and move it around, dropping it in yet another new location. As you will see, the Canvas is updated as the new position information ripples through the system from the Viewer to the sequence, finally to be displayed in the Canvas.

Controls, the next tab in the text clip's Viewer window, is unique to text clips. It replaces the audio tracks that accompany most video clips. The Controls tab is where we customize the text we use in this clip.

STEP 26

Click on the Controls tab in the Viewer window (Figure 6-18). At the top is a field to enter your text. Type in the word "see-thru" and then hit the Tab key to apply the text change. Hitting Enter would be a mistake here, since it would just result in a hard return in the text field. Either hit Tab to get out, or click the mouse pointer elsewhere in the Control tab.

The change in the text will occur immediately. To watch your text appear in the Viewer as well as the Canvas as you work in the Control tab, simply grab the Control or Video tab, tear it away from the Viewer window, and drop it in an unused area of the Desktop.

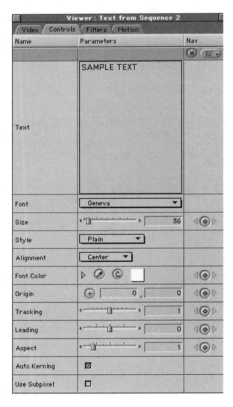

Figure 6-18

The remaining parameters in the text Control tab are standard fare for text editors. There are options for changing the font, color, size, alignment, and a host of other settings. If the list of fonts seems small, this is because Final Cut Pro can only use TrueType fonts, of which there may not be a great many on your machine. One of the great benefits of using applications such as Photoshop for creating text images is that they can usually access both Post Script and TrueType fonts for generating text images for use in Final Cut Pro, whereas our Final Cut Pro text generator is limited to the TrueType set. Many fonts are also available or can be converted to TrueType fonts, if you are limited to Final Cut Pro's text editing capabilities.

Remember that we are not creating text to be viewed as text; we are only creating a stencil for matting in a layer of video. Therefore, we need to keep in mind a couple of things. We will need a fairly large fat font, such as Verdana. If the letters of the word are too thin, there will be little video viewed through them. We also need to make the letters substantial enough to be seen at all, and the tracking—the space between the letters themselves—will need to be kept quite tight.

Figure 6-19

STEP 27

We will be able to adjust this again once our imagery is in place, but for now, set the Font as Verdana, Size at 125, Style as Bold, Alignment at Center, Color at its default White, Tracking at 1, and the Aspect at .8. Finally, reposition the text based on its new shape and size, using either the Viewer or the Canvas window Wireframes (Figure 6-19).

Setting the Travel Matte-Luma

After the text has been repositioned in an area of the screen that appeals to you, we will edit in the video clip we want to matte into the text.

STEP 28

Select a clip from the Browser window that you want to use. Make sure that you use a different clip from the one in the Video 1 track and that this clip contains footage that is generally lighter in detail.

If both clips are exactly the same or if they both contain footage that is equally bright or dark, it will be difficult to see that matting has taken place, just as if we used a stencil to paint green letters on a green wall. The text clip that we use will disappear completely, so don't expect to see the white letters following the next operation.

STEP 29

Edit the clip into a third video track, either by creating the track from the Sequence drop-down menu Insert Tracks command and then editing the clip into that target track, or by simply dragging the clip into the sequence where the third track should be and then trimming it down so that it fits your stack of clips (Figure 6-20). Either way, you should end up

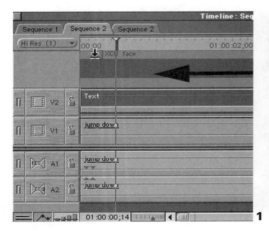 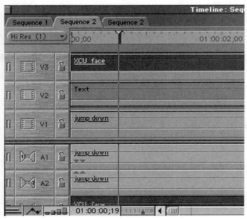

Figure 6-20

with a darker video clip on the Video 1 track, the text clip on the Video 2 track, and a lighter video clip in the Video 3 track. The clips should be stacked vertically so that for the moment, the video clip in the Video 3 track obscures the clips in the lower tracks.

STEP 30

Select the clip in the Video 3 track. Go to the Modify drop-down menu, scan down to the Composite Modes, and select Travel Matte-Luma (Figure 6-21). When you do this, the Canvas window, which was previously showing only the Video 3 track, will suddenly reveal most of the Video 1 track, with the exception of the text, inside of which is video matted from your Video 3 clip.

What is occurring here is that Final Cut Pro is using the luma values of the text clip to determine what should be used as the stencil. Anything white in the text clip is stenciled and allows the Video 3 layer clip to show through onto the background clip. Anything that is black blocks the stencil and doesn't let the Video 3 layer clip show through onto the background clip. It is performing the simple act of matting.

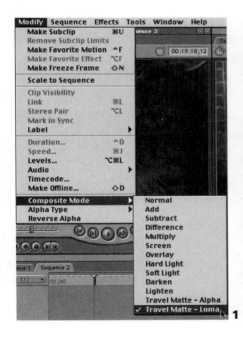

Figure 6-21

But unlike a normal stencil that either blocks or doesn't block, our Travel Matte performs more like a fine-tunable screen. We can adjust how strong or weak the matte is based on how close to white an element in the clip is. White text such as we created in the Controls tab produces a complete matte, allowing the Video 3 clip to show through totally. Black, on the other hand, blocks the matte, and all we see is the background Video 1 clip.

The implication of this is that all the values in between black and white—the grays—also have luma values that are not quite here nor there. These can be used to yield mattes that are partial mattes. And since color can be keyframed within the Viewer window, you can fade your stenciled mattes in and out, just as you would with any text, but with much more pleasing results.

STEP 31

To see this difference, double-click the text clip in the sequence to load it into the Viewer again (Figure 6-22). Go to the Control tab and scroll down to the Font Color parameter. Click the little triangle to open the color selection tools.

There are sliders here for adjusting Hue, Saturation, and Brightness. Hue adjusts the actual color type, whether the color is blue, red, green, or whatever color you wish it to be. Saturation controls how intense the color is. Brightness controls the luma values, or how close to white or black the color is.

Since our Travel Matte is a luma matte, changing the Hue and/or Saturation will have only a limited ancillary effect, so we really want to adjust the Brightness.

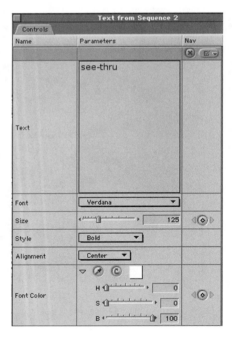

Figure 6-22

Figure 6-23   1   2

STEP 32

At present, Brightness should be set at 100, or white. Click the slider and drag the value down to around 50 (Figure 6-23). Although the Viewer window will show that the text has merely become darker, the Canvas window will update to show that the matte has become partially transparent, based on the closeness of the text color to black. Our luma matte is becoming transparent. Bring the Brightness value in the Control tab back up to 100, so that the matte stands out for the next exercise.

The Travel Matte-Alpha

What is an alpha channel?

The other type of Travel Matte is capable of delivering equally interesting results, but is based on a different method of contributing transparency information to the matte. This is called the alpha channel. The alpha channel is a unique system of including transparency information in an image, not by using colors within the image (although that is possible as well). Instead, alpha transparency information is included in its own special channel that accompanies an image. There, it performs like a carry-along matte for the image. Any application that can access alpha channels uses the information from the alpha channel to determine what should and should not be matted out when displaying the image.

Some explanation of the RGB system of digital images is necessary here. Digital images generally use a system called RGB to store information about each pixel in the image. R stands for red, G stands for green, and B stands for blue. All the colors that can be displayed by a computer (which includes most of the colors visible to the human eye) can be expressed by some mixture of these three colors. The amount of each of the three colors determines which color is displayed. The scale of possible variations on each primary color in RGB is 256. Therefore, in every pixel, there is a separate channel each for red, green, and blue. Each channel carries a value of between 0 and 255 for the amount of that color. The mixture of the three channels gives the pixel its color appearance.

Although this doesn't seem like much variation, the combination of 256 colors for each of the three primary colors yields over 16 million different possible colors, from which the term "millions of colors" originates. Each pixel of an RGB image carries bits of data describing what the red, green, and blue values for the pixel are. The reddest possible red in the RGB color scale would carry the value "255, 0, 0" zero being the value when none of a particular color is being used in a channel (blue and green in this instance).

The alpha channel was developed as a system that would not intrude on the original three color channels but could be used to carry useful transparency information about the image. A separate channel was devised that is limited to variations of strictly luma information. The alpha channel uses 256 levels of gray to communicate whether or not a pixel should be transparent, regardless of its color data. Because this alpha information can be expressed as grays as well as completely black or white, a pixel can also be partially transparent, depending on how close the alpha level is to black or white.

This may sound very similar to the luma matte that we just performed earlier, and in fact, the same process is used. But the difference is that we needed two images to complete the luma matte, the text clip for the matte and the actual image we were matting (not counting the background the matte was being stenciled onto). But an image that contains an alpha channel carries the matte inside of it as a channel. Thus, an image with an alpha channel is like two images in one: our color information that we see as image detail, and then an unseen channel of grayscale information that determines only what parts of the image are transparent. Instead of needing a foreground image, a matte image, and a background image, the foreground image actually contains the matte image as a unique channel.

It should be noted that not all digital images can carry an alpha channel, and that some video codecs can support alpha channels while others cannot. Most applications require you to save an image in a format that supports alpha channel data. You can tell if a format supports alpha channels if the format you are saving in allows you the choice of "millions of colors +", the plus indicating that an alpha channel is present (Figure 6-24). Most applications can read and utilize alpha channels, Final Cut Pro included, and many use alpha channel transparency information for processing transparencies as we are doing without informing you that the process is taking place.

Unfortunately the Apple DV codec does not support alpha channels. Although we are working with them in Final Cut Pro, the media files on our Scratch disk actually do not contain alpha channels. Final Cut Pro is adding them to the clips only for use in the application. If you need to export a video file for use elsewhere in another application and you

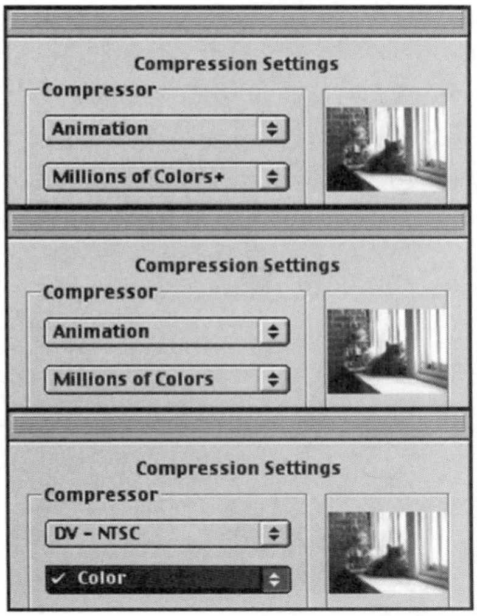

Figure 6-24

want to send an alpha channel with that file, you need to export it using a codec that does support alpha channels (e.g., the Animation codec).

Although an alpha channel does also use grayscale luma information to carry transparency data about the image, it is not the same sort of matte as a luma matte. We will use our text clip again as an example of the way that alpha channels can show information that does not display when a luma matte is being used.

Adding a drop shadow to the text for accentuation

Although our Travel Matte-Luma does its job, it looks a bit flat and one-dimensional. We will add a drop shadow to the image for the text matte. First, we'll attempt to follow the intuitive method of simply adding the drop shadow to the luma matte already in place.

STEP 33

Double-click the text clip and bring it up in the Viewer window. Go to the Motion tab and look for the Drop Shadow parameter (Figure 6-25). Click the checkbox to enable the drop shadow effect.

As we adjust the settings here, make sure that the Viewer and Canvas windows are in sight to keep up with what your changes are doing across the application.

STEP 34

Set the Offset value at 5, and drag the Brightness value for the HSB Color control down to 0 to describe a substantial, believable shadow. Although our matted shadow will not be black, the strength and density of the shadow must be accurate for the drop shadow to appear real. Set the Softness value at 15 and the Opacity value at 75 percent (Figure 6-26).

Figure 6-25

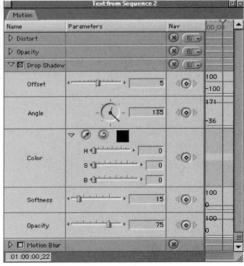

Figure 6-26

Set your background in the Viewer window to Checkerboard 1 or 2 so that you can accurately gauge the strength and depth of the drop shadow. The drop shadow as it appears in the Viewer window should look nice and robust. The contrast between the very dark shadow and the very light text gives the text a multidimensional feel.

Unfortunately, a glance at the Canvas window will reveal that the drop shadow doesn't even register. The luma matte we created to stencil the video clip is based on the white luma values of the text. Shades of black that dip below that level of white simply aren't matted in at all. This unfortunately includes that nice healthy drop shadow.

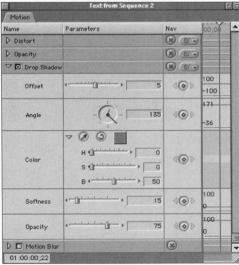

Figure 6-27

STEP 35

Adjust the Font Color of the drop shadow in the Motion tab, bringing the Brightness level up to 50 (Figure 6-27).

In the Canvas window, the drop shadow is now becoming a part of the matte, as the color of the drop shadow approaches the white that the luma matte works with. But it also looks unacceptable. The matte is very rough-edged, and it is difficult to distinguish between the matte of the letters, which should be solid and crisp, and the drop shadow, which should be darker and softer, mimicking a change in focus as well as opacity.

This is where the alpha matte comes to the rescue. When you created the text clip, Final Cut Pro discreetly created an alpha channel in the text clip to allow the sequence to translate the empty areas of the clip as transparent. This alpha channel is just sitting there, coming into play only when called for.

Viewing the alpha channel

The normal view for these windows is in RGB, which we have described as the ordinary color range of digital images. But Final Cut Pro offers two additional viewing modes to let you see what is in the alpha channel as well—RGB, Alpha, and Alpha+RGB.

Figure 6-28

STEP 36

Load the text clip in the Viewer window and revisit the Wireframe drop-down bar. Below the Wireframe mode selections, you will find three choices for viewing the clip: RGB, Alpha, and Alpha+RGB. Switch the Mode to Alpha (Figure 6-28).

As we said, the normal mode for watching a clip is RGB mode. But if we switch it to Alpha, the view changes to a black and white image. These are not the colors you will see. This is the exact transparency information for the clip; it is the mask unmasked. Any areas that are white in the alpha, such as the text, are completely matted. Any areas that are black will be totally transparent. The bonus here is that with the alpha channel, there are 256 levels of gray for defining partial transparency of the image or matte. The drop shadow appears in the alpha channel with exactly the strength it carries in the actual RGB image, even though the color is reversed here. When you are using colors other than black and white, it may not even be related on the color wheel.

Viewing in alpha mode can be difficult if your clip contains imagery that is not directly related to the alpha channel. In this example, the imagery is the text, which is also where the alpha channel is. But since the alpha channel can carry transparency information about any area of the image, you might need to see the alpha channel and the RGB channels at the same time so that you can determine the relationship between the alpha channel and the RGB channels.

STEP 37

Revisit the Wireframe drop-down bar. Switch the Mode to Alpha+RGB. Switching back and forth between RGB and Alpha will not give you this sort of perspective, so Final Cut Pro offers a third viewing mode—Alpha+RGB. When viewing in this mode, the RGB data is visible, making the image appear as it would in RGB mode. But any areas that contain alpha channel matting information, such as the white and drop shadow areas of our text clip, show as pink. The brilliance of the pink areas describes the gray level to which transparency will be applied to the area. Areas that are not pink are areas that are fully transparent.

To fully appreciate the difference between the Travel Matte-Alpha and the Travel Matte-Luma, viewing on a black background is best. We will temporarily turn off track visibility for the Video 1 track, which is presently our background video clip.

STEP 38

Go the sequence and click on the green track visibility light for the Video 1 track at the extreme left-hand side of the Timeline window (Figure 6-29). This does not affect the contents of the sequence at all; it only hides the Video 1 track from view. You can turn track visibility back on at any time to make that track visible again.

Now that you have a black background, look at the Canvas window and view the text clip with the present luma matte, taking one last glimpse at the hard edges and lack of drop shadow in the Viewer window view of the clip. Now we will switch the matte from luma values to alpha values.

Figure 6-29

Setting the Travel Matte-Alpha

STEP 39

Select the video clip in the Video 3 track, the clip we are matting into the text layer (Figure 6-30). Go to the Modify drop-down menu, choose Composite Modes, and select Travel Matte-Alpha.

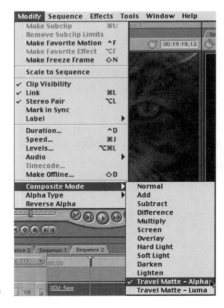

Figure 6-30

Instantly, the drop shadow smoothes out behind the text matte, this time with the soft, blurred edges that resemble the drop shadow of the text clip. This is possible because the alpha channel can more accurately assign levels of transparency based on the level of gray in the alpha channel, although it does not have to interact with the actual levels of gray in the RGB image. Since the luma matte must make its transparency matte from the gray levels in the RGB image, adjusting the transparency amount will affect what you see in RGB mode, and therefore out to video. The alpha channel simply makes whatever it covers transparent, no questions asked.

Turning the visibility of the Video 1 layer back on reveals that even on a video background that has less contrast than the black one, the difference between the two drop shadows is sharp (Figure 6-31). A proper understanding of the alpha channel system will allow you to extend this to other areas of the application, pulling off matting stunts that would otherwise require much more complex workarounds to be done at all.

As you work with other effects, keep in mind that most of them are simply combinations of alpha channels automated as plug-ins by Final Cut Pro. When you choose a color to key out or you create a garbage matte, you are only giving Final Cut Pro a different set of parameters for establishing the transparency of areas within a clip. Once again, there is more than one way to skin a cat, and sometimes you will get more acceptable results by constructing your own alpha channel mattes than by using a plug-in from the Effects Filters set. And remember that you can construct matte clips in Photoshop and other image-editing applications. As long as it carries an alpha channel, any shape or image you create can become the Travel Matte-Alpha clip.

Figure 6-31

Special effects: motion effects

Motion for clips

In the preceding chapter, we experimented with keyframing in the Timeline window, as well as adjusting the physical properties of a clip from within the Viewer window's Motion tab to create the drop shadow effect. But we have yet to fully explore the functionality of Final Cut Pro with regard to the many other physical parameters for a clip or the ability to keyframe them to achieve interesting effects that change over time. We will now move a little further with each of these ideas, learning to fully manipulate a clip within a sequence. If you can see it in your head, you can do it in Final Cut Pro.

To begin with, we will use the text clip we used in the previous chapter, but this time without the Travel Matte applied from the Video 3 layer clip. You can either click the green track visibility light to hide layer 3 or just select the Video 3 layer clip and delete it from the sequence. When you do so, your text clip will pop back into the Canvas as an actual graphic text element rather than a matte for a video clip (Figure 6-32).

The main reason we will be using a graphic element like a text clip is that some of the subtleties of the physical properties we are about to adjust are difficult to see in a full-frame, full-motion video clip that also contains motion in itself and covers so much of the viewing area that it obscures what you have adjusted. This does not mean that the physical properties we will toy with here have no effect on video clips; the physical properties we will alter in the text clip are simply easier to distinguish than if applied to a video clip. Once you understand how the properties work, you can apply changes wherever you want throughout the system, even to an entire sequence at once.

Figure 6-32

STEP 40

Double-click the text clip and load it up into the Viewer window. For the moment, we will leave the Control tab alone and leave the text formatted as it is. Click and drag the Motion tab away from the Viewer window so that you are able to view it, the Viewer Video tab, the Canvas window, and the Timeline window simultaneously (Figure 6-33). This may take some shuffling of your Desktop space and re-sizing of windows; once again, you become aware of the benefits of having multiple monitors in your editing setup.

This is one of those times when you will really appreciate the Macintosh's ability to easily integrate two computer monitors into the workflow. Although you can get by with one monitor, you can wear out your wrist moving windows around to get at other windows.

Figure 6-33

The clip motion attributes

Let's take a look at the various physical attributes in the Motion tab of the Viewer window that we can work with in a clip. Each attribute has a set of specific controls or parameters which relate only to that attribute. To get at the parameters for an attribute, click the little triangle to see its drop-down parameter list. From the top, the attributes are Basic Motion, Crop, Distort, Opacity, Drop Shadow, and Motion Blur. We will cover each of these attributes separately, adding up their combined effects to the text clip until we have a text flyover that really flexes Final Cut Pro's compositing toolset.

STEP 41

Basic Motion; Scale

The first attribute is Basic Motion. Click the drop-down triangle and review the parameters. Examine the first parameter, Scale (Figure 6-34).

Although Scale does not seem directly related to motion, bear in mind that motion is a basic concept in dimensional space. Motion as we see around us in the real world takes place in three dimensions: side to side, up and down, and also closer and further away from our position as the viewer. But the video frame we are compositing with exists only on a two-dimensional plane. That is to say, there are only left and right and up and down.

To simulate the third dimension of real-world movement, we have to control the scale of the object we are working with. When an object is closer to us, it appears to be larger than when it is further away. Thus by adjusting the Scale parameter, we can make an object seem to be closer than other objects or approach the viewer over time.

Next to the word Scale, there is an adjustment slider. The default value for this will always be 100 percent, since Final Cut Pro always assumes that we intend to use the object at the original size it was created. For precision purposes, there is also a numerical entry field next to the slider.

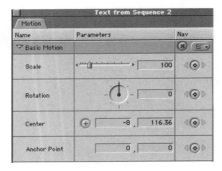

Figure 6-34

STEP 42

Click on this field and enter 400, increasing the scale of the text to 400 percent. When you hit Enter, the text size will jump to a much larger size, extending past the edges of the video frame in both the Viewer and the Canvas windows (Figure 6-35).

Do not be concerned that the text outside the frame is not visible; it is still intact, only outside the viewing area of the frame. Changing the value back to 100 percent will reveal that the text is still there. Likewise, if you drag the slider down to zero, the text disappears, since it is now being displayed at 0 percent of its normal size. The range of scale goes from 0 percent to 1000 percent.

STEP 43

Take the slider up to around 1000 percent and take a look at the text. If you see no text in the frame, it is because you are looking at the space between the letters, which has also been magnified along with the text itself. Moving the text clip around using the Selection tool will bring some of the text back into view.

One thing you will notice is that the text no longer looks presentable at 1000 percent. The edges that you can see have taken on a pixellated appearance that is reminiscent of what we saw when the Viewer window or Canvas window was viewed at percentages higher than 100 percent. But unlike those instances, this time the effect will be seen in the resulting video out to Firewire, since this is an effect we are manipulating in the clip with rather than a video preview window on the computer screen.

Once again, Final Cut Pro is being asked to interpolate or make up information about the object, information that was not originally present. When you created the text clip, you

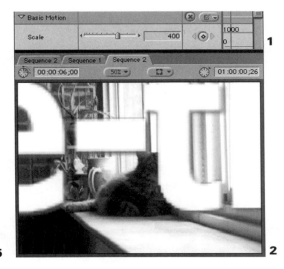

Figure 6-35

chose a font size and other settings for the text that determined how large it would appear when viewed at 100 percent in the Viewer and Canvas frame windows. But when you increase that size using the Scale attribute, Final Cut Pro has to make up the information that was not included, resulting in pixellation and blockiness.

Unfortunately, there is no way around this. You have to plan ahead when you create your text or other graphic images for the largest possible size they will be required to display in the frame. The good news is that you can down-sample (i.e., decrease the scale of the text) without any perceptible loss of detail. So if you create your text at the largest necessary size, you can later scale the image down to the specific size you need. The lesson here is that Final Cut Pro can throw away data and reduce the scale to yield acceptable results, but increasing the scale above 100 percent usually results in unacceptable pixellation.

Since the Controls tab of the text clip in the Viewer window is conveniently available, we can easily test the correct method for getting large acceptable text.

STEP 44

Pull the Scale in the Motion tab back down to 100 percent, so that when we bump the size of the text, we don't explode it out to 100,000 percent of the present size (Figure 6-36)! Next, tear away the Controls tab from the Viewer window and place it on the Desktop so that we can adjust it and watch it update in the Viewer and Canvas windows. In the Size field, enter the number 400.

Instantly, the text in the Viewer and Canvas jumps to a much larger size that does not display the pixellation of the previous scale increase.

But once again, there is a problem that may not be immediately apparent. When you enlarge the size of the text in the Viewer window, it extends past the edge of the clip's video frame, as you would expect, since it is now too large to fit within the frame. If you move the Wireframe around in the Viewer or Canvas window, however, you will see that the edge of the frame from the Viewer window now cuts off the text that extends past the edge of the Wireframe. Although the text is still intact, as pulling the font size back down in the Controls tab will reveal, we cannot see beyond the edge of the clip's frame size.

The Video frame, in either the Viewer window or the Canvas window, is like a mask for all the elements that are in the project. It can only show you items that fit inside the mask of the frame. Enlarging the Scale of the clip made the entire frame size larger, thus

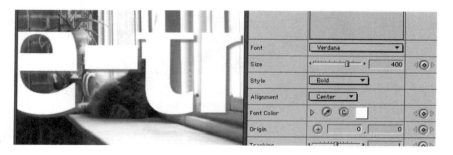

Figure 6-36

allowing us to move the clip around the smaller Video frame without encountering an edge. Even though we got pixellated results, we could move the frame around because the frame had been scaled up the same amount as the text.

But when you only increase the size of an object that lies inside the video frame such as the font of our text, rather than the frame size itself, you encounter the edge of the video frame. The frame stayed the same size, while the text extended past its edges. Bring the Font Size back down to 100 before moving on.

Although this may not seem problematic now, it will severely hamper your ability to use large good-looking text as a moving object in your sequences. You will need either to sacrifice some image quality to get the larger text size or to sacrifice heavy scale and movement adjustment of the text clip. Generally, the best trade-off is to sacrifice some image quality rather than the ability to adjust scale and movement. If your technique and timing are good, you can use other elements, which are explored subsequently to hide the image degradation.

Pan and Scan for large images

A still better solution is to use another dedicated image-editing application that is not restricted to the video frame size that Final Cut Pro needs to work in. This is the primary reason for the popularity of Photoshop among Final Cut Pro editors and its resulting tight integration into the application by the fine people at Apple who wrote the code.

Using Photoshop, you can generate text and other graphics images that are not restricted to the 720 x 480 frame size that Final Cut Pro's text editor conforms to, up to a limit of 4,000 x 4,000 pixels. Such images are easily integrated into the sequence as clips just as the internal text clip was, and in the formats described in Chapter Four on importing media. Where the internally generated text clip showed its edges when moved around the frame, a Photoshop-generated image with a frame size of 1440 x 960 could be moved easily around the 720 x 480 video frame eliminating the edge problem encountered using the Final Cut Pro Text Generator clip.

This technique is sometimes referred to as Pan and Scanning, since you can move a larger image around a smaller frame to focus on what you want, just as if you were panning a video camera to select what you want in the frame as you shoot. Pan and Scan is often used in converting motion pictures to video that have a wider aspect ratio than the television's 4x3 rectangle. If letterboxing is not used to help fit the entire wider image on the screen at once, filmmakers sometimes resort to showing only part of the frame at a time. If they want to show the other part of the frame for some reason, they can Pan and Scan to the other side of the frame.

If you do not yet use an image-editing application such as Photoshop, but you plan to work with graphic elements a great deal, you may want to start considering your options. Using a dedicated application to generate your graphics has a profound effect on the quality and aesthetics of your finished products, besides opening up the motion and scale possibilities previously described.

Rotation

The second parameter for Basic Motion is Rotation. The setting can be entered either using the directional dial, or by numerical entry. Either way, the value entered is that of degree of rotation between 0 and 360. If you type in the value 180, your text layer will flip upside down as it is rotated 180 degrees from the present position (Figure 6-37). Entering 0 will return it back to its proper orientation in the frame.

There is a further dimension of rotation that can be understood by looking at the dial. If you enter a value higher than 359, you are actually entering a degree of rotation that includes a complete rotation and begins the second one. So entering the numerical value of 540 will land the clip upside down, having rotated the clip one and a half times.

STEP 45

Enter 540 into the Rotation field to demonstrate this behavior and then look at the dial (Figure 6-38). A small red hand on the dial will indicate that more than one rotation has occurred.

It may seem less than obvious why one would want to rotate an image more than 360 degrees, since the end result will still be a simple degree of turn for the clip. This is because when we keyframe the rotation, we will be able to specify how many times it completes a full rotation on its axis between two moments in time, as well as the actual degrees of rotation that it begins and ends with. The flexibility of being able to keyframe the number of complete rotations as well as beginning and ending values will define our ability to speed up or slow down the spin of an object, based on the number of times it spins over a specific duration.

Figure 6-37

Figure 6-38

Center

The next parameter, Center, is perhaps the most obviously vital component of motion (Figure 6-39). Center defines the two-dimensional physical position of a clip in the frame, based on the standard geometric coordinates of the X and the Y axis. The numbers on the X axis determine horizontal position, and therefore motion to the left and right. Numbers on the Y axis determine vertical position, and therefore motion up and down.

Since the X and Y axes work from geometric standards, the coordinates 0, 0 are for the geometric center of the screen. Positions to the left of center on the X axis are described with negative numbers; similarly, positions above the screen center on the Y axis are described by negative numbers. When a clip is moved, the X and Y values of the Center parameter in the Motion tab change based upon the new position inside the video frame of the Canvas window.

Although this was useful in a static sense, as it was in the correct placement of our text clip in the last exercise, it takes on primary importance in keyframing motion paths. It is the Center parameter that defines where the clip is at any time during its playback. If you keyframe that position over time, you will be able to create a basic motion path.

Figure 6-39

Anchor Point

The Anchor Point parameter offers an unusual flexibility that changes the clip's relationship with all the other parameters in the Basic Motion attributes (Figure 6-40). The Anchor Point works more or less as a central point or axis for each clip. All other parameters function based on what the clip has assigned for its true center, or Anchor Point. This Anchor Point does not affect anything outside the clip, and, unless changed, it will always default to the exact center of the clip, which is why it is often sorely underused by editors who should know better.

Its influence is most plainly seen when combined with Rotation values. For example, a round plate whose Anchor Point is directly in the center will spin in place like a wagon wheel when rotated. But when the Anchor Point is moved out to the edge of the plate, the rotation will make the plate itself rotate around a point in space on the edge of the plate. This makes it easier to mimic the real world motion of lop-sided objects whose center of gravity is not defined by the center of their surface. The Anchor Point is, as usual, keyframeable, meaning that you can alter the Anchor Point of a clip over time to yield interesting new effects based on the other clip's parameter's reaction to the Anchor Point change.

Crop Settings

The next attribute in the Motion tab is Crop Settings (Figure 6-41). The crop settings are interesting in that they allow you to hide a section of a clip from view without having to insert a matte of any kind. This can be convenient for operations such as displaying only a section of a larger graphic image or video clip inside the frame. Rather than change the scale of an image to make it smaller, which would still include detail you may want to leave out, the Crop attribute simply limits how much of the frame of the clip is visible.

The parameters for this are adjustable as a square or rectangle, based on the left, right, top, and bottom edges of the clip.

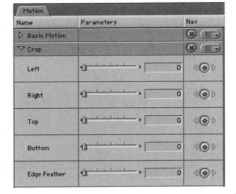

Figure 6-40

Figure 6-41

STEP 46

Click and drag the slider for either the left or right edge and drag it to around 25 percent (Figure 6-42). You will see that half of the frame of the clip has been obscured. The other half is not resized, as it would if be you had changed the actual width or scale of the clip as you will be able to do with the following Distort attribute.

The last parameter for the Crop tool is Edge Feather. This allows you to soften the edge you are cropping the clip with so that the image being cut off with the tool fades into the background now visible on the other side of the crop line. This can be displayed by moving the crop line directly over one of the letters in the text clip were are adjusting. Drag the left or right slider such that one of your letters is seemingly cut in half. Set the Edge Feather at 50 percent and you will see the cutoff soften such that the letter appears to disappear into the background.

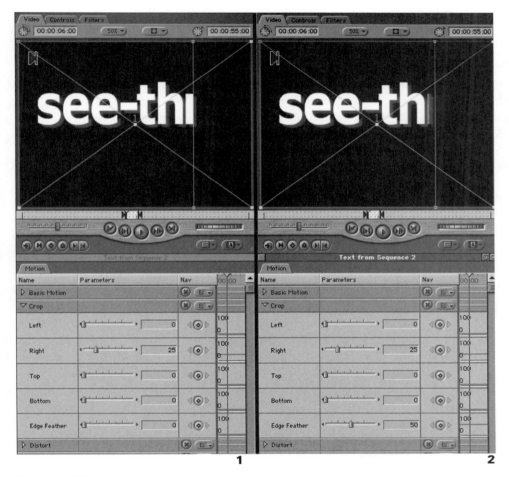

Figure 6-42

All transparencies, including the Crop tool, are an effect of unseen integrated matting. The soft, feathered edge you are witnessing is the result of Final Cut Pro allowing some pixels to be seen and others to become transparent along the edge of the crop. The value you enter into the Edge Feather field simply informs Final Cut Pro how far outside of the crop edge to apply the gradual mixture of visible and matted out pixels. The effect is rather pleasing, having a soft edge, and when the Crop parameters are keyframed, it can be used for a gentle, gradual revelation of imagery in the composited sequence.

Distort

The third attribute, Distort, is a powerful tool for altering the very shape and aspect of a clip (Figure 6-43). Unlike the Crop attribute previously described, the Distort parameters change the actual size and shape of the clip in the frame by altering its physical dimensions rather than by covering them up. The effect is rather like having an image imprinted on a rubber balloon. As you stretch and release the rubber, the image changes its appearance as well, resulting in very precise distortions based on the amount of stretch applied to the edges of the material.

The first four parameters of the Distort attribute are Upper Left, Upper Right, Lower Right, and Lower Left. These parameters correspond to the corner points of the clip's frame. There is no slider adjustment to these parameters because the points exist as X and Y coordinates and, therefore, can be assigned to any location in the frame. Interestingly, they can even be assigned between the other points, so that the clip appears to double back on or fold up into itself.

STEP 47

To display this type of clip distortion, enter 0 and 0 as the new X and Y coordinates for the Lower Right parameter (Figure 6-44). The Lower Right corner of the clip will jump to the center of the video frame, which carries the coordinate value of 0,0. Not only has the shape of the clip changed, but the text within the clip now appears curved and scrunched as a result of its position in the frame and based on the amount of virtual squeezing of pixels to adjust and fit them all inside the new frame shape.

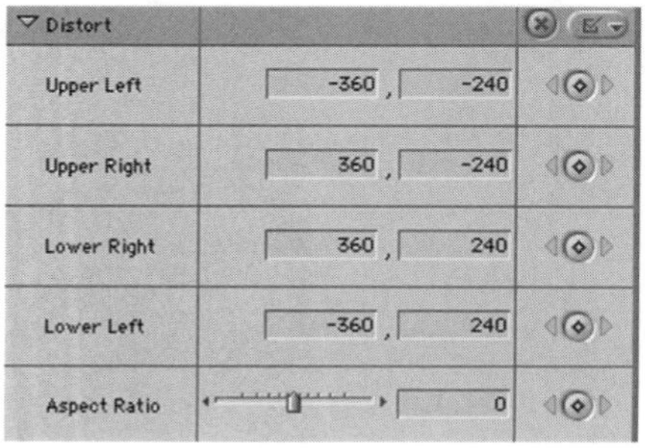

Figure 6-43

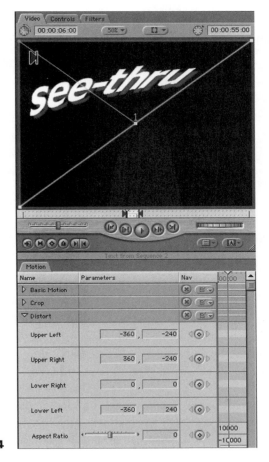

Figure 6-44

Notice that although we are still looking at a two-dimensional text image in the frame, the simple change in shape and curvature suggests three-dimensional depth. This is another example of using the alteration of two-dimensional shapes to suggest natural depth perspective. The Distort parameter allows us to mimic the real-world perception of change in size and shape that tells our brain that an object is nearer or farther away.

STEP 48

To remove the Distort effect and to return any coordinates of any attribute or parameter back to their original positions, hit the red X parameter reset button just to the right of the attribute's name (Figure 6-45).

The last parameter of the Distort attribute is called Aspect Ratio. It is adjusted by a slider, with a default value and unaltered setting of 0. The Aspect Ratio changes the degree to which the clip in the frame is distorted two-dimensionally. Dragging the slider to the left reduces the Aspect Ratio of the clip, making it appear shorter and fatter evenly across the

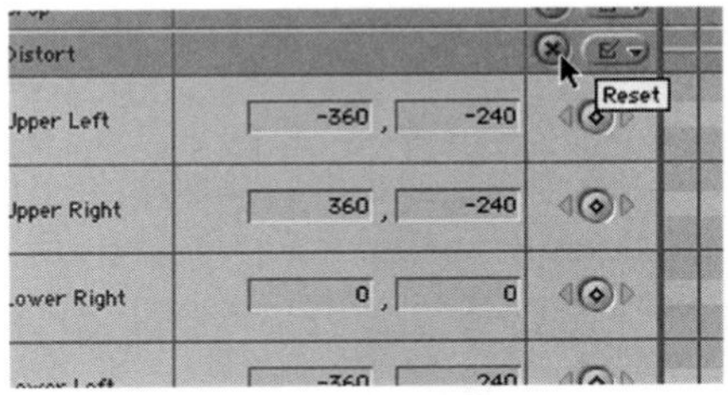

Figure 6-45

horizontal plane. Dragging the slider to the right increases the Aspect Ratio so that the clip appears taller and skinnier once again, evenly distributed across the vertical dimension (Figure 6-46).

The reason this parameter is referred to as an Aspect Ratio instead of a Short/Fat–Tall/ Skinny tool is that it was initially most useful in taking footage shot anamorphically to yield a widescreen, cinematographic look with letterboxing. Many cameras allow you to shoot in a 16x9 mode, which introduces letterboxing at the top and bottom of the frame. The difference in screen shape can yield interesting effects, so many videographers use it when shooting.

Unfortunately, older versions of Final Cut Pro could not decipher the difference between regular footage and 16x9 Widescreen. After all, anamorphic 16x9 footage is simply footage that is stretched in the camera. Final Cut Pro could not read the file data that the cameras inserted in 16x9 footage specifying that the footage had been anamorphically recorded. The result was tall, skinny images that fit within the video clip frame size but were obviously distorted.

Although many of us would love to maintain the skinny figure that anamorphic footage gives us, we needed a way to get the clip back to an accurate relationship between the clip's height and width, otherwise known as the Aspect Ratio. 16x9 and 4x3 are common Aspect Ratios for video, and we saw that there are others that were available when we set up the Audio/Video Settings earlier in this book.

The Aspect Ratio determines not only what shape the frame size of a video or film is, but also whether or not the video or film frames displayed in the frame are shaped correctly. In film and some video cameras, the Aspect Ratio is stretched up by the lens, which determines the shape of the image recorded on the film stock or video CCDs. With other video cameras and software such as Final Cut Pro, the Aspect Ratio is determined by how the system organizes the digital bits of each frame. Either process results in an *anamorphic* image, from the Greek *ana*, or Up, and *morphë*, or Shape.

With Final Cut Pro version 1.2.5, the application gained the ability to automatically detect and correctly reshape 16x9 footage without using the Aspect Ratio parameter of the Distort attribute. The tool can still come in handy, however, to change the Aspect Ratio for other reasons, such as correcting typographic elements, reworking improperly processed footage that has been printed or telecined at the wrong ratio or for any other situation that requires an adjustment to be made evenly across the frame on either a horizontal or vertical plane.

Figure 6-46

Opacity

The next attribute, Opacity, has been explored in the discussion of the sequence Timeline, and we have already set keyframes for it there (Figure 6-47). If the clip for which we set keyframes for Opacity were presently loaded into the Viewer window, this is the location in which those keyframes would be displayed. Alone among the Motion attributes, Opacity can be set from the Timeline.

Even though it may seem easier to set Opacity keyframes in the Timeline, fine-tuning them should always occur in the Motion tab of the Viewer window. This is because the timing of Opacity keyframes frequently coincides with that of other keyframes (e.g., Basic Motion and Distort) to produce profoundly three-dimensional effects that really seem to move from far away to near at hand.

Figure 6-47

Why is this so? Although your eyes are generally too sensitive for you to notice it, you detect very definite differences in the amount of light something reflects. This difference tells your brain about how far away an object is. An object very far away reflects less light into your eye. Although you do not think about this when viewing objects, the part of your brain that does the work of perception does.

This is not really news to anyone who works with light for a living. Photographers, videographers, and cinematographers all know about the fall-off of light as you move away from its source. It follows a definite mathematical formula, referred to as the *inverse square law*. This physical law states that as the distance between you and a light source doubles, the amount of light making it to you is quartered. The amount of light you receive from a light source drastically decreases as you move away from it.

Those who live by the camera must deal with this in order to get the proper exposure when they shoot. But it is also valuable information to us when attempting to create believable three-dimensional distances between objects in our video animation. When an object should be closer to the viewer, it should be very bright and near 100 percent opacity. As it gets further away, its reflectivity should drop, leading to decreased opacity. Put simply, further away should be harder to see, and not just because the object is smaller.

As you change the scale and position of an object, you will likely change its opacity as well to increase the perception that not only is the clip getting larger as it approaches the screen, it is getting brighter. The touch of realism that adjusting opacity gives is subtle but every bit as important as any other attribute you adjust in the clip.

Drop Shadow

The next attribute, Drop Shadow, is a fast, convenient tool for adding depth and dimensionality to an image with no muss or fuss (Figure 6-48). It must be enabled by checking

Figure 6-48

the box next to the name, otherwise it will not be present. Drop Shadows are like an unspo-
ken signpost that an object stands out closer to the viewer than the ones behind it. It is a
truly paradigmatic symbol of three-dimensional images, because even though most drop
shadows are really very primitive and out of place in the context of the image they accompany, they still give the sense of depth they are used for.

We have already used the Drop Shadow in our discussion of Composite Modes and
Luma and Alpha Mattes. But it is worth mentioning that the controls for the Drop Shadow
are keyframe-able as well, allowing some flexibility in the Drop Shadow over time. Why
leave the ubiquitous Drop Shadow simply sitting there in one position, when you can
imply that time is passing by having the shadow change position as if it were being cast by
the sun moving rapidly across the sky? Remember that no effect need remain static over
time unless you have chosen for it to do so for reasons of aesthetics. The choice to leave a
Drop Shadow alone should be made because the Drop Shadow shouldn't move, not
because it can't.

Motion Blur

The final attribute in the Motion tab is for Motion Blur (Figure 6-49). Like the Drop
Shadow attribute, Motion Blur must be enabled by checking the box next to its name.
Motion Blur is a unique effect that works only with elements in a composition that move.
Motion Blur affects only the movement of items within Final Cut Pro. A clip moving
across the Canvas window can display Motion Blur, but a clip that is stationary in the cen-
ter of the window cannot.

Items such as film and video footage that display natural motion blur in their own
original frames will not be affected by motion blur settings unless their entire clip moves in
the larger video frame. Motion blur is applied to entire clips rather than any separate part
of a clip's frame. Therefore we will be able to add motion blur to our text image when we
move it across the screen, although we could not add it to any object inside the static video
clip in the background.

But what is motion blur? Motion blur is a phenomenon that occurs when an object
moves faster than the perceptual frame rate of the viewing camera or eye can gather still
images. As we said early in this book, video and film are simply a series of still images that
update so rapidly that the viewer perceives continuous moving objects rather than a series
of images of still objects. If the frame rate slows down too far, we see the still images indi-

Figure 6-49

vidually, and the illusion of continuous motion is broken. Speed up the frame rate, and the illusion becomes stronger and sharper.

At around 24 frames per second, which is the standard frame rate of film, the illusory motion becomes fluid, each frame being nearly indistinguishable from the previous one. Slightly faster, PAL video uses a standard frame rate of 25 frames per second, very close to the film frame rate. NTSC video uses an even faster frame rate, increasing it up to 29.97. Since both PAL and NTSC video are interlaced formats that scan each frame as two separate fields sequentially, the true frame rate should really be regarded as double this number, as fast as 59.94 individual fields per second.

What does this have to do with motion blur? Motion blur occurs whenever an object passing in front of the camera or eye moves too fast for the viewer to actually record each increment of the movement. If you pass your hand in front of your face quickly, you will see a blur of your hand, because your eyesight is gathering image information at a relatively slow frame rate. In the fraction of a second that your eye gathers an entire frame's worth of visual information, your hand has actually moved several inches or several feet, depending on how fast you move it. For those few fractions of a second, you see your hand in all the places it has traveled since the beginning of your eye's gathering of the visual frame.

For our eyes, this is nothing new, and aside from some interesting optical illusions based on spinning wagon wheels and such, we are quite used to what our eyes can and cannot perceive sharply. But where motion blur really takes on significance is in the field of image gathering. Photographers and cinematographers must take care that when they record images, the shutter speed, which is the part of the camera that determines how long the image gathering period lasts, is fast enough to overcome excessive motion blur. If the shutter speed is too low, there will be so much movement between the beginning and the end of the exposure that there will be nothing but blur in the frame.

Still photographers have to be more careful than the rest, because the shutter speed for still cameras is a choice to be made by the photographer based on many factors, the most important of which is the desired effect. The photographer can add or take away motion blur by changing the shutter speed. For still photographers, motion blur is an important effect that can be used to deliver a sense of duration and motion in a still image.

Film and video, as we have said, have standard frame rates, making this less of an issue. Although shutter speeds can still be customized to add or take away motion blur, generally speaking there is a normal, definitive amount of motion blur that occurs in footage shot with a shutter speed that matches the frame rate. With video, which is composed of two interlaced fields for each frame, the shutter speed is a fiftieth of a second for PAL and a sixtieth of a second for NTSC, such that the shutter speed more or less doubles the frame rate and matches the field rate. For film, which is noninterlaced and does not use field doubling, this is 24 frames per second (Film motion blur can be further increased or decreased by the shape of the shutter itself. An iris-shaped shutter allows the cinematographer even more control over motion blur. This is not a function of the frame rate though, and does not concern us here).

As with Drop Shadow, Motion Blur must be checked ON to be applied to a clip. There are two parameters for Motion Blur, the percentage of blur and the number of samples. The percentage of blur lets you tell Final Cut Pro how intense the motion blur should appear. Final Cut Pro creates motion blur by taking a certain number of frames before and after the frame in question and blending them together, just as we said your eye gathers image information about where you hand was over the entire period it gathered a single "frame" of vision.

The result is that you see a mixture of where the object is in the frame now with an image of where the object was just before and after the present frame. Since the blend across the frames is smooth and is reblended with each new frame, you get an effect not unlike what real motion blur does: you are kept from getting a tight focus on each position of the object within each frame.

Figure 6-50

Although it would seem like a bad idea to obscure focus, motion blur can add realism to moving objects that otherwise appear not to be visually integrated into the frame. We can't help it; when we see motion blur, we assume that an object really moves through space, even if that object is a two-dimensional text image crossing a video screen. Since our job in compositing is to mimic reality on a two-dimensional plane, motion blur is an essential part of any movement we add to our compositions. Some experimentation will prove this point; although until we have a moving clip as in the next example, Motion Blur will not appear to have any effect.

Percentage of blur increases or decreases blur by adjusting the range of previous frames that the blur blends into the frame when Motion Blur performs the act just described. The scale of percentage ranges from 1000 percent, which blends ten frames of the clip into the blur, to 0 percent, which blurs none. 100 percent blends only one frame. The default value of the percentage is 500 percent, which gives a healthy "standard" blur (Figure 6-50).

Sample refers to how detailed the blending process is in calculating the motion blur. More samples will result in a much more fluid blend between the individual frames that are being joined to create the blur. Fewer samples means that the blend may look a little rough, since very few positions between the number of frames being sampled are figured in. The bottom end of fluid samples is 4, which looks pretty rough. A setting of 32 looks fantastic. The trade-off is that the larger the number of samples, the longer it takes Final Cut Pro to calculate and render the frame. With 32 samples, this can take several seconds. With 4, the calculation is quick. A setting of 8 or 16 usually gives the best balance between speed and quality.

Making the motion

Now that we know what each of the attributes does, let's go through and use them on our text clip in the sequence, this time including keyframes to get a sense of how values can change over time in our sequence, as well as in society at large. First, make sure that you have a clean text clip trimmed out to at least six seconds in length edited into the first frame of a sequence. You don't really need a video clip in the background for this example, but you may find that it adds a dimensionality to motion and other attributes that is difficult to achieve with monotone backgrounds. Feel free to add another video clip to the sequence, although you may want to disable its visibility while setting the motion keyframes for the text clip.

We will create a bit of motion with this text that uses all our attributes. The text will start dimly seen and far in the distance. As it approaches the viewer, it will spin around. When it reaches nearly screen size, it will spin around one more time, although this time with a different axis. Finally it will come to a rest in the center. At any time you can check on the way your keyframes affect the sequence with a quick render using the selection tools we discussed in the preceding chapter.

If a setting doesn't make sense, clear off your keyframes for that attribute by Control-clicking the keyframe and selecting Clear (Figure 6-51). Once all the keyframes are cleared

Figure 6-51

for an attribute, lay them out again correctly and you will get the results you want. If you still have trouble, just enter the numbers offered here and work backward through the process, advancing frames and looking at the effect that each keyframe has on a particular frame.

Preparing the project for the motion settings exercise

STEP 49

Move the sequence playhead to the beginning of the text clip. Now double-click the text clip so that it loads into the Viewer window. Make sure that the title line in the Viewer window reads "Viewer: Text from "nameofyoursequence" (Figure 6-52). If it does not, you may be keyframing the wrong clip! Remember that Final Cut Pro will allow you to keyframe any clip or sequence that is loaded into the Viewer, regardless of whether it is the one you want to work with in the sequence. Keep your eyes open.

STEP 50

When your text clip opens in the Viewer, its playhead will be located on the same frame of the clip that the playhead in the sequence and Canvas are parked on. Do not move this just yet, because we are going to set our first keyframe on this frame. Grab the Motion tab from the Viewer window and drag it to the side so that you can make your settings there while watching the update in the other three windows.

STEP 51

After you tear the Motion tab away from the Viewer window, reshape the Motion tab window into a short, wide window by tugging on the reshape tab at the bottom right-hand corner of the window, so that you have complete access to the Motion tab's own Timeline (Figure 6-53).

Not surprisingly, this Timeline uses the same navigation conventions that the sequence Timeline window uses. The difference of course is that this Timeline shows only the clip loaded into the Viewer window. Instead of layers, this Motion Timeline is divided into the various attributes and parameters we just finished going through.

Figure 6-52

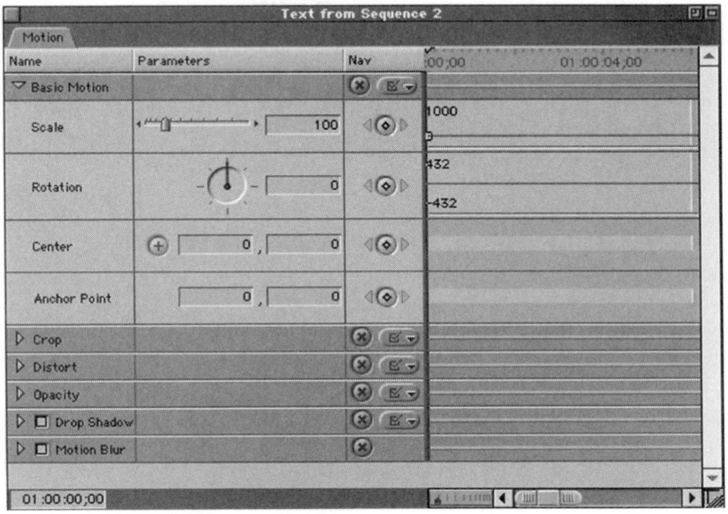

Figure 6-53

Notice that there is a light gray and dark gray area of this timeline (you may have to adjust the time scale of the Motion Timeline to see this). Light gray indicates any area in the sequence Timeline window in which the clip is present.

STEP 52

Hit the Down Arrow key to move the playhead to the end of the clip, and you will see that the playhead jumps to the end of the light gray area (Figure 6-54).

Why would Final Cut Pro not chop the Motion Timeline off at the end of the clip? The answer is functionality. Final Cut Pro will allow you to set keyframes for a clip in an area beyond the actual duration of the clip. This gives you even more options when determining the keyframe values and their positions, as well as keeping your keyframes safe should you accidentally trim off the end of a clip where you had previously inserted keyframes.

Figure 6-54

Figure 6-55   **Figure 6-56**

Setting Scale keyframes

The first keyframe we will set is for Scale.

STEP 53

Return the playhead to the first frame of the clip by hitting the Up Arrow key. In the Motion tab index, just to the right of the Scale percentage field, is a small button containing a diamond. This button is for setting and removing keyframes. Click the button and watch as the diamond changes to green (Figure 6-55).

Whenever the playhead is parked on an attribute or parameter keyframe in the Motion Timeline, this diamond will turn green. Hit the Right Arrow key to move to the next frame, and the green light disappears. Hit the Left Arrow key to return back, and it lights back up.

The keyframe we have just set has the value of 100 percent scale, which was the default value for scale in the scale percentage field next to the keyframe button. As described previously, a single keyframe is no different from having no keyframes at all. Keyframes have an effect only when there is a change in value, and that requires two keyframes.

To set the second keyframe, let's move to a new position later in the clip.

STEP 54

With the Motion tab window active, type +400 and then hit Enter (Figure 6-56). Just as with the sequence Timeline, this moves the Motion Timeline playhead up four full seconds.

STEP 55

When your playhead is parked at the fourth second, click the keyframe button again. Once again, it inserts a keyframe on the position of the playhead with a value of 100 percent (Figure 6-57). You now have two separate keyframes on the Motion Timeline in the Scale parameter.

But this does not change anything in our clip, because both keyframes carry the same value. In our example here, we want the text to start out very small and grow larger over time so that it appears to approach from a distance. We need to return to the first keyframe and assign it a different value.

We could hit the Up Arrow key to move the playhead to the beginning of the clip where our first keyframe is, but there is a faster way, especially if the keyframe you are going to is not directly on the beginning or end of a clip. If you have more than one keyframe in

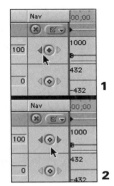

Figure 6-57

Figure 6-58

the Motion tab timeline, the keyframe button will display a sideways triangle pointing in the direction of the next keyframe from where the playhead is currently parked (Figure 6-58). If the playhead is parked between two keyframes, triangles will appear pointing in both directions. If there is no keyframe on a side of the playhead, the triangle will be grayed out. Clicking on the triangle will move the playhead to the next keyframe in that direction, eliminating the need to scrub around with the playhead and wait till the keyframe button turns green.

STEP 56

Hit the left triangle to return the playhead to the first keyframe of the Scale parameter. There are three primary ways to change the value of an existing keyframe. The first is simply to change the value in the field.

STEP 57

With the playhead parked on the first keyframe in the Scale parameter of the Motion Timeline, type 10 into the Scale field, then hit Enter (Figure 6-59). When you do, the green line between the keyframes slants, showing in a linear fashion that there is now a difference in value for the two frames separated by four seconds.

The second method to change the value is less precise, but it allows you to change the Timeline position of the keyframe in time as well as its value.

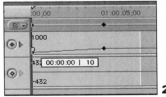

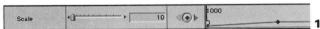
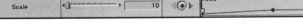

Figure 6-59

STEP 58

Hit Command-Z to undo, or simply type the value 100 back into the field to return it to its previous state. Move the mouse pointer directly over the keyframe until you see the arrow turn into a small crosshair. Click the keyframe, and drag in either direction. Up and down will change the value of the scale, and left to right will actually reposition the keyframe earlier or later in the timeline.

Using this method, you can reposition the keyframe in time as well as change its parameter value. If the numbers move a little too fast for your taste, simultaneously pressing the Shift and Command keys while clicking and dragging the keyframe will "gear" down the adjustment, allowing somewhat more precise control.

The third way is to manually adjust the scale using the Wireframe controls in the Canvas or Viewer windows. It is very imprecise, and should be used only when you are in such a hurry that visually estimating the scale change is the best you can do.

STEP 59

Make sure that the Canvas window is set to Image+Wireframe and that the text clip is selected in the sequence Timeline, so that you see the Wireframe of the text clip visible around the text in the window (Figure 6-60).

STEP 60

Set the Motion Timeline playhead on the keyframe you are adjusting. Then click and drag at either of the corner points of the Wireframe box in the Canvas window.

As you drag, the box will grow or shrink depending on the direction you are dragging. Also notice as you drag the corner point and change the scale of the Wireframe in the Canvas window, the value in the Scale parameter of the Motion Timeline in the Viewer win-

Figure 6-60

dow changes as well. Although you are making the adjustment in the Canvas window, you are actually adjusting the clip in the sequence which is loaded into the Viewer. Thus the Scale change is selected in the Viewer's Motion tab. Such is the nature of global settings.

As you will be able to tell, this method is very imprecise, which is not a good situation when working with complex keyframes.

There are two important exceptions. If you want to set the clip's scale nonuniformly (i.e., you want to make the height different from the width), perform this hands-on action with the Shift key pressed. This allows you to set height and width by dragging up and down and left and right respectively. Final Cut Pro accomplishes this by changing the Scale parameter and also the Aspect Ratio parameter of the Distort attribute at the same time. Of course if you want to keyframe this nonuniform shape change, you need to set an initial keyframe for both Scale and Aspect Ratio.

The other exception is for when you want to quickly set new values for Scale and Rotation simultaneously. This is great for creating animations of texts and other graphics that zoom and spin into the frame. Such animations require very tight coordination between the Scale and Rotation settings. If you set a keyframe for both Scale and Rotation parameters and then Command-click a corner point of the Wireframe box, you will be able to set the two values together and achieve natural spinning animations.

For the moment, set the first Scale keyframe's value at 10 percent, using either method of entry. We will also be doing some extra interesting effects using the other attributes that will require a couple more Scale keyframes. The action of the other Scale keyframes will become understandable when all the keyframes for the other attributes are in place.

Our final spin in the animation requires that the last jump in Scale occur faster than the first. Therefore we will also set a keyframe at the third second, this one for a value of 50 percent. This means that for Scale, the text will grow from 10 percent to 50 percent over the first three seconds, and then from the third to the fourth second, it will grow from 50 percent to 100 percent.

This is to mimic the real-world perceptual phenomenon in which an object farther away appears to move much slower at high speeds than an object close by. As the text gets larger and closer to the viewer, its speed must seem to increase the way that a car's speed appears to increase as it approaches us, even though the car's speed maintains a steady 50 miles per hour.

STEP 61

Move the playhead in the Motion Timeline up to the third second by typing "+300" (Figure 6-61). Add a keyframe by entering a Scale value of 50 percent. Notice that after you

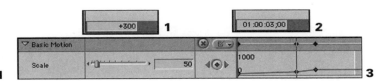

Figure 6-61

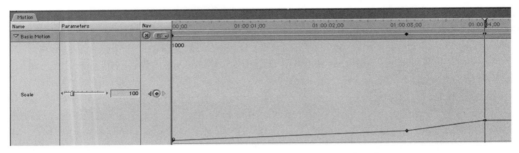

Figure 6-62   Correctly keyframed scale parameter motion timeline

create the first keyframe for a parameter, simply changing the value of it on a frame for which there is no current keyframe creates one.

When your keyframes for Scale are properly set, they should look like Figure 6-62.

Setting Rotation keyframes

The second parameter is Rotation. Although we won't be making our text spin in more than one full rotation, we want to give the sense that as the text approaches us, it spins in from the side as well. We will set an initial keyframe at the beginning of the clip with the value of –90 degrees. Negative values for Rotation simply imply counterclockwise rotation.

STEP 62
Reposition the Motion Timeline playhead at the beginning of the clip on the same frame as the first Scale keyframe. Make sure that the Scale parameter's keyframe button diamond is green to indicate the playhead is currently on that frame.

STEP 63
Click the Rotation keyframe button to create an initial keyframe, and then enter a value of –90 into the Rotation field (Figure 6-63).

With the initial keyframe set and the value entered, the Canvas window should now show the clip as a very small sideways rectangle near the center of the frame. Of course, as the text grows in size, we expect it to also straighten out.

STEP 64
Move the playhead up to the three second mark either by typing "+300," or by hitting the right-side triangle next to the keyframe button in the Scale parameter. Our next Rotation

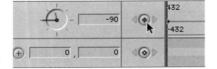

Figure 6-63

keyframe will be on the same frame as the Scale one, so that would be the fastest way to navigate to it without searching.

Once at the three-second point, we want to add a second keyframe. We could do this as we did the last three, by clicking the keyframe button. But the easiest way is to simply change the value of the parameter. If there is already an initial keyframe anywhere on the Motion Timeline for a parameter, simply changing the value of that parameter with the playhead parked on a frame that does not currently have a keyframe will automatically create a new keyframe there.

This does not work across parameters. There must be an initial Rotation keyframe, for instance, for this to work in the Rotation parameter.

So the process of setting a lot of keyframes for a parameter is to set the first keyframe with the button, and then go about your business setting new values at new positions by positioning the playhead and then entering new values by any method.

STEP 65

To set our new Rotation keyframe at the three-second point, type 0 into the field and hit Enter (Figure 6-64). A new keyframe will appear. Now if you scrub back through the

Figure 6-64

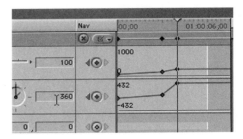

Figure 6-65   Correctly keyframed rotation parameter motion timeline

frames by using the Left and Right Arrow keys, you will see the text grow and rotate at the same time.

We still have one more keyframe, which will give one full rotation. This rotation will be aligned with the Scale keyframes that jump from 50 percent to 100 percent.

Since we have our keyframe at the third second, we need to add another at the fourth second so that our second full rotation completes as the scale reaches 100 percent.

STEP 66
Move the Motion Timeline playhead to the fourth second by entering "+100" and enter the value 360 into the Rotation field, which will rotate the text object fully once. When you add this value, the Rotation values should look like Figure 6-65.

Setting the Center parameter

The next parameter is perhaps the most important parameter for Motion work. It is the Center parameter, and how well you can use it over time really says a lot about how good you are at keyframing in general. Many parameters can be set almost by chance and still yield acceptable results, but the Center attribute and Motion Paths demand precision. The more time you spend working with clip positioning and keyframing, the better your motion paths will look. There is no substitute for practice.

Then again, the tools are easy and intuitive. Even the term "Center" itself simply means "where the central pixel of the clip is positioned on a given frame." This is where the math comes in, because X and Y coordinates apply to everything in motion, and if you can shuffle the X and Y coordinates around well enough, you'll be able to convince the viewer that you've been working with the Z, or depth, coordinate as well.

What is a Motion Path? A Motion Path is simply a trail of position values that change over a range of frames. When an object has a Motion Path, it has a new position value for each frame within the Motion Path. As the frames are played back, the object moves to the new position demanded by its Center keyframes. This movement over time is motion, and the line describing these changing position values over time that appears in the Canvas and Viewer when Wireframes are visible is called the Motion Path.

Intuition tells us that we'd want to set our first keyframe in the first place that the clip appears. But unfortunately intuition does not usually involve strategic thinking. In the present example, we know where the text will end up—right smack in the center of the screen at its present 0, 0 coordinates. Our text may zoom in from the left, right, up, or down, but it will end up in a specific place based on the design of the graphic.

STEP 67

Since we know that we want our text to end up in the center of the screen, advance the playhead to the third second, where our motion will stop, and click the keyframe button to create our first keyframe that has the default X and Y coordinates of 0 and 0, the center of the frame (Figure 6-66).

STEP 68

Once these values have been set into the keyframe, return the playhead to the first frame of the clip by typing "-300" and hitting Enter, or by hitting the left triangle next to the Keyframe button in either the Rotation or Scale parameters (Figure 6-67).

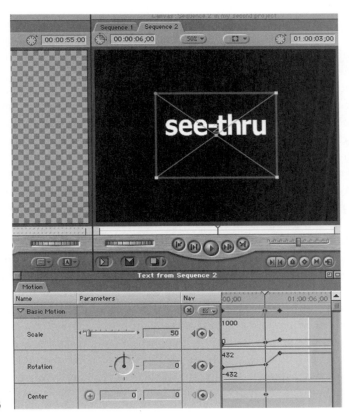

Figure 6-66

Figure 6-67   **1**   **2**   **3**

When setting position keyframes and Motion Paths, handling the clip manually in the Canvas window is usually the way to go, at least to get the keyframe quickly set in the neighborhood of the position you want to place the clip. You can more easily see the physical position of the clip as you drag and drop it, rather than trying to Figure out the X and Y coordinates of the new position.

STEP 69

Make sure that the Toolbar is set to the General Selection tool, return to the Canvas window, and click and drag the clip to the top, left-hand corner of the frame. If you refer back to the X and Y coordinates of the Center parameter in the Motion tab, they should be somewhere in the area of −300 and −150.

Since we might have to do mental math quickly at some point to adjust it again, choose a rounded value near the ones you have now and enter those values. For instance, if you got −291 and −142 from clicking and dragging, enter new round values of −300 and −150. Although these values are not substantially different in position, they will be much easier to work with if you need to figure pixel distances later. If the position must be exactly as you determined from the drag and drop, then leave the values you have.

A new Center keyframe will appear on the first frame of the clip as soon as you drop the clip the first time. If you didn't get it right the first time, keep moving the clip until you get the desired position. The keyframe will change value every time you adjust the position. If you can't get the X and Y values exactly right by manually dragging, you can simply type the X and Y values into the Center fields.

As you drag and drop the clip to create its second keyframe, you will notice that a small line appeared in the Viewer and Canvas windows between the initial keyframe at the third second and the next keyframe at the first frame of the clip. The thin line, which has many bumps along its length, is called the Motion Path. Each bump along the line is the position value for a frame in between the two keyframes. As we will see a little later, these bumps will tell us how fast the clip is moving and even how it speeds up or slows down.

A simple useful shortcut to remember when moving the clip manually with click and drag is how to constrain the movement to right angles. If you hold the Shift key down while clicking and dragging the clip in the Canvas or Viewer windows, the movement will be constrained to either horizontal or vertical straight lines. This is very convenient if you want to make sure your text doesn't move diagonally at all. Simply set your first or final keyframe, move the playhead to the Timeline location of your next keyframe, hold the Shift key, click, and drag. You will see that the movement is possible only in vertical or horizontal directions.

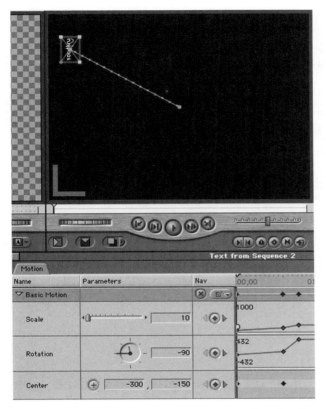

Figure 6-68   Correctly keyframed center parameter motion timeline

After you have set the beginning keyframe for –300 and –150 and at the third second, the clip is positioned at the Center, your keyframing for the Center parameter should look like Figure 6-68.

Looking at the initial motion

STEP 70

In the sequence Timeline, scrub through the frames and watch in the Canvas as the text clip now zooms up, spins 90 degrees, and moves from the top left corner to the center of the frame simultaneously over the first three seconds. Render the sequence, and play it back.

Sometimes, it is difficult to separate your eyes from the two-dimensional world of Final Cut Pro keyframes and see the faux Z dimension you are creating, but try to see how the combination of scale and size create a silent third value that perceptually equals depth. Also notice that at each frame you park the playhead on between the Center keyframes, the clip is centered on a bump on the Motion Path. As you scrub up the Timeline, the clip moves down the Motion Path.

After enabling the Clip Keyframes display button in the Timeline window, the area underneath the text clip will be occupied by a blue line. The blue line is an indicator that there has been a change in the default Motion tab settings for the clip in the track. The Clip Keyframe display bar will also reveal the existence of any Effects Filters applied to the clip with an additional green bar. Both of these indicators are valuable, particularly because after you render a clip, it may be difficult for you to remember that effects were applied to it. Knowing that effects were applied and then rendered (which you would know because of the absence of the red render bar) may one day keep you from accidentally unrendering a section of your work.

Setting the Anchor Point keyframes

The final parameter for the Basic Motion attribute is the Anchor Point. As we described earlier, the Anchor Point is not an obvious physical value. Changing the Anchor Point alone will not make the clip smaller or turn it upside down. The Anchor Point is rather a value that alters all other physical values. The Anchor Point determines the central point from which all other activities are performed.

This can be confusing, because changing the coordinates of the Anchor Point often has the opposite effect from the one you were hoping for. Successful use of the Anchor Point depends on abstract thinking and simple concrete logic. Just drawing out on a piece of paper what you intend to do makes it clear what settings you need. Above all, do not avoid the Anchor Point just because it is sometimes counterintuitive. Try this exercise, then watch television and start looking for examples of its use in animation.

For our example, we want the initial zoom out and rotation to proceed normally, as if the text were turning itself as it approaches the viewer. For this effect, the text will have an Anchor Point value of 0, 0 for X and Y respectively. This means that the axis of the rotation will be at the exact center of the text clip. The clip will spin like a wheel, with the text at the center.

STEP 71

Create the first keyframe for the Anchor Point at the first frame of the text clip with the default value of 0, 0. This keyframe should be lined up with the keyframes of the other parameters (Figure 6-69).

Until we give the parameter a new keyframe value elsewhere, the Anchor Point will remain in the X and Y center of the clip. But in our description of what we want to end up with, we proposed to have the text perform another quick rotation after reaching the end of the first zoom-out/rotation, this time with the rotation from a different axis in the clip. Now we will need to add more keyframes.

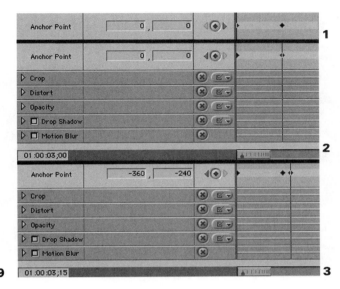

Figure 6-69

Since our first zoom/rotation ends and the next one begins at the third second, it would seem apparent that our new Anchor Point value to create the different rotational axis should occupy that keyframe. Unfortunately this is not so, but let's make the mistake anyway and see why.

STEP 72

First make sure that your Canvas and Viewer windows are set to Image+Wireframe so that the Motion Path we will be affecting is shown. It may also be beneficial to lower the screen size of each window to 25 percent to aid in screen redraw times and to allow you to manually control Center parameter values if they should end up far off-screen.

STEP 73

Move the Motion Timeline playhead up to the third second. Enter new values −360 and −240 for X and Y for the Anchor Point keyframe (Figure 6-70).

When you enter this value, the physical position of the clip and the shape of its Motion Path will change drastically in the Canvas and Viewer windows, even though your Center position values have not. The clip appears to act as before for a scant few frames, and then begins to spin wildly in a circle long before it reaches the keyframes at the third second.

This is because the Center and Rotation values are now being figured based on having the center pixel of your text clip in the extreme upper left corner rather than in the center. When we set this new Anchor Point keyframe, we told Final Cut Pro that Center and Rotation values should be computed from there instead of the 0, 0 center of the clip.

To make matters worse, our new Anchor Point value is the second, not the first, keyframe value, so that the Anchor Point axis and position are changing over time. This makes

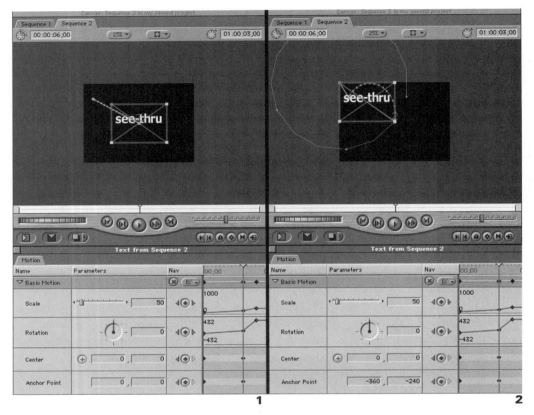

Figure 6-70

it even more difficult for us to track and predict the motion path we are creating. Scrub through the frames and try to isolate the Anchor Point change as it occurs and observe the effect it gradually has on the rotation and position values.

In the Wireframe view, the Anchor Point progression line, which obviously is changing positions between the two Anchor Point keyframes, can be distinguished from the Motion Path created by the two Center keyframes. The Motion Path always contains bumps that indicate the Center position of a given frame between two keyframes. The Anchor Point progression line is simply an unbroken line between the relative position of the Anchor Point in the first keyframe to its relative position in the second keyframe (Figure 6-71). This can be a little tricky to see, since the Anchor Point's effect is to change the frame's X and Y position on which the Center and Rotation values are based.

STEP 74

Position the Motion Timeline playhead at the first keyframe of the clip. Using the Left and Right Arrow keys, scroll through the first three seconds of the text clip and back to examine

Figure 6-71

the behavior that is being created by simultaneously shifting the values of the Anchor Point and the Center and Rotation. Keep an eye on the Anchor Point's shift from the center of the clip to the left top corner of the clip. Then observe how this results in the spiral of Center and Rotation as these parameters also change based on their own keyframes.

Now, the effect we really wanted was to have the text clip first rotate and zoom based on an axis at the center of the text clip. Then after the clip reached the three-second point, we wanted it to perform another rotation, this time rotating from its top left corner rather than from the center axis. What happened instead is that we gave the parameter one keyframe too few. In order to perform what we are describing here, the Anchor Point value needs to remain the same all the way until the third second, and then change positions to the new value.

With only two keyframes as we have it now, the parameter values interpolate between the first frame and the third second; as a result, the axis slowly shifts over the length of the clip. This results in a spiral as the Rotation and Center values shift consistently with the Anchor Point. We need another keyframe positioned at the third second to keep the Anchor Point keyframe value at 0, 0 up until that point.

Just after the three-second point, we want the Anchor Point to change to the top left-hand corner at −360, −240 so that the rotation occurring between second three and second four will use the new axis for the rotation. To make this change in Anchor Point a little more natural, we will keyframe it so that the Anchor Point shifts over time.

STEP 75

To make sure that you have these keyframes set appropriately, make sure the values and timing match the following:

- The first frame of the clip should have an Anchor Point keyframe value of 0, 0.
- The third second of the clip should have a keyframe value of 0, 0.
- One half second later (15 frames for NTSC and 12 frames for PAL), there should be a third keyframe value of −360, −240. These keyframes should look like Figure 6-72.

STEP 76

Scrub through the clip on the Timeline window using the Left and Right Arrow keys, and observe the different effect of our new Anchor Point keyframes.

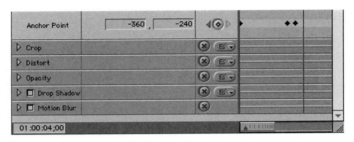

Figure 6-72 Correctly keyframed anchor point parameter motion timeline

Right up to the third second, the rotation spins obediently in its quarter-turn using the 0, 0 as the Anchor Point axis. But as soon as it hits the keyframe at the third second, it encounters an interpolation of Anchor Point X and Y coordinate values covering one half second.

It begins the second rotation, but does so as the rotational axis is shifting. By the time the rotation is complete, so is the axis shift. The end result, as you will see, is that when the clip appears far away, its rotation appears to be centered on its own axis. But as it approaches the viewer, it appears to gradually shift into an interesting spiraling change in rotational axis from center to extreme corner.

This little performance could have been keyframed entirely from within the Rotation and Center parameters, but it would have taken an enormous amount of tweaking and mathematical work to get the keyframes positioned and coordinated. Instead, our Anchor Point adjustment required only three keyframes. Clearly, this tool can be useful in creating complicated motion paths and rotations with far fewer critical Center and Rotation keyframes. Learn to use the Anchor Point, and it will save you hours of work one day.

Setting the Distort attribute

The next attribute we will set is Distort. Technically, there is no need for the Distort attribute here, but we want to take advantage of its power to shape our text so that it also appears to unfold from facing up to facing forward in its initial rotation. We will also use it to add a slight distortion curve to the text itself so that it appears to have some depth.

We will set the first four parameters first, these being the X and Y coordinates that determine the position for each corner of the clip. Remember that the values for these coordinates are based only on the clip itself, not its sequence. As we will see, changing the X and/or the Y value will merely stretch the clip to compensate. Stretching it beyond the limits of the clip's frame would result in part of the clip becoming masked in the Canvas window, just as our text clip was when its Font Size setting was too large to fit in the window.

Setting the Distort parameters, like setting the Center parameter, is a job that is best initially handled by visual spec in the Canvas or Viewer window. There is a special tool for Distort effects within the Video windows.

Figure 6-73

Figure 6-74

STEP 77

To access the Distort tool, go to the Toolbar and select the second toolset from the bottom (Figure 6-73).

This toolset contains the Crop and Distort tools. The Crop and Distort tools set values for the like-named attributes within the clip, based on the parameters as discussed above. Although we will not be adjusting the Crop parameter, its keyframing and adjustment are similar in form to the Distort we will work with.

STEP 78

Select the Distort tool and move to the Canvas window. We do not need to set any initial keyframes for the first four parameters, because the values assigned to them will remain constant throughout the animation. Move the Motion Timeline playhead up to the fourth second, where the Scale keyframe is set to 100 percent, making it much easier to visually spec as we adjust.

Since we are directly adjusting the shape of the clip using the Canvas window's Wireframe, having the clip scaled up to 100 percent will facilitate the operation.

STEP 79

Make sure that the window is set to Image+Wireframe, and then use the Distort mouse pointer to grab the bottom left-hand corner and move it up to around a quarter from the bottom and the extreme left-hand side of the frame. Then grab the bottom right-hand corner of the clip and drag it down about an equal amount outside the boundaries of the frame, but vertically aligned with the right edge of the frame (Figure 6-74).

The effect of these two adjustments will result in a pinched left-hand side of the text and a stretched right-hand side.

STEP 80

If you cannot get the Distort tool to perform this properly, simply enter the following coordinates for these two parameters: Lower Right 360, 440, and Lower Left –360, 100. Leave the upper two parameters, Upper Left and Upper Right, and the Aspect Ratio intact.

Setting the Opacity attribute

Earlier in the text, we set Opacity using the Pen tool in the sequence Timeline, but we can just as easily do it here. The keyframing we do for Opacity will have the same effect on the clip that it did from the sequence Timeline. In fact, any keyframing you do at the sequence window appears here. In addition, any keyframes you enter here will appear on the sequence Timeline if you have the Clip Overlay button enabled there.

Rubberbanding on the sequence Timeline is just a quick shortcut for the Opacity parameter. But when you are trying to coordinate many different visual effects into one grand motion, being able to line up keyframes is essential. Therefore, it's usually best to set

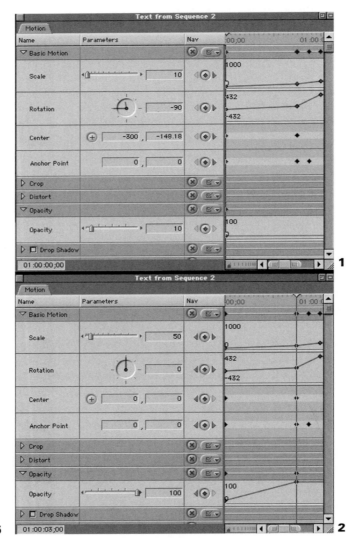

Figure 6-75

Opacity keyframes within the Motion tab whenever you want to group changes in visibility with other Motion properties.

Since our text clip is supposedly far in the distance at the beginning of the animation, it stands to reason that its Opacity should be dramatically lower.

STEP 81

Return the playhead to the beginning of the clip, create a new keyframe for Opacity and enter its value as 10 percent (Figure 6-75).

We don't want to start the clip as completely transparent, because initially our clip does not start out infinitely small. Part of the believability of this animation is that it looks real. From the viewer's perspective, the video frame does not extend into infinity, and starting from an infinitely small and dark setting would suggest that it does. Straining believability is not a good idea when performing sleight of hand; ask any magician. Therefore, our clip will appear very small and semi-opaque at first, but with definite substance, rather than as a pin prick on the horizon.

STEP 82

After you have set the first Opacity keyframe, advance to the third second. At that point, the clip will be in full visual range, and there will be no further reason to attempt to hide it. Enter a value of 100 percent (Figure 6-75). This keyframe will bring the clip up to full visibility.

If you Left and Right Arrow scrub through the clip on the sequence Timeline, you will see the accentuation that change in Opacity gives the clip as it is changed by the other attributes.

Setting the Drop Shadow attribute

Next we will set a drop shadow for the text clip. We will be keyframing the change in the drop shadow so that it appears that a real stationary sun to the left of the screen is causing it. We will start with the three-second mark as our principal shadow, and then keyframe the changes from there.

STEP 83

Click in the checkbox to enable the Drop Shadow attribute, and then move the Motion Timeline playhead up to the third second (Figure 6-76).

STEP 84

At the three-second mark, insert an initial Offset parameter keyframe value of 3. Offset is the distance between the object and its drop shadow. At the three-second mark, an Offset of 3 should look pretty nice. The problem of course is that when the Scale changes, the Offset needs to change values as well, since when the text is further away and smaller, its shadow will also appear smaller.

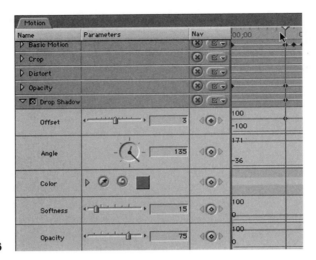

Figure 6-76

STEP 85

Move the playhead to the beginning mark and enter a new keyframe value of 1. Now as you step forward through the frames from the beginning of the clip, you should see the shadow hug the text a little more tightly.

The Angle parameter is especially important if our shadow is to look like it was cast by another object. Since the drop shadow is actually projected from the clip and not a stationary light, we need to keyframe the shadow's changing position in relation to the imaginary light. Doing so is easy. The area of real importance for the drop shadow is the giant rotation at the end of the animation.

Our initial shadow angle will be set for a light in the top left-hand corner.

STEP 86

Create and set the initial keyframe for the Angle parameter at the beginning of the clip at 145 degrees.

STEP 87

Advance the Motion Timeline playhead to the third second. This is where our text begins a large rotation and where we need to correct the Angle keyframes so that the shadow appears to remain stationary.

STEP 88

Add a new Angle keyframe that maintains the value 145 at the third second (Figure 6-77). This second keyframe exists solely to keep the Drop Shadow Angle value of the first three seconds locked in at 145 degrees. The next keyframe will actually cause the rotation of the Drop Shadow Angle in the opposite direction of the clip's rotation, which from the viewer's perspective means that the light casting the shadow is stationary.

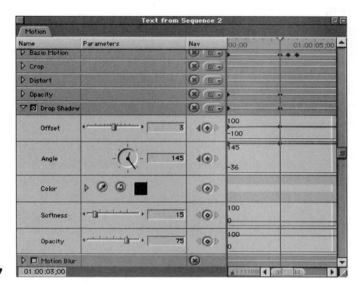

Figure 6-77

STEP 89

Advance to the fourth second and enter a new Angle keyframe value of –215 (Figure 6-78). The importance of entering your values numerically here is supreme. It is easy to accidentally turn the Angle wheel in the wrong direction, resulting in a sun that reels across the imaginary sky while your text reels about the frame.

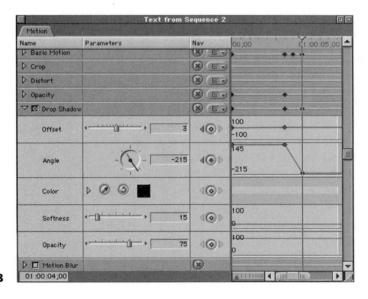

Figure 6-78

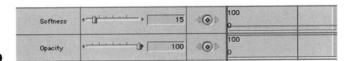

Figure 6-79

Negative 360 degrees equals one counterclockwise turn. Since our text is turning clockwise and we want our shadow to remain in the same direction it was, our shadow angle should turn the exact same amount of rotation but in the reverse direction. This will cancel the travel of the shadow, leaving its imaginary light source appearing to be in the same position.

STEP 90

The primary reason for the drop shadow is to accentuate our text, so enter a low value for Softness of 15 and a high value for the drop shadow's opacity of 100 (Figure 6-79). Since these values will remain consistent for the entire animation, there is no reason to keyframe them.

Applying nonlinear interpolation techniques

Render the animation, and take a quick look at it as it plays back in real time. You may be surprised to see that, despite all our precise settings, something still looks slightly unnatural about the movement. There's nothing wrong with the settings, except that we haven't gone quite far enough with them. Before going further, we have to explore the idea of linear versus nonlinear interpolation.

All the keyframes we have applied so far have used linear interpolation: when the value changes between one keyframe and the next, the change in value is evenly spread over the frames in between the keyframes. For example, if at frame 1 there is a keyframe whose value is 1 and at frame 10 there is a keyframe whose value is 10, at frame 2 the value will be 2, frame 3 the value will be 3, and so on. Although there are no keyframes between frames 1 and 10, Final Cut Pro evenly interpolates the exact value for the frames in between.

When it interpolates those values, the process is called linear, because there is no variation in the rate of the value's change. But we know that the real world does not operate along such rigid lines. Not even a robot could begin moving at exactly ten miles an hour, travel an entire distance at the same speed, then stop on a dime. In the real world actions start slowly, speed up, slow down and stop, and not necessarily in that order.

So Final Cut Pro allows you to perform what is called nonlinear interpolation. This means that between two keyframes, there can be variations in the values between the keyframes. For instance, between frames 1 and 10 of the above example, frame 5 could have a value of –5, or 20, or anything else, as long as it gets back to 10 on the tenth frame.

Of course the values that come into play for nonlinear interpolation depend on the attribute or parameter involved. This nonlinear interpolation of keyframes is divided into two types: Ease In/Ease Out, which is the method for Motion Paths, and Smooth, which is the method for everything else. Ease In/Ease-Out allows us to vary the speed of a moving clip across the frame, slowing down and speeding up between keyframes. Smooth allows us to introduce curves into the interpolation of attributes and parameters such as Scale, Rotation, and other keyframeable effects.

Adjusting the motion path using Ease In/Ease Out

STEP 91

To set Ease In/Ease Out for the Motion Path, select the text clip on the sequence. Move the Motion Timeline playhead to the first frame. Make sure your mouse pointer is set to the General Selection tool. In the Canvas window, change your view mode from Image+Wireframe to Wireframe (Figure 6-80).

Ease In/Ease Out must be set on the Motion Path of the object itself in the Canvas or Viewer window, and it is much easier to see this in Wireframe mode in which the clip's imagery is not visually distracting.

STEP 92

Once in Wireframe mode, you will see your text clip's Wireframe in the top left-hand corner. If you move the mouse pointer over the Center of the clip, the pointer will change into a crosshair. When it does so, Control-click, and you will receive a contextual menu with the options Make Corner Point, Ease In/Ease Out, and Linear (Figure 6-81).

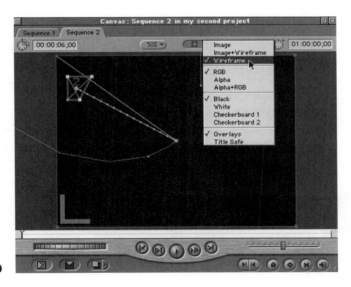

Figure 6-80

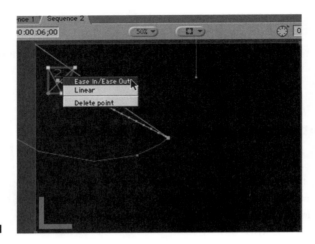

Figure 6-81

STEP 93

The interpolation is presently Linear; select Ease In/Ease Out on the contextual menu to change this to nonlinear interpolation.

When you do so, you will see two new points appear, a small one and a large one (Figure 6-82). You may need to zoom in on the Motion Path to see the points, since they look similar to the Motion Path's frame bumps, but slightly larger in size. These are referred to as Bezier handles, and they allow you to adjust the Motion Path of your clip based on a direct manipulation of the curves created by tugging on the Bezier handles themselves.

Of the two, the larger handle is for creating curved shapes in the Motion Path. Grabbing either of the two larger Bezier handles drags the formerly straight Motion Path out along a curved path. Although it's difficult to be entirely precise using this tool, it can be a quick way to create smooth, curved motion when you need it.

STEP 94

Grab the end of the larger Bezier handle and drag it slightly up so that you introduce a gentle curve to the first leg of the Motion Path that looks like Figure 6-83.

The Motion Path has been slightly altered by the Bezier handle so that the path of the text clip curves somewhat as it moves to the center of the frame.

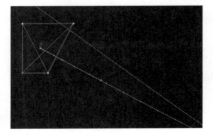

Figure 6-82

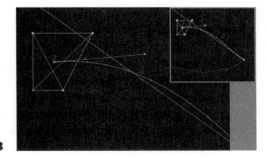

Figure 6-83

STEP 94

Float the crosshair mouse pointer above the smaller bump on the Bezier handle and drag in either direction along the handle (Figure 6-84). Unlike the larger Bezier bump at the end of the handle, it will not create a curve in the Motion Path and its adjustment is restricted to the distance between frame bumps inside the present Motion Path line. As you drag the handle, the many other smaller frame bumps on the Motion Path become closer together or farther apart.

This is the essence of speed change, or velocity. Each small bump on the Motion Path indicates the position of the clip during that particular frame. Each bump is a position in time as well as the X and Y coordinates of its position in space. As you tug on the Ease In/Ease Out Bezier handle, the bumps get further apart or closer together, meaning that the clip will move a further or shorter distance between the two bumps or frames. When two bumps are very close together, it means that the clip is moving very slowly, because it is only traveling the small distance between the bumps over a single frame. When the bumps are far apart, the clip is traveling very fast, because it is covering far more distance between the frames.

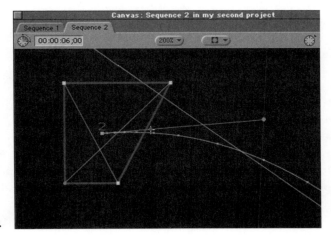

Figure 6-84

STEP 96

Float the mouse pointer over the smaller Bezier handle. When it changes to a crosshair, click and drag the small handle toward the center of the clip. This will cause the frame bumps to be much closer together near the beginning of the Motion Path and further apart near the three-second mark where the second Center keyframe is located.

Thus, the clip will appear to move slowly at first as it grows in size and then it will speed up slightly as it approaches 100 percent Scale. Using your imagination, you can make fast-moving clips rush to a stop, start up slowly, speed up, and slow down naturally or whatever other combinations of Ease In/Ease Out you can think of. The key is to look at the material and the Motion Path and decide what looks natural.

It should be noted that the Ease In/Ease Out nonlinear interpolation is far from extreme in its adjustment of velocity. While you can change velocity, as we have just done, the change will be gentle. The purpose in Final Cut Pro is to smooth out the motion of clips moving along Motion Paths, particularly as they move around corners. Other applications such as Adobe After Effects and Pinnacle System's Commotion have far more comprehensive and intuitive tools for this sort of nonlinear interpolation. If you really want to get into realistic inertia-affected motion, you might consider looking into one of these animation packages. The theory of motion and keyframing is the same, but the tools are more robust.

Adjusting the other attributes and parameters using smooth curves

We can also adjust the other attributes by using Smooth curves in the Motion Timeline. Nearly all attributes and parameters that have interpolated keyframeable values can be treated with nonlinear interpolated Smooth curves. If there is an interpolated value line between two keyframes in the Motion Timeline, Control-clicking one or both of the keyframes will allow you to apply Bezier handles to them, giving you control over the rate of the change in value occurring between the keyframes.

A direct and clear example of the power of Smooth curves with Bezier handles can be seen by adjusting the Rotation attribute of our text clip.

STEP 97

Go to the Motion Timeline, scroll up to the Rotation attribute, then resize the scale of the Timeline such that you are looking at all three keyframes of the parameter.

STEP 98

Float the mouse pointer over the middle keyframe located at the third second. When you do this, the mouse pointer will turn into a crosshair, indicating that it is ready to adjust a keyframe.

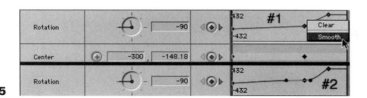

Figure 6-85

STEP 99

Control-click the keyframe, and hold the mouse button to receive a contextual menu offering two choices: Clear and Smooth. Select Smooth and release the mouse button (Figure 6-85).

Once you release, the keyframe should sprout two blue Bezier handles pointing in either direction from the keyframe. These two handles are linked to produce an equal and opposite curve when adjusted; if you drag up and out with one handle, the handle on the other side of the keyframe will move down and out to produce a smooth curve from the keyframe preceding it to the keyframe following it. If there is no keyframe preceding or following the keyframe that has been set to Smooth, it will show only one or no Bezier handles, since one or zero keyframes on a timeline imply no change in value and Bezier handles exist only because there is a change in value.

STEP 100

Grab the left-hand Bezier handle and drag down and out to produce a curve shaped like a trough (Figure 6-86).

STEP 101

Now move the playhead to the beginning of the timeline, make sure that your viewing mode is set to either Wireframe or Image+Wireframe, and frame scrub (depress Right Arrow key) through the interpolation. Keep an eye on the actual rotation in the Canvas window and match it to the changing values in the numerical field in the Rotation attribute line of the Motion tab.

As you frame advance through the clip, you will notice that the rotation value in the numerical field far exceeds the initial keyframe's value of –90, resulting briefly in rotation in the wrong direction! But the clip still ends up at a value of 0 by the time it makes it to the Smoothed second keyframe. Remember that negative rotation values simply mean rotation in the counterclockwise direction, so making the trough deeper simply forces the rotation even further in the counterclockwise direction before sending it back to a value of 0.

Figure 6-86

Also notice the smooth gradual change in rotational angle values as you frame advance both visually in the Canvas and numerically in the attribute field. Visually, the rotation slows down as it reaches the bottom of the trough, then speeds up as it returns back in the direction it came from, ascending the second curve of the trough. Also notice that in the numerical field, the angular values for the interpolation are expressed to two decimal points. Both linear and nonlinear interpolation are this accurate, far more accurate and speedy than the average editor would be in figuring out the curve manually. This accuracy is most valuable in determining the curve values for this and other attributes and parameters. Since curves are mathematically determined, it's easy for Final Cut Pro to crunch the numbers for a nice smooth curve value that looks very natural.

One feature that may seem problematic is the fact that adjusting a Bezier handle in one direction automatically adjusts the handle in the other direction as well. That may be fine in many situations where the value change should be smoothed along the entire timeline. But what if we want to set a curve for only one side of the keyframe, leaving the other interpolation intact? There are two options available.

STEP 102

The first is to Command-click the Bezier handle you want to adjust (Figure 6-87). When you do so and drag the handle out, you will see that this does increase the extremity of the other side's interpolation without affecting its curvature. Performing this on the right-hand Bezier handle of our example text clip will show the left-hand trough increasing or decreasing in depth as you adjust, but only relative to the change in the right-hand Bezier handle. The shape of the trough is the same, but the values become more or less extreme relative to the amount of drag you create on the right-hand side.

STEP 103

While that can be valuable, many times the Bezier handles must be adjusted completely independently. To accomplish this, Shift-Command-click either handle and adjust its curve to your satisfaction. Using the Shift-Command click, straighten the interpolation between the first and second frame, but now Shift-Command-click the right handle and create a curve for the rotation between keyframes two and three. This will cause the rotation at the beginning of the second keyframe to occur more quickly and give the appearance that the right edge of the text clip is leading the spin.

Your Rotation keyframes should look like Figure 6-88 when set correctly.

Render your timeline and take a look at the effect this has on the rotation in real-time motion. Also go in and experiment with other attributes and parameters that offer nonlinear interpolation through Smooth keyframing. You'll quickly learn how to exploit curves for smooth natural effects.

Figure 6-87

Figure 6-88 Correctly keyframed rotation parameter motion timeline

Applying motion blur

The final attribute to be set for the text clip has been left to the last for good reason. Motion blur, as described earlier, is a very processor-intensive operation that slows down rendering enormously. While certainly worth it for the realism it affords motion, the slow-down in processing a single frame can really cramp your style when you are busy setting complicated motion keyframes in place. As such, it should generally be disabled until you are ready to use it.

Unlike other attributes, motion blur is not keyframeable. Motion blur is a constant effect that you set into place. It's somewhat similar to the physical phenomenon of gravity. Although gravity causes the weight of an object and gravitational pull is the same for all objects, not all objects weigh the same amount. Motion blur is the same for the entire clip, but its actual visual effect depends on what the frames being blended by the Blur are doing at any given time. Fast movement will equal much motion blur; slow or no movement will result in little or no motion blur.

The key is to make sure that you aren't setting the percentage, or number of previous and future frames included in the blur, too high or low. If the motion is only 30 frames in length, 500 percent motion blur will likely be too high a percentage.

STEP 104

Enable Motion Blur for the clip by clicking in the checkbox for the attribute. Since our motion is very short and very fast, set the percentage to 100, so that it is blending only the frame prior and frame following (Figure 6-89).

STEP 105

Set the Samples to 8, which will result in a rather smooth blend between the blurred frames, but won't take an eternity to render.

STEP 106

Return to the Canvas window, park the playhead somewhere within an area of movement and you should see a nice blur. The blur should not be so extreme as to make the text completely unreadable, but should suggest motion, even on a parked frame.

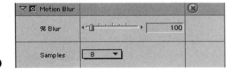

Figure 6-89

Render the sequence once again and observe the blur in action. Depending on factors such as the size of your text clip, the keyframing you've set and the values for motion blur, your results may vary slightly. Experiment with the settings and you will arrive at the perfect setting for the clip you are working with.

Speed and duration

Another effect that you can apply to video and audio clips is Speed. It is easy to change the speed of a clip directly. Because of the flexibility of the Speed control dialog box, it's also easy to coordinate the new speed with a specific duration so that you can slow the clip down to match another element in your sequence.

To change the speed of a clip, we want to load it into a sequence first. Although you can change the speed of a clip by simply selecting it in the Browser or loading it into the Viewer, this will actually apply the speed change to the original clip, which is more than likely your Master clip. Clips that have had their speed adjusted must be rendered to play back correctly, and rendering can only occur in a sequence.

STEP 107

Edit a clip into a fresh empty sequence. Adjust the Time Scale of the sequence so that the entire clip is visible on the sequence timeline.

STEP 108

With the clip selected, go to the Modify drop-down menu and select Speed (Figure 6-90). When you do so, you will receive a dialog box that allows you to customize the resulting clip generated by the Speed command.

The first field is for percentage. 100 percent, the default, is normal speed. Lowering the percentage will decrease the speed; raising it will increase the speed. Below this is a Dura-

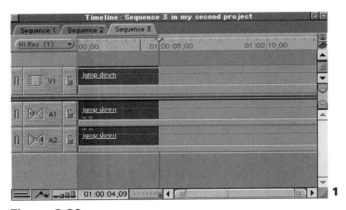 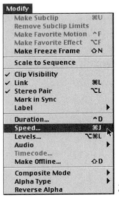

Figure 6-90

tion field. Notice that if you hit the Tab key after entering a new value into the Speed field, the Duration field updates, revealing the new duration of the clip. Reducing speed to 50 percent will result in the duration doubling, while doubling the speed to 200 percent will result in halving of the duration.

STEP 109

Type 50 percent into the Speed field, and watch the Duration field double in length (Figure 6-91).

By the same token, changing the Duration field will give a new speed percentage value. This is probably the most valuable tool for wily B-Roll video editors who want to make the most of every frame they have, and who aren't afraid of adjusting the speed of a clip to make it fit just right into its edit slot. Although Fit to Fill, one of the Canvas' editing overlay boxes, also performs this function automatically, you can't specify the speed change before committing to the edit. With the Speed command, you enter the space you have for the clip, and you can immediately see the speed change your clip will be subject to when processed before committing to it.

The Reverse checkbox simply applies a negative value to the percentage. This means that in addition to any speed changes made, Final Cut Pro will also reverse the order of the frames in the clip, making it run backwards.

Frame Blending should always be enabled. When you speed up or slow down a clip, you are removing or creating frames of video to make the video appear faster or slower. This can cause strobing or stuttering if the new set of frames do not appear to be as continuous or sequential as the original set were. Frame Blending creates new intermediary frames that are a blend of the two new frames left over from the speed change. The result is smoother and cleaner speed changes.

STEP 110

Enable Frame Blending, leave Reverse disabled, and hit OK. The clip in the sequence will have doubled in length and require rendering. If you render, you will observe the very smooth slow motion.

1

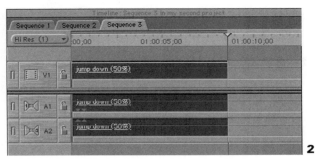

2

Figure 6-91

Bear in mind that any change in speed will require a render, since no video media file yet exists for the speed-changed clip. Interestingly, if you change the speed and then change it back to 100 percent or normal, the speed change will not require rendering, since the second speed change simply refers the clip back to the original clip media. When you change the speed, the clip in the sequence will be marked with the percentage of the speed as a handy reminder that the clip is modified.

Effects filters

What are effects filters?

You can also apply effects to a clip that are based on the use of filters, otherwise known as plug-ins. In reality, filters are special scripts that are applied to clips to produce specific effects. Rather than adjusting one physical element of a clip at a time, filters usually change many different aspects of a clip at once. Instead of being designed to change the raw physical attributes of a single clip, filters usually change several at once to produce a desired effect.

The effects filters can be applied in a manner similar to the application of the Transition. You can simply drag and drop the effect from the Effects tab of the Browser window onto the clip, or you could apply the filter by selecting the clip and accessing the filter from the Effects drop-down menu. In this example, we will use the latter process although there is no functional difference between the two. Use whichever seems more effective in your personal workflow.

Applying a color effect; the invert filter

Let's apply a couple of effects filters of varying sorts to see how they affect the clip.

STEP 111

Edit a fresh clip into the sequence and double-click it to load it into the Viewer window. As with the Motion tab settings, we will be able to see our work update in the Canvas window, but the Filter parameters and their keyframes themselves must be set in the Filters tab of the Viewer window.

STEP 112

Once the clip is loaded into the Viewer window, click on the Filters tab to bring it to the front.

STEP 113

Go to the Effects drop-down window, select the Video Filters submenu, and then select Channel and Invert (Figure 6-92).

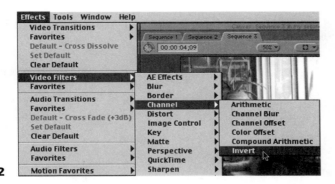

Figure 6-92

Your clip should immediately and radically change in the Viewer and Canvas windows. The Invert filter works its visual magic by changing the values of chroma, luma, and alpha equally and opposite within a clip. Whites become blacks, yellows becomes blues, all colors invert based on their opposites in the RGB color wheel. But you are not limited to this default appearance. By adjusting the parameters of the Invert filter, you can choose exactly which chroma, luma, or alpha channels are inverted, as well as the extent to which they are inverted.

STEP 114

Go to the Filters tab in the Viewer window, and tear it away (Figure 6-93). We want to be able to watch our effect change in the Viewer Video window as we adjust the parameters of the filter. If you tear the tab away and find no Invert filter present in the tab, you have either misapplied the filter to the wrong clip, or you have loaded the wrong clip to the Viewer window.

For a standard captured clip in Final Cut Pro, the filter tab will have two sections: a Video Filters section and an Audio Filters section. Any filters applied to the clip will appear here. If there are no audio clips linked to the video, there will be no section for audio clips on the filters tab. The reverse is also true for audio-only clips such as imported audio CD tracks.

After applying the Invert filter, you should see the Invert filter in the Video section of the Filters tab. There is a checkbox for enabling and disabling the filter. This can be very handy if you've worked hard to get the settings for a filter correct, but you need to make it disappear for a moment. Rather than deleting the filter by selecting its name and clicking Delete, you can simply turn it off at the checkbox and then re-enable it when you are ready for it again. Be warned that disabling the filter with the checkbox removes any render files associated with it, so you'll have to re-render if you do so.

In the parameters for the Invert filter, there is a drop-down bar allowing you to choose the image data that you want to invert. You can choose from many arrangements that will have different effects. Even though you know that the actual frames of DV are digital RGB images, you can invert based on luma (or Y), any of the specific RGB channels individually,

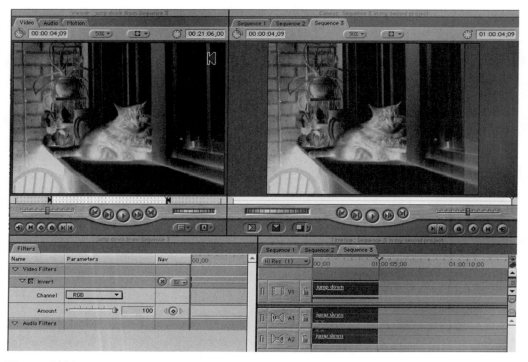

Figure 6-93

an alpha channel if present, or UV. For the moment, leave this set to the default RGB (Figure 6-94).

The other parameter you can alter is the Amount. This parameter is a percentage based on the degree to which pixel values are inverted. A value of 100 percent means that pixels are completely inverted, and 0 means they are not inverted at all (Figure 6-95).

STEP 115

To illustrate the idea, type 50 into the Amount field and hit Enter to apply the change. The screen should be completely gray, since all pixels are being inverted exactly halfway to the

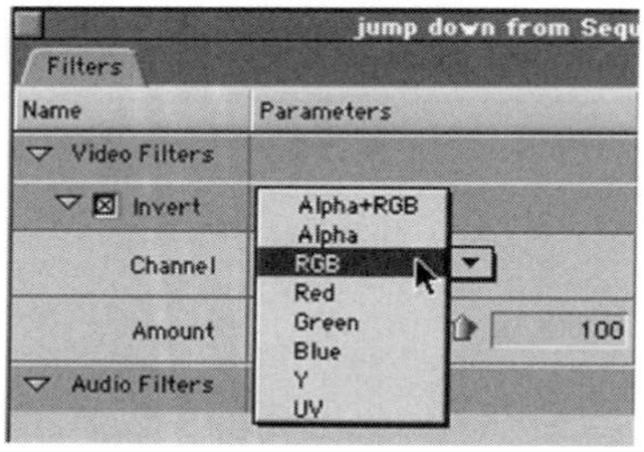

Figure 6-94

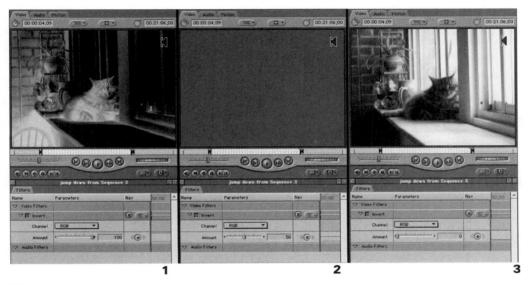

Figure 6-95

opposite. The halfway value for any inverted color is always exactly middle gray. Type in the value 100 again to return to the full Invert.

You will see that this parameter is keyframeable, as are many parameters in most Effects Filters. The keyframing lesson of the previous section applies to this keyframing, and most parameters can be adjusted with Smooth curves as well.

Applying an effects filter matte; the garbage matte

STEP 116

Make sure the Viewer window with the Filters tab is active, and go to the Effects drop-down menu (Figure 6-96). Select the Video Filters submenu, then Matte and, finally, Four-Point Garbage Matte.

When we discussed matting in the section on Composite Modes, we talked about using luma values or alpha channels to create transparency mattes, or areas of an image that you could see through to the next layer. The approach there was to block or allow transparency in the frame using the luma or alpha values of another layer as a mask. With the garbage matte, though, we can create areas of transparency from directly within the clip itself.

The garbage matte is exactly what its name implies. It is used to matte garbage that you don't want seen out of the frame. But instead of building complex alpha channels or being constrained to an exact rectangle or square, as the Crop tool would require, you can arbitrarily set the positions of the four corner points of the matte. If the shape you need to matte is more complex than is possible to matte with four points, an Eight-Point Garbage Matte is also available in the same Effects filters Matte submenu.

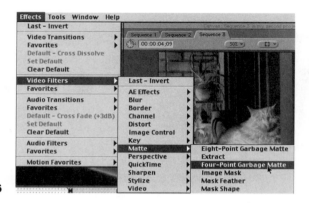

Figure 6-96

We're going to make an interesting framing effect using the inverted colors and the garbage matte to focus attention on one area of the frame.

STEP 117
Move the clip with the Invert and Four-Point Garbage Matte filters up to the Video 2 layer (Figure 6-97).

STEP 118
Set the target track to the Video 1 layer, and then line up the sequence playhead with the first frame of the clip in the Video 2 layer.

STEP 119
Now, select the Video 2 layer clip and copy it, using Command-C. Hit Command-V for Paste. Paste will make an exact copy of the Video 2 layer clip in the Video 1 layer, using the playhead as the insertion point. Alternatively, hold the Option and Shift keys down, and

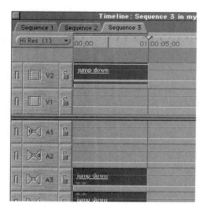

Figure 6-97

Figure 6-98

then move the clip back to the Video 1 layer (Figure 6-98). Final Cut Pro will automatically Copy and Paste it there for you.

You now have two copies of the exact same clip lined up frame for frame, one above the other. We will remove the Invert and Garbage Matte filters from the bottom background clip. Copying and Pasting will have included the filters, but they are necessary only in the Video 2 layer clip.

STEP 120

Double-click the Video 1 layer clip, and bring it up into the Viewer window. Select the Filters tab, then select and delete each of the two filters by clicking on the name and hitting the Delete key. Make sure that you are deleting the filters from the Video 1 layer clip and not the Video 2 layer clip.

STEP 121

If you hit the Track Visibility Green Light in the sequence to turn off the Video 2 layer, you should see that you now have a background clip that exactly mirrors the clip above it with the exception of the Effects filters applied to the latter (Figure 6-99). Turn the Video 2 layer

Figure 6-99

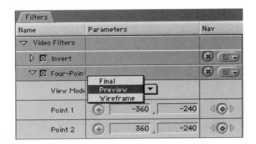

Figure 6-100

track visibility back on, and you should see the layer we are going to use our garbage matte on.

Let's set the parameters for the garbage matte.

STEP 122

Double-click the Video 2 layer clip so that it loads up into the Viewer window. Pull the Filters tab away from the Viewer window so that you can watch each window update as you work.

The first parameter on the Garbage Matte filter is the View Mode. Your choices are Final, Preview, and Wireframe (Figure 6-100). Each mode has its benefits. The Final view allows you to see the matte as it stands by itself. The matted out areas will appear as transparent, just as the image will appear in the Canvas above the background layer. The Preview mode gives the same view but also includes small numbers indicating the positions of the corners of the matte you will assign in the following parameters, making it easier to tell which corner needs to be adjusted for the desired effect. The Wireframe mode shows no transparency and displays a Wireframe shape representing the present shape of the matte. This mode allows you to set the positions while looking at the entire image. For our example, set the mode to Preview.

Underneath the View Mode are four point assignment parameters that correspond to the four corners of the matte. When the View Mode is set to Preview, the four points are identified in the Video tab of the Viewer window. You can enter precise X, Y position data into the fields for each point, but you can more easily set the positions by clicking the Crosshair button and then clicking a position in the Video window of the Video tab of the Viewer window.

STEP 123

Click the Crosshair button for Point 1, and move the mouse pointer to the Viewer window (Figure 6-101). The pointer will convert to the crosshair. Click on a position that you want to establish as the top left corner of the matte. You will see that the matte closes in and the Point 1 corner now occupies that position. In the Canvas window, you will see the matte taking shape.

Figure 6-101

STEP 124

Follow the same process for each of the other three points until you have created a four-cornered matte that crops the clip, revealing only the section that you want to reveal. Remember that the Four- and Eight-Point Garbage Mattes are valuable because they do not have to follow a rectangular or square shape like the Crop tool in the toolbar, so you can get a little wild with the shape (Figure 6-102).

Be careful that you don't overlap the lines of the matte, because the garbage matte is a geometric entity. It can create transparency only between the points. If you cross the matte lines, you will end up with unpredictable results.

The next parameter is for Smoothing the corners of the matte. There is no Bezier control of the corner points, so to soften them, you need to adjust this parameter. Unfortu-

Figure 6-102

nately, you cannot apply a different corner softening effect for each corner; garbage mattes are rarely used for very tight matting operations but are useful for quickly eliminating material from the frame.

STEP 125

Set the Smooth value to 40 to get a nice round edge to the matte (Figure 6-103). The next parameter is for Choke, which shrinks the matte based on the value chosen. Choke is often misunderstood to refer to the choking of the image you see. Choke really refers to the invisible matte you are creating. Positive values of Choke will choke (i.e., reduce) the matte size, revealing more of the clip. Negative Choke values will increase the size of the matte, hiding more of the clip. It may seem pointless to set a Choke value when you could easily have just set the matte position correctly, but remember that this parameter is keyframeable. Since the Choke value expands or shrinks the matte size equally for all four points, Choke can be an easy way to adjust the size of the matte without having to keyframe four corner points individually.

The next parameter is Feather Edges. This parameter we have encountered before; it simply adds a gradient transparency edge to the matte so that the edge appears to fade into the background.

STEP 126

Give the Feather Edge parameter a value of 20 to create a wide fade between the inverted foreground and normal background clips.

The following two parameters are checkboxes for Invert, which just reverses which side of the matte causes the transparency, and Hide Labels, which causes the Point numbers to disappear from the Viewer window when in Preview mode.

Take a look back at the Canvas window now, and you will see that the Garbage Matte isolates the small section of the top inverted clip and the edge of the clip fades off into the normal uninverted background clip. The effects must be rendered to play back correctly,

Figure 6-103

and you can always go back in and change the matte's parameters. If you find that four points are not enough to create the complex matte shapes you need, the Eight-Point Garbage Matte doubles the number of corner points available for the matte. If you still find this too limiting, then you have promise as a compositor. You have stretched Final Cut Pro's rotoscoping abilities to the limit and may want to look into a more robust 2D animation application such as Pinnacle Commotion or Adobe After Effects.

This is only an example of two of the many effects that come bundled with the Final Cut Pro nonlinear editor. A full discussion and demonstration of each Filter is far beyond the scope of this book. You have the tools to master the Effects filter interface and its keyframing convention. Experimentation, as well as referring to the descriptions of the Filters contained in the Final Cut Pro manual, is the key to mastering them all.

Nesting

Up to this point, we have only discussed effects work with regards to single individual clips. This is fine, but of course situations will arise where you need to apply the same Effects filters equally to all the individual clips in a sequence. Or perhaps after completing an edit of a sequence, you want to apply Motion tab–type effects to a sequence in its entirety rather than its individual constituent clips. This would be impossible to do with the sequence's clips on a one-to-one basis.

The answer lies in Nesting. Nesting simply refers to placing a sequence inside another sequence so that the first sequence is treated as a clip. You can manually move one sequence into another by simply dragging and dropping it just like you edit in a clip. You can load a sequence from the Browser window into the Viewer window just as if it were a clip and then edit it into another sequence. And you can apply effects to a "clip" sequence as if it were a clip. About the only thing you can't do with nesting is to nest a sequence inside of itself.

There is another benefit to Nesting that is less obvious. It can retain your render files, even in editing situations where ordinarily the render files would have been thrown out. Whenever you render, Final Cut Pro creates render files for the clip in the Render Folder of the Scratch Disk. It organizes these files according to the sequence where the rendered clips are in use. When a rendered clip is played back, Final Cut Pro goes into the Render folder and finds the clip based on which sequence it was rendered in. That's why if you render a clip in one sequence and then copy it onto another one, it has to be rendered again. The sequence you copy it into has no render file associated with it.

To examine this, take a look at the Render Manager. The Render Manager is a special tool that allows you to eliminate old render files from your disk to free up drive space. To get to this tool, go to the Tools drop-down menu and select Render Manager (Figure 6-104). When the Render Manager opens, you will see a window containing folders bearing the name of your projects. If you drop down the triangles next to the project name, you will find subfolders bearing the name of each of the sequences inside the project. Drop the sequence subfolder down and you will see the various oddly-named render files associated with the sequence.

You can remove files, subfolders or even a whole project's render files by clicking a checkmark in the Remove column of this window. Although render files are not identified by name or even by the name of the clip they are associated with, you can usually tell which files are irrelevant based on the Last Modified column. If you knew that you re-rendered sequence material after such and such a date, it's pretty easy to tell which files are still around but unused and wasting hard drive space.

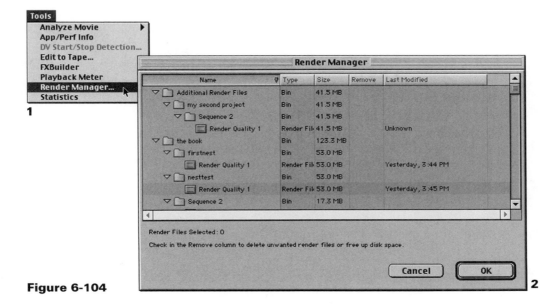

Figure 6-104

Render Manager demonstrates the idea that rendered clips are connected to their render files through the sequence. And since they are connected through the sequence, this makes the sequence a way of protecting render files. By placing the rendered clips into their own sequence and then including that sequence in another sequence, or in short, nesting, you are not breaking the initial link between the rendered clip and its render file through the nest sequence. If the rendered clip is nested into a sequence, the render files stay connected to the rendered clip, even if you edit the sequence in ways that would ordinarily endanger the render files.

STEP 127
To illustrate this point, take the previously-rendered example using the two clips and the garbage matte on separate video layers. Render the clips so that your footage plays back correctly.

STEP 128
Using the Razorblade All tool from the Toolbar, make a cut in both clips on the same frame (Figure 6-105).

STEP 129
Use the Roll Edit tool to trim the newly created edit points in one direction or the other (Figure 6-106).

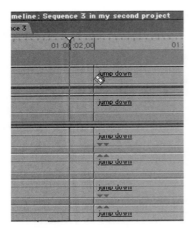

Figure 6-105

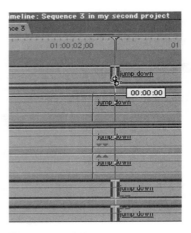

Figure 6-106

STEP 130

Now if you re-trim the edit points back to where they were, you will see that the red line has returned above the re-trimmed section of the clips, meaning that these sections must be re-rendered (Figure 6-107). The action of cutting and trimming the clips caused the render files linked to them to become unlinked.

In theory, we haven't altered the clips at all; we've merely cut on a frame, trimmed the clip, and then re-trimmed the clip back to its original length. Unfortunately, Final Cut Pro regards this as change enough and forces a re-render.

STEP 131

Return the two clips back to their normal rendered status (you can leave the razorblade-cut edit points, just Command-Z to undo the steps back to the point before you razorbladed).

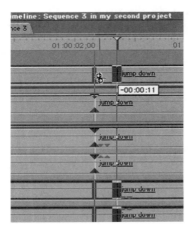

Figure 6-107

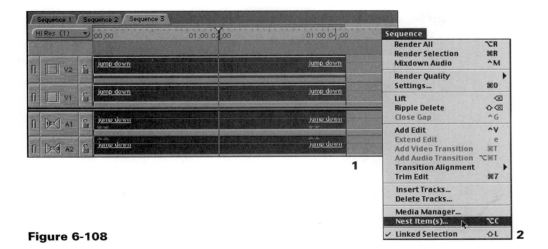

Figure 6-108

STEP 132

Select all the clips, go to the Sequence drop-down menu and choose Nest Item(s) (Figure 6-108).

When you do so, you will be presented with a dialog box allowing you to name the new nested sequence. Do not leave the confusing default; name it something that makes sense to you (Figure 6-109).

You will be able to change the dimensions of the frame and the aspect ratio if you wish, but remember that you will have to render it if you do so, just as you would if you put the clips into a sequence that did not have the same frame size conventions of the original clips. This is essentially what you are doing, since the Nest Item(s) command creates a new nested sequence and includes the selected items in it.

Below this are options for retaining effects markers and audio levels with the nested items. This is important since Final Cut Pro will keep any adjustment you have made intact with the nested clips. The other checkbox included is the Mixdown Audio. This option may be useful in some cases. Remember that there is a limit to the number of audio tracks that Final Cut Pro can play back in real time. When you nest clips together, you also nest any linked clips together. Since two audio tracks are frequently linked to any video

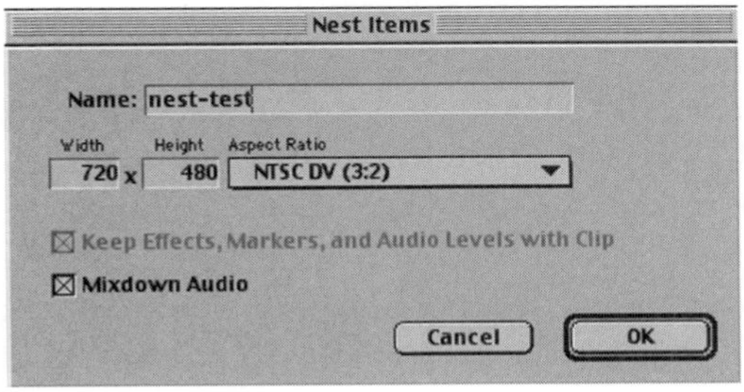

Figure 6-109

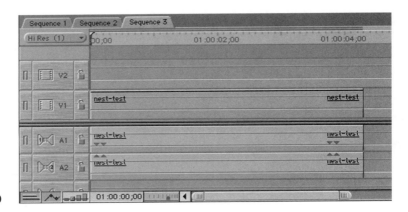

Figure 6-110

track you've captured, nesting repeatedly can lead to a situation where you have more than eight tracks playing back simultaneously, even though it appears that only two are.

Mixdown Audio fixes this situation by quickly and discreetly rendering a Mixdown of the multiple audio tracks of the nested items. The sound will be the same, but you will remove the possibility of accidentally forcing Final Cut Pro to play back audio tracks beyond its capacity. You will still be able to go back into the nested sequence and change anything about the audio tracks if you wish after nesting.

STEP 133

Hit OK, and you will return to the sequence to find that the clips have disappeared and been replaced by what appears to be a single clip that bears the name of the nested sequence you just created (Figure 6-110). A glance over to the Browser will reveal that there is a new sequence icon in the project; it is the nested sequence you just created. Grab it and drag it over into the sequence and you will see that it is treated just as if it were a simple clip.

When you select Nest Item(s), all the selected items are bundled into one sequence that appears in your Project tab. All these items also seem to disappear in the sequence, but in fact they are just wrapped inside the nested sequence that appears on the sequence Timeline after the Nest Item(s) command.

STEP 134

Once again make a Razorblade cut, this time in the center of the nested sequence that contains the clips you just previously cut and re-rendered. Perform the same trim action on the nested sequence (Figure 6-111).

When you perform the same trim out and trim back in, you will find that the nest does not have to be re-rendered. This is because a sequence is being edited, not the nested clips inside the sequence. Since the sequence itself has no render files to become unlinked, no re-rendering needs to occur, even though the sequence's contents have render files.

Do not become too confused by the concept of nesting items. It is a useful feature that becomes clear as you use it. Nesting can protect render files, as well as allow you to apply

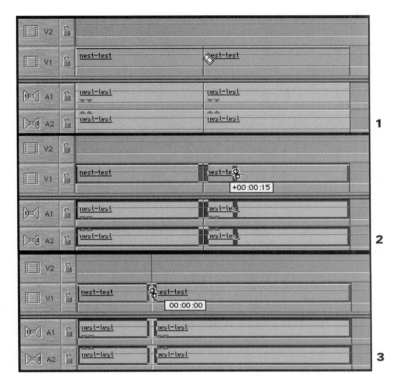

Figure 6-111

Effects Filters and Motion Effects to an entire sequence as a whole object, just as if it were an individual clip. If you understand these two functions, you can utilize the Nesting functions of Final Cut Pro in very productive ways.

STEP 135

Double-click on the nested sequence in the sequence Timeline. Instead of opening up in the Viewer window as a clip would, the nested sequence opens up as yet another tab in the sequence Timeline window (Figure 6-112).

After all, a nest really is a sequence, not a clip, regardless of how it is being treated on the video tracks of the sequence. Double-clicking the nested sequence here, or in the Browser window will open the sequence up in the Timeline window as its own sequence tab. This allows you to go back into a nested sequence and make changes to the nested items if you must.

You can also load this, or any sequence, into the Viewer as if it were a clip in a couple of ways. You can grab the sequence and drag from either the Browser or parent (sequence the nested sequence sits in) into the Viewer window. You can also open the nest (or any sequence or clip) into a new Viewer, leaving the previous Viewer window open and loaded with its current clip.

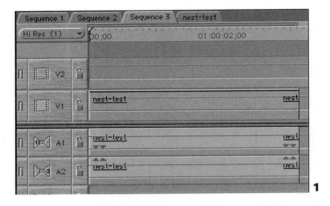

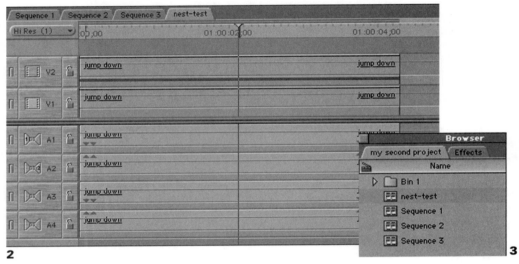

Figure 6-112

STEP 136

To create a new Viewer window instead of replacing what is in the current Viewer window, hold down the Option key and double-click the sequence (or clip) icon in the Browser window's project tab or bin. This will bring it up in its own new Viewer window. Incidentally, you can always create a new Viewer window for any individual clip in the Browser window by this same method. With individual clips, you also have the option of Control-clicking the clip in the Browser window and selecting Open In New Viewer (Figure 6-113). This option does not exist for sequences however, and you should rely on Option double-clicking for those. Clips in sequences on the timeline must be loaded into a Viewer window through the drag-and-drop method, or by selecting the clip and choosing Clip In New Window from the View drop down menu.

Figure 6-113

Once the sequence is loaded into the Viewer window, you will have access to the same tabs that you worked with on the individual clips. You may want to go through and perform the same exercise on the sequence that you did on the clips earlier so that you get the hang of using effects filters and Motion effects and experience the ease with which entire sequences can be processed as individual clips.

7 Getting Your Project out of Final Cut Pro

Getting your project out of Final Cut Pro

There are three basic ways of sending your material back out to the DV deck to record onto DV tape. We call such copies *master copies*, because they are copies coming directly off of the editor in the first generation of a dub. Actually the term originated with analog recording systems and is less relevant for Firewire DV editors these days, because the Firewire master copy can be duplicated with no loss in generation. But we still cling to the term, because it means that this is a finished copy that you want to archive and preserve, regardless of its lack of susceptibility to generation loss. It is your master copy.

Once you've finished cutting your video production, you're ready to output the results to tape via the Firewire connection. The process is the easiest of all those described in the previous chapters, because you've been sending the video and audio out to Firewire all along for preview as you edit. Recording the material to tape is simply a matter of deciding how much you want to automate that process.

The first and simplest way to record your finished production out to Firewire and DV tape is to put the tape in the deck, park the playhead at the beginning of the sequence, hit record on the deck, and then hit play in Final Cut Pro. There is no real difference between this method of Firewire output, which you've been using all along to preview the edit, and the following methods, aside from the varying degrees of automation. This simpler method may prove best for editors using A/D converter boxes or those who are using cameras and decks not yet supported or controlled by Final Cut Pro.

Output to Firewire part 1: Print to Video

There are, however, two other methods that may be useful in customizing your master copy that gets recorded to DV tape. The first method is Print To Video. Print To Video is a very simple yet powerful tool that allows you to customize what goes to tape. You can define leader material such as color bars and tone, slate information, countdowns, and the amount of black that occurs before your sequence begins playback and after it ends. You can specify whether the entire sequence or just a portion of it is played back. You can have Final Cut Pro play the sequence back as a defined number of loops, inserting black space between each copy of the sequence. You can also add trailer black onto the end of the recorded sequence so that your recorded product doesn't end on blank video resulting in a timecode break. And you can keep tabs on exactly how long the final recorded piece will be, based on the amount of extra time you've added on to include the features just described.

Preparing for Firewire output

STEP 1

To start the process, make sure that your deck or camera is connected to the Macintosh's Firewire port and powered on. Select a sequence in the Project tab or the Timeline window that doesn't require rendering or has already been rendered and open it up in the Timeline window. Now go to the File drop-down menu and select Print to Video (Figure 7-1).

Customizing your Print to Video

When you select Print to Video, a dialog box will appear named Print to Video. This is the window that lets you customize the output. The top section of the box contains leader, or any elements that you may want to include prior to playback of the actual video in your master tape. You enable or disable each element using the checkbox. For some features, you customize the element based on your individual needs.

Color Bars and Audio Test Tone

The first Leader element is for Color Bars and Tone, essentially the same clip we used in the first exercise in this book. Here Final Cut Pro will automatically output them to tape before your project. Color bars are the reference colors that are always included in any tape used in the broadcast industry or in any situation in which calibrating video equipment is necessary. Once your video leaves the DV tape, it is no longer digital and exact. The color settings on almost all televisions and video monitors are different. Many times, these settings are so far apart that the color relationships in the video footage are completely wrong. This means that although a color on your videotape is blue, it may be closer to green or red than it should be on the video monitor that it is passed to.

This does not mean that the blue color cannot exist on your video monitor; it merely means that most video monitors are not *calibrated*, or set to display blue exactly as yours is. Thus the blue may appear as a different hue or saturation. This is where calibration comes in. Color bars contain each of the colors necessary for establishing the correct color balance of a video monitor. Since the color bars are digital and are preconfigured to be correct, when piped through a video monitor they reveal how far off the monitor's color settings are. Using color bars, it's easy to see that a color is not being displayed correctly and an adjustment must be made in the monitor's color and brightness controls.

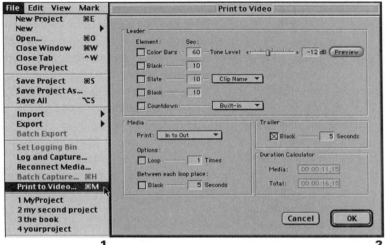

Figure 7-1   1   2

Color bars are not only useful for calibrating your video monitors, they are necessary in the broadcast industry for making sure that your video fits within the broadcast specification for video. Such specifications are referred to as "broadcast legal" video. When you send a tape to a television station, the video engineer passes the tape through two devices—the Waveform Monitor, which measures the luma values, and the Vectorscope, which measures chroma values—that check to make sure that the luma and chroma values of the video signal are not too extreme for broadcasting.

Some luma and chroma values can exceed this "broadcast legal" specification. The engineer can then arbitrarily clip the chroma or luma to make it fit the specification, resulting in loss of fidelity to the original. A better alternative for the engineer is to adapt the luma and chroma more gently so that it fits within the specification. But the engineer cannot do this adaptation without a reference to what is "legal" on your tape. Changes can be made based on the Waveform Monitor and Vectorscope assessments, resulting in legal video that more closely resembles what you wanted it to look like.

Without reference color bars on your tape, you are always at the mercy of having your video not displayed correctly wherever it is played back. Including them is always a good idea, even if you think you may never need them. Most stations require a minimum of 60 seconds of color bars prior to slate information on the tape. Remember that tape is always cheaper than embarrassingly bad-looking video, even if your family is the only group that will ever see it.

Final Cut Pro also includes a reference audio tone with the color bars. In essence, the reference audio tone performs the same function as the color bars. The reference tone is a standard 1 KHz tone you've probably heard a thousand times when a television station finishes its programming for the night. This standard tone lets the engineer check to make sure that what your tape thinks is 1 KHz agrees with what the station's equipment thinks is 1 KHz.

It also allows the engineer to match the audio levels with whatever the station is using. If your reference audio tone is set at –12dB, the engineer will know how to adjust it when the signal appears on the station's equipment as much higher or lower. It will be possible to boost or lower the signal with great precision so that your audio is reproduced faithfully.

Even if you never intend to have your material pass through a television station, the –12 dB reference tone allows you to set the volume on the monitor you are using at the optimum level before playback. Just play the tone back through the system it will be played on. When you hear the tone on the tape playback, set the volume on the video monitor or television accordingly, and you can be sure that the audio levels you worked with in your project will be in the same intended neighborhood when played back.

This system of matching audio levels between the recording side and the playback side using a reference tone is referred to as *gain staging*. It has far more important benefits than those listed above, especially with respect to avoiding distortion and clipping when the audio is rerecorded through analog or digital systems later on down the road. Just remember that anything you record without reference color bars and audio tone leaves you operating in the dark about the state of your video and audio. You will have no way of gauging how the product looks or sounds when reproduced on another video monitor or when rerecorded later.

Although the color bars are reasonably standard, the audio tone that an engineer wants can vary in different stations and localities. –12dB is a popular standard, but if you are preparing tapes for a specific venue or purpose, make sure to ask in advance which standard is preferred, then make sure to include the information about the reference tones with the tape case and slates.

STEP 2
For the Leader settings, specify 60 seconds of color bars and tone, with the tone set for –12 dB (Figure 7-2).

Below this is an option is for black. This option is offered between each leader option for the purposes of making your leader smoother, more predictable, and easy to use later on. You can customize the amount of black in seconds here.

STEP 3
Enable the Black checkbox and insert 10 seconds of black.

Figure 7-2

Slate

The third option is for a slate. Slates are simply text images that appear before your video piece plays back. They can give information about the format of the video and audio, the title of the piece, who produced it, whom and what it is for, information about the reference bars and tone, and anything else you want to include. Often, production facilities have specific policies about what they require in the slates before a submitted piece of video. Make sure you know what should and should not be there, as well as how long the slate should appear.

There are three options for the slate (Figure 7-3).

You can use the Clip Name—the default setting, which includes just the name of the object being recorded. If you are recording a sequence, the name of the sequence will appear in the slate. If you are outputting only a single clip, its name will appear. Clip Name is the default setting because the other two choices, Text and File, require you to enter information or select a file, whereas Clip Name will just use the name of the object you are printing to video.

If you click the Slate drop-down bar and select Text, a small text box appears to the side of the bar. You can type all the relevant information here and it will appear as the slate during the Print to Video. This is the fastest way of getting information into the slate prior to recording, since it requires no external file preparation, but is capable of including far more information than the simple clip name.

Be careful when using this, though. It's easy to forget to change this text, and you will find yourself using slate information from a previous recording session if you do not change it, since the text remains the same until you disable or alter it.

Finally, the Slate drop-down bar gives you an option for File. If you choose this, a small folder icon appears next to the bar. Clicking on the icon allows users to navigate to the file they have created for a slate. This option is very useful if you wish to use a standardized company logo image or you have a specific design you want to use that cannot be achieved using the limited text box of the previous bar option.

The File option allows for the use of standard PICT files, Photoshop images, or any of a number of QuickTime still image file types. Remember that the non-square pixel rules

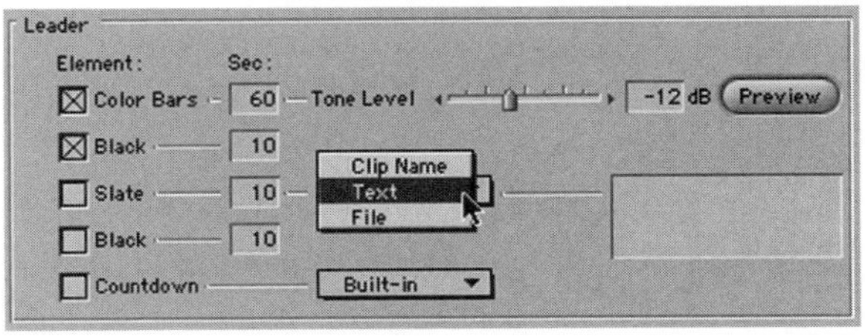

Figure 7-3

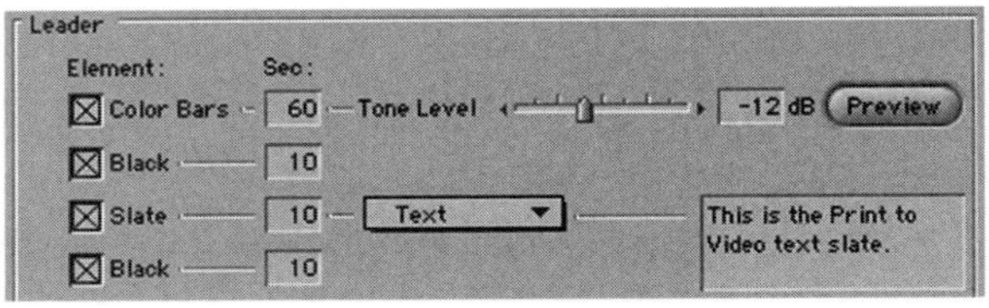

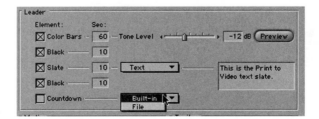

Figure 7-4

are still in effect here as discussed earlier in the book and to prepare image files such that they appear correctly without appearing squeezed or stretched.

If you choose to include a slate in the playback, there will be a slight render delay before playback for recording is possible. Since the slate will be played back as video out to Firewire, it must be rendered. The length of the render delay depends on the duration of the slate, as well as the type you use.

Clip Name and Text are very fast renders because they are using a standard black and white text format that processes and compresses very quickly. If you use a strange file format with the File option, the render may take longer as Final Cut Pro converts then processes and compresses the video. In either case, the render of still images in Final Cut Pro is rather fast and is well worth the wait if you want the benefit of sharp-looking hand-made graphics that say more about your product than what its name is.

STEP 4

Enable a slate, set it for 10 seconds, and then choose Text from the bar. In the text box, type in "This is the Print to Video text slate" (Figure 7-4).

Below this is another optional Black as referred to in the Slate option.

STEP 5

Enable the black checkbox and insert 10 seconds of black.

Countdown

The final option of the Leader options is for Countdown. This option also contains a drop-down bar for selecting the source of the countdown. The two options are for the Built-in or File (Figure 7-5).

Figure 7-5

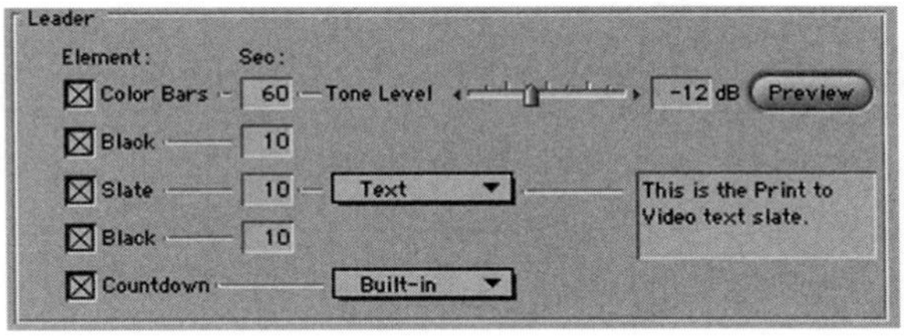

Figure 7-6

The Built-in countdown is a standard countdown file that comes prepackaged with Final Cut Pro. It is precisely clocked for 10 seconds, in which each second is displayed with a clock wipe showing the passage of the second. The end of the countdown results in a "2-pop." When the countdown reaches 2, there is a "pop" that makes it easier for audio to be synced up with the video should there ever be the need. Countdowns are imperative for precise editing situations and broadcast facilities that must cue up tapes with a great degree of accuracy.

You can also create your own customized countdown QuickTime Movie file and utilize it by changing the drop-down bar to File and then navigating to the new countdown file. You can use Final Cut Pro itself to create your custom countdown file; just create and export a suitable file and save it in a safe place for use when you need it. But remember that using a custom-made countdown file does not guarantee that it is accurate to a 10-second countdown. You have to arrange that yourself when you create the file. The Print to Video settings box will not constrain your file to fit a 10-second countdown. It will simply play the file you select.

STEP 6
Enable the countdown and select Built-in from the bar (Figure 7-6).

Media

In the bottom left corner of the Print to Video window, there is a box labeled Media. This box lets you customize what gets sent to tape through the Firewire tube. Although the practice of sending media through Firewire is no different than playing from the Timeline, the Print to Video window allows us to automate the process and create seamless master tapes based on our individual needs. This can include looping the media, only recording a specific portion of the sequence and including pauses between the loop repetitions.

The first option is a drop-down bar named Print. The two choices on the bar are In to Out and Entire Media (Figure 7-7).

In to Out allows you to create a range in your sequence or clip defined by In and Out points on the sequence Timeline or Viewer window. Then choosing In to Out in the Print bar will restrict playback to those edit points. This can be very useful if you are only backing up sections of your sequence to tape and are not yet mastering the entire sequence.

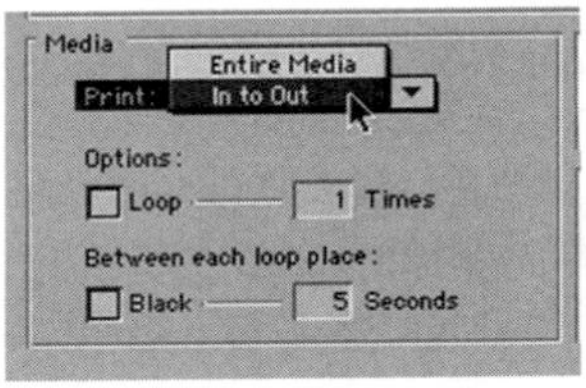

Figure 7-7

Choosing Entire Media plays the whole clip or sequence from the first frame to the last frame of the last clip. Although this is a good quick way of printing your finished sequence, it can be messy as well. If you have forgotten about orphaned clips late in the sequence that are separated by gaps from the parts that you do want to record to tape, Entire Media will play all the way until it has played these orphaned clips as well.

Of course, none of this is really ever out of your control. You can always stop the Print to Video by stopping the recording at the deck and hitting Escape on the keyboard. Print to Video does not involve Device Control. The only difference between Print to Video and simply playing from the Timeline and hitting the record button on your deck is that Print to Video allows you to customize the process as defined here. Print to Video will not start or stop the recording on your deck.

Below the Print bar are two options. The first checkbox is for Loop, which is accompanied by a numerical field for establishing the number of looped repetitions. Beneath this is a checkbox for a black pause between repetitions and a field for determining the length of the pause.

STEP 7

For the Media settings, select Entire Media from the bar, then disable Loop and Black between Loop checkboxes (Figure 7-8).

Trailer

To the right of the Media box is the single option for Trailer. You may include black following the end of your recorded product, and you may define the length of this black trailer.

Do not be fooled by the seeming unimportance of this one box. It is very important that you insert black trailer at the end of your recording. Always tack on at least 10 seconds of recorded black trailer at the end of your program. At the end of the Print to Video, the Firewire output will switch from the Print to Video back to displaying whatever frame the

Figure 7-8

playhead was parked on when Print to Video began. Thus at the end of your emotionally touching video, the recording could instantly switch to a video still image of your crazy uncle doing something embarrassing.

This is no big deal, and you could certainly just park the playhead over a black area of the sequence before Printing to Tape, but it is much better to get into the habit of manually including black frames to the end of your productions. This feature will take on the added significance of timecode issues when we move to the Device-Controlled version of Print to Video called Edit to Tape in the next section.

STEP 8

For the Trailer settings, enable Black and enter 10 seconds into the numerical field (Figure 7-9).

Duration Calculator

Finally, the box underneath Trailer is the Duration Calculator (Figure 7-10). The two Timecode fields refer to Media (the length of the actual sequence or clip) and Total, which is the sum amount to the frame of all the leader, media, and trailer footage, including any looping of the media. This is just a short idiot check to make sure that you have enough recording space on the tape for the duration that you have prepared to Print to Tape. It will also keep you from accidentally looping or including far too many seconds of leader if you make a mistake entering those parameters.

STEP 9

Your Duration Calculator should show the length of your sequence in the Media field and that same amount plus 1 minute and 50 seconds in the Total field, that being the sum total of the leader and trailer features you have tacked on to the piece.

Print to Tape

Once these settings are correct, hit the OK button to begin rendering the Slate and the Countdown. Since the Built-in Countdown is not a still image, it takes a bit longer to render up. While the render process is completing, cue up a DV tape in the deck, or, if you are using an A/D converter box or other pass-through device, prepare whatever deck you are using to record.

Figure 7-9

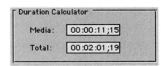

Figure 7-10

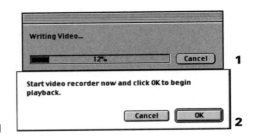

Figure 7-11

STEP 10

When the render process is completed, you will be greeted by a message box reading, "Start the video recorder now and click OK to begin playback." (Figure 7-11) This is your cue to manually begin recording on the deck or camera. Let the tape record for five seconds of pre-roll, then click OK.

The video will start feeding down the Firewire tube, showing you the Leader, your product and finally the Trailer black footage. Near the end of the Trailer black, press "Stop" on your deck, and you will have completed your first mastering-to-tape process.

Remember that you can stop the process at any time if you realize that you have made a mistake. Pressing the Escape key always aborts Capture or Print to Video functions immediately. You may also notice that the computer screen went black when the actual Print to Video began. That is because we specified it should do so in the View External tab of the Audio Video Settings, which was discussed early in the book.

If you want to have the video playback on the Desktop as well as stream out to Firewire during Print to Video, go back to the View External tab in the Audio/Video Settings and enable Mirror on Desktop during Recording for the View During Recording Using field. If you enable this, be warned that producing full-frame full-motion video in both places can put an enormous strain on the processor and you may encounter Dropped Frames warnings, particularly with older Macintoshes. The performance hit you take may not be worth it; especially since you should be more concerned with watching the playback on the deck or camera's video monitor as it records rather than the low resolution computer screen version.

Out to Firewire part 2: Edit to Tape

What is Edit to Tape?

Edit to Tape performs the same essential function as the other two methods of getting your project out to DV tape. It differs only in the precision and automation of that process. The most significant difference is that Edit to Tape uses Device Control and Timecode to automate the process. In many ways, this follows the same rules as Logging and Batch Capturing from DV tape through Firewire did. This means that some configurations will not have the option of Edit to Tape and will have to rely on Print to Video.

If your editing system restricts you to Capture Now because you do not have access to Device Control and Timecode, Edit to Tape will not function for you. Users with DV converter boxes and/or those using a DV deck or camera as a pass-through device will not be using Timecode and will not be able to access the Edit to Tape feature. On the other hand, there is at least one deck in common use that requires Edit to Tape because it does not have any recording buttons on its interface. Edit to Tape is also a function that may or may not function correctly on some DV devices, depending on the specific device's adherence to the Firewire protocol and Final Cut Pro's ability to control it accurately. Consult the Apple Final Cut Pro Web site for specific details regarding your DV device.

The reason that Edit to Tape is unique is that it allows you even more flexibility in recording your project to tape. The Print to Video and directly recording from the Timeline use an edit maneuver politely referred to as "crash recording." This means that you press Record on the deck and begin recording wherever the playhead of the deck happens to be at the time. Although you can perhaps get close to the frame on the tape you want to start recording from, it is impossible to start recording with any degree of frame accuracy.

As we discussed earlier in the book when talking about Logging and Timeline editing, the essence of editing is getting the appropriate material to follow each successive cut as you work. This is what establishes a linear timeline and tells a story. Crash recording does not let you select the frame you will begin recording from on the tape, and so recording directly from the Timeline or Print to Tape do not fully qualify as editing.

Edit to Tape, on the other hand, allows you to select an In point frame on your DV tape to begin recording to. This means that you can perform edits onto DV tape directly from Final Cut Pro rather than assembling the entire project in a sequence and printing its entirety out to tape. The term "assemble editing" is an old term in the editing industry. It means to edit material by tacking on each cut to the end of the last cut, thus "assembling" the edit.

It may seem somewhat crazy to go to the trouble of working in Final Cut Pro and then record pieces of a project onto DV tape one section at a time. But there are definitely instances where this can be beneficial. Let's say, for example, your project is a very long-form one that you are completing in 20-minute sections at a time. As you finish each sequence and print it to tape, you can clear off the previous section's media and free up drive space. But instead of having to recapture them all individually at the end, sequence them together into one piece, and print the whole thing out again as one continuous piece, you could simply assemble edit the pieces one at a time onto a final master as you complete them.

Edit to Tape can yield still other conveniences and benefits. It is likely that your master tapes will contain more than one project. To avoid wasting tape and potentially creating timecode breaks as a result of recording in areas that do not already have timecode, use Edit to Tape and you can select the exact frame that you begin recording from. No more guess work or hand cueing tapes using manual deck controls. The Edit to Tape window, which uses many of the same keyboard navigation conventions as the Log and Capture window lets you select your Assembly Edit In point with ease and accuracy, while allowing you the same Leader, Media, and Trailer flexibility that the Print to Tape function offered.

The only real catch is that there must be timecode on the DV tape prior to engaging the Edit to Tape tool. Like the linear tape Assemble Edit system, Edit to Tape requires that you tack the clip you are editing in onto the end of a previous cut. That does not mean that the previous cut need be video of any importance. It merely means that there needs to be timecode available for the Edit to Tape function to establish as an In point for the edit.

There need not even be continuous timecode after the In point established in the Edit to Tape window, although a few seconds of post-roll might be wise. The bonus of Assemble Editing is that it behaves according to the rules of what is called *three-point editing*. With three-point editing, you need to provide the system with both the In point and the Out point for the cut you are editing in, but you only need to include the In point of the tape you are recording to. When the system begins recording the edit to the tape, the Out point will be automatically figured by the length of the cut being edited in.

There's no reason to calculate the Out point when it will be recorded anyway at the length you established when you set the In and Out points for the incoming cut. You've been doing three-point editing all along in the previous editing exercises in this book, although there has never been a timecode issue because you weren't editing onto a tape. Any time you perform an edit without establishing an Out point for the edit, you are performing a three point-edit.

After you complete an assemble edit to tape using the Edit to Tape window, Final Cut Pro continues to record an extra number of video frames for you to tag the next assemble edit to. With DV equipment, this can be up to 45 extra frames of black. For this reason, editing in sequences to tape out of order is not a real possibility. Although the In point where you begin recording to tape would be accurate, the Out point would overwrite a small portion of any video following the edit. For this reason, you cannot perform what is referred to as Insert editing.

Firewire DV devices cannot perform Insert editing, and the option will be disabled in the Edit to Tape window. Other professional editing configurations using serial Device Control are not limited to assemble editing. If Insert editing techniques become an important function for your production needs, you may want to look into developing such a solution. Final Cut Pro can and does support frame accurate Insert editing, but not when using the Firewire DV protocol. Thus, you could integrate such a solution into your Final Cut Pro life, but it would require some equipment and configuration adaptations.

Although you don't need to have an entire tape of continuous timecode to use the Assemble Edit to Tape function, it is usually a good idea. The process of covering an entire tape with continuous timecode is called *striping* a tape. When you stripe a tape, you record timecode all the way through its length, accompanying that timecode with video black. With most formats, timecode must have video to relate to in order to record correctly. Once the timecode is striped onto the tape, you can rerecord whatever video you want where the black video is and the timecode numbers already on the tape will remain intact.

This process of striping a tape with timecode and video black is called Black and Coding, and is a function conveniently available in the Edit to Tape window as well.

Why is Black and Coding a good idea, when we have established that it is not technically necessary for Edit to Tape? There are a couple of reasons. Since the process of assemble editing simply picks up the timecode of the frame where you start recording from and continues it (called *regenerating timecode*), you need never fear timecode breaks again. There will always be continuous timecode on a tape that has been prestriped with Black and Code. Remember that even starting and stopping recording in the camera is a form of assemble editing, and so already having continuous timecode prestriped eliminates the possibility that you might accidentally start recording in an area of the tape that did not already have timecode, resulting in a timecode break.

Another benefit of Black and Coding your tapes is that it naturally performs a technique referred to as "unpacking your tapes." Videotape is a physical material stretched between two rollers in the cassette casing. On rare occasions, the tape can be rolled too tightly in the factory, resulting in uneven tension on the tape as it rolls from one side to the other. This can create havoc on the recording process. Remember that the servo lock function in a deck performs to keep the tape moving at an exact speed. Uneven tension in the tape stock can cause this speed to alter, resulting in any number of problems.

This problem can nearly always be avoided simply by winding the tape all the way to the end and then winding it back, a process referred to as unpacking your tape. The bonus of Black and Coding is that it naturally performs this function while also striping the tape with continuous timecode. Although Black and Code must be performed in real time, meaning that Blacking and Coding an hour-long tape takes an hour, it may save you the misery one day of finishing a complex shoot only to find that your tapes are hopelessly artifacted and have numerous timecode breaks.

Setting up Edit to Tape

STEP 11

Go to the Tools drop-down menu and select Edit to Tape (Figure 7-12). You will find that opening the Edit to Tape window has the same effect as opening the Log and Capture window; when you do so, the Edit to Tape window takes control of the Firewire connection. As with the Log and Capture window, while the Edit to Tape window is open, you cannot preview sequences or clips out to Firewire. If you want to edit, you have to close the window. Both Log and Capture and Edit to Tape assume full priority over the Firewire connection when enabled, locking out all other windows from the Firewire pipe.

You will see three tabs at the top of the Edit to Tape window: Video, Mastering Settings, and Device Settings. Now that you've progressed through the bulk of the application, the conventions these three tabs utilize should be instantly familiar.

Figure 7-12

The Video tab

In the initial Video tab, you will see the predictable Timecode fields in the top left- and right-hand corners of the tab. These represent, on the left, the duration of the proposed edit, and on the right, the current frame of the DV tape.

  Prior to inserting a DV tape, the message at the bottom of the tab will read "Not Threaded," just like the Log and Capture window (Figure 7-13). Once your tape is loaded up, the message will change to "VTR OK," indicating that Final Cut Pro is communicating with the deck and recognizes a DV tape is present. The Timecode field in the upper right hand corner will show the frame number that the DV device's playhead is parked on the tape, although you may have to hit Play and then Pause to get the frame number to report in the box.

STEP 12

Pop a DV tape in your DV deck or camera that contains footage and therefore timecode to assemble edit to. (For this set of exercises, employ a DV test tape that you don't mind

Figure 7-13

Getting Your Project out of Final Cut Pro

recording to. Since Edit to Tape involves an automated recording process, you may accidentally record over footage if you are not careful.)

Setting the Assemble Edit In and Out Point

Here, as with the Log and Capture window, the J-K-L and I and O keyboard shortcut conventions function. Use the J-K-L keys to navigate around your tape. When you find a likely In point frame to begin recording your sequence or clip material, hit the I key (Figure 7-14). In the bottom left hand corner of the tab, the In point Timecode field will update, with the frame you selected. As we said, Firewire DV users will be unable to perform Insert editing, so adding the Out point to the tape is not necessary.

The Mastering/Editing bar

At the top and center of the tab are two items of importance. The first is a drop-down bar. Initially, the bar should read Mastering (Figure 7-15). Clicking on it will reveal the other option on the bar, Editing. The difference between the two has a lot to do with the difference between Assemble and Insert editing, although you can use the Editing option for either one. The key is that each setting is dedicated to a different sort of recording to tape that depends on the job you are doing.

When the bar is set to Mastering, it is assumed that you are performing the same function that you were with Print to Video, namely that you are recording an entire project, complete with Leader and Trailer materials. The only difference will be that instead of hitting record on the deck, you will select an In point on the tape and Final Cut Pro will begin the record process automatically for you on that frame. With Mastering selected, the Mastering Settings tab, which is exactly the same as the one in the Print to Video window, is enabled, allowing you to customize the Leader and Trailer materials. Like Print to Video, Mastering will record the entire project or a range of In to Out as specified in the Mastering Settings tab.

When the bar is set to Editing, the assumption is that you are performing specific edits of clips or sequences to your tape. Since in many cases this would involve Insert editing into the middle of a project, Mastering Settings are disabled, the reasoning being that you wouldn't want slates popping up between cuts in your finished project. In addition, the Editing option only records out the single clip or sequence selected at the time and depends on three-point editing for a choice of Insert or Assemble edits performed. With Editing chosen on the bar, it is possible to provide Final Cut Pro with an Out point as well as an In point, which cannot be done with Mastering.

Figure 7-14

Figure 7-15

Of course, with a Firewire DV device, the Insert edit could not be performed anyway. So choosing and entering an Out point is liable to result in an area of discontinuous video over the timecode already present on the tape, since Insert editing does not overwrite timecode and inserts only video or audio tracks. In other words, unless your editing configuration is capable of performing Insert editing to tape, do not select an Out point. If you wish to Assemble Edit using the Editing function, prestripe the tape with timecode, then select only an In point. Better yet, use only the Mastering option, disabling the Leader and Trailer options if you are tagging a cut with frame-accurate Assemble Editing.

Black and Code

The feature next to the Mastering/Editing bar is the Black and Code button (Figure 7-16). As mentioned earlier, this feature allows you to conveniently automate the process of pre-striping continuous timecode across your entire tape. Simply pop a blank tape (or any tape you want to erase and stripe with continuous timecode) in the DV device, click the button, choose a sequence preset from the same list in the Audio/Video Settings you use for projects, and then click OK. Final Cut Pro will rewind the tape to the beginning and begin striping the tape, recording video black and timecode.

You can always stop the process of Black and Coding at any time by hitting the Escape key to abort. This is not a very complicated action. It is only an example of Final Cut Pro's inclusion of one-click automation for a function that we have been performing for years by prerecording tapes with the lens cap on the camera. There is no difference in the results. Timecode is timecode and video black is video black.

Mastering Settings tab

The second tab, the Mastering Settings, is functionally the same tab as in the Print to Video window. You have the same options available there, including the Duration Calculator. The only difference is that the process of Edit to Tape does not allow preloading the selected edit clip or sequence into the window, so the Media Timecode field will remain zeroed out, while the Total field will reflect only the total amount of Leader and Trailer material added to the Assemble edit about to occur. As mentioned before, Mastering Settings are called into play only in the recording process when the Video tab shows Mastering in the drop-down tab.

Figure 7-16

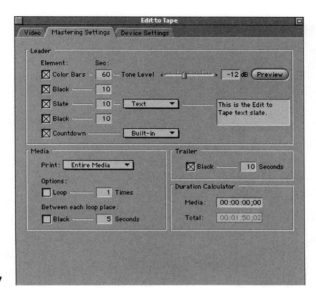

Figure 7-17

STEP 13

In the Mastering Settings Leader section, choose 60 seconds for color bars and tone (Figure 7-17).

STEP 14

Choose 10 seconds of black.

STEP 15

Choose 10 seconds for a slate. Switch the Slate option bar to Text and type in "This is the Edit to Tape text slate."

STEP 16

Set 10 seconds for black following the slate.

STEP 17

Enable the Countdown and choose Built-in from the Countdown option bar.

STEP 18

In the Media box, select Entire Media and disable Loop and Black Between Each Loop.

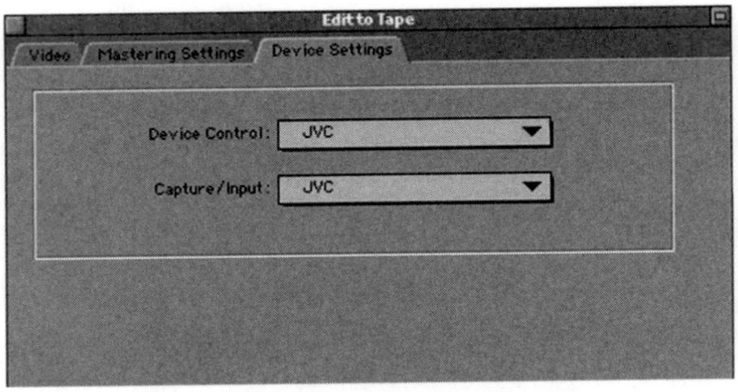

Figure 7-18

STEP 19

Then select 10 seconds of Black for the Trailer section. The Duration Calculator should show the Media Timecode value as zero and the Total Timecode value as 00:01:50:02. (Two extra frames are included into the Total record length to accommodate for NTSC Drop Frame timecode. PAL users will register 00:01:50:00.)

Device Settings

The last tab in the Edit to Tape window is the Device Settings (Figure 7-18). Just as in the Log and Capture window, the Device Control and the Capture/Input drop-down bars allow you to change the device presets without leaving the Edit to Tape window. You should set them exactly as you have them for capture unless you are using different equipment for the Edit to Tape process. If you have them set to disable Device Control because you have been working without a DV device while editing, you may need to change them back to get deck communication active again.

STEP 20

Return to the Video tab. Navigate through your destination DV tape to locate the In point you want to begin recording to, then click the I key to establish the In point. Make sure that there is adequate pre-roll time prior to the In point, since the recording device will need to attain servo lock, just as if you were capturing from the tape instead of recording to it.

Committing the Edit to Tape

To begin the actual Edit to Tape, simply drag and drop the item you want to record into the Video tab video window (Figure 7-19). The Edit process is very similar to the drag-and-drop edit process using the Canvas window.

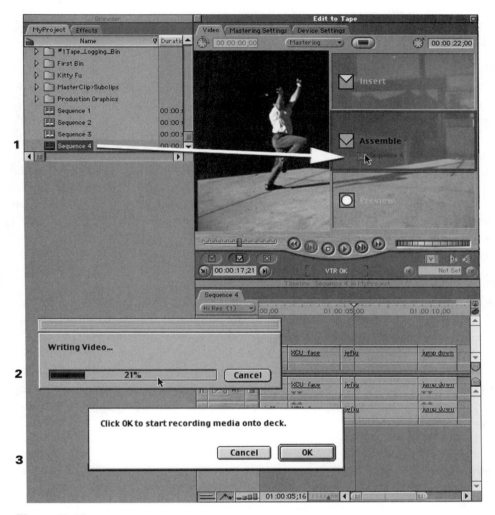

Figure 7-19

STEP 21

If you want an entire sequence or an In to Out point of a sequence, click and drag the sequence icon from the Browser window Project tab to the Edit to Tape window. If you want a single clip, you can drag it from either the Browser window Project tab or the sequence Timeline into the Edit to Tape window.

Bear in mind that if you do so and you want to use specific In and Out points from the clip, you need to select the appropriate clip location. Remember that master clips in the Browser window may not contain the same In and Out points as the version of the same clip in the sequence. Also bear in mind that you may need to set In and Out points in the sequence itself if you have selected In to Out in the Mastering Settings, since setting it that way without In and Out points will result in Entire Media.

STEP 22

When you drag the item over to the Edit to Tape window, a set of overlays will appear, similar to those seen when dragging and dropping edits into the Canvas window. The three overlays that appear are Insert, Assemble, and Preview. Since you are using a Firewire DV device, Insert editing is not technically possible, so that overlay will be grayed out and unavailable.

At the bottom of the three overlays is a Preview overlay, which applies only to the unavailable Insert edit. Since Insert editing actually records over material on the tape as part of its functionality, Final Cut Pro offers an idiot check function to make sure that you are editing in what you want. This Preview is unnecessary for Assemble edits, because if you make a mistake in recording, you can always simply rerecord the cut without fear of having harmed any cuts occurring after it.

STEP 23

In between Insert and Preview is the Assemble box. Drop the item on the Assemble box and let go. Final Cut Pro will immediately begin rendering up the Leader and Trailer material we specified in the Mastering Settings.

STEP 24

When this is complete, you will receive a dialog box telling you to press OK to begin the recording process. Do so, and the deck will leap into action, pre-rolling to the edit point, and your item will automatically master out to tape.

Exporting

You don't have to restrict yourself to printing to tape. Although DV is a fantastic way to make video for full-frame, full-screen viewing, you can also use the Export function to produce QuickTime movie files for use in Web, CD-ROM, and many other functions. To explore all the possible variations of export is far beyond the scope of this book, but stepping through a few of the options should get you through the initial stage and give you the tools for further experimentation.

We have already explored the Export process briefly in the preparation of audio CD files for use in Final Cut Pro. There, the idea was to use the Export process to convert the 44.1 KHz sample rate of the CD audio file so as to match the 48 KHz sample rate of our sequence and ensure trouble-free integration into the project. Our process here will involve some of the same steps and the same reasoning, although in pursuit of a different result.

There are three primary purposes for exporting media from Final Cut Pro: (1) to create media files optimized for such specific purposes as the Web or CD-ROM, (2) to create media files that are unchanged from the source for external manipulation in another digital application, and (3) to create metadata files that serve to organize or document the media you are working with in a project, such as creating a Batch List for a paper log. Each of these purposes requires a different strategy, and, predictably, Final Cut Pro has developed a convenient workflow for dealing with each option.

Export for distribution: Optimizing the QuickTime movie

The first option we will discuss is exporting to optimize a media file for use in specialized distribution. The most common use for this is to create multimedia files for use in Web and CD-ROM or DVD distribution. In this instance, we are actually creating a new media file that contains the content of the media we are exporting from Final Cut Pro but uses different parameters to allow it to function elsewhere.

There is much you can change about a file while exporting it. Depending on the codec (Apple DV, Cinepak, Animation, MPEG, etc.), the frame size, aspect ratio, and frame rate, you can change the media to suit your needs specifically. Which option you choose when exporting will determine what you can change about the media as you export it as a file. But to export the file as we want it, we need to first specify the file format, which will determine what media will be exported into the new file.

Choosing the File format

STEP 25

Select a clip that you want to export (Figure 7-20). Make sure the Viewer is active and the clip is loaded into it. For this exercise, we are exporting a single clip, although you could also set In and Out points on the sequence Timeline and export as an In to Out range.

STEP 26

Next go to the File drop-down menu and select Export. When the submenu pops up, you will see a list of Export options.

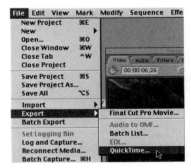

Figure 7-20

STEP 27

For this first exercise, select QuickTime. An Export dialog box will pop up asking you to specify the settings for the QuickTime Movie.

Remember that QuickTime is an architecture, not a file type. Just as a skyscraper and an outhouse are both buildings, there is a world of difference between the two. And there are just as many variations between the different sorts of QuickTime export possibilities.

STEP 28

Underneath the File Location box and the Export As field, locate the Format option drop-down bar (Figure 7-21).

In the audio CD exercise, we selected AIFF from this bar in order to export the audio tracks we were working with. We could also select AIFF for the current clip, but if we did so, we would generate only the audio files of the clip, because the AIFF format supports only audio.

Other options on the list have their specific purposes, and independently purchased codecs can add even more flexibility, such as the ability to export MPEGs suitable for DVD authoring. The format we want to work with is QuickTime Movie. The QuickTime Movie can support a video track, stereo audio tracks, and even timecode, if the chosen codec supports it. Remember that the media files we captured from DV tape and stored on our Scratch Disk are also QuickTime Movie files that use the Apple DV codec.

STEP 29

After selecting QuickTime Movie from the Format list, look for the Options button in the dialog box. Underneath the Format bar is a Use bar with Export presets for the Format you've chosen, but it is rarely a good idea to settle for presets where you have the option of customizing the file manually. The Option button will take you to a fresh dialog box where you can refine the export to exactly what you need.

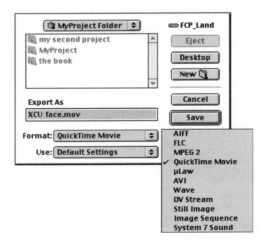

Figure 7-21

Choosing and setting the compression

The main issue at stake for creating media files for Web and multimedia functions is to keep the data rate low. We know that the data rate of the Firewire DV codec we are using in Final Cut Pro is a steady 3.6 MB/sec. Unfortunately for users accessing CD-ROMs, the data rates shouldn't exceed a maximum of 128 KB/sec, and for most low-bandwidth Web surfers, the data rate should be even lower. This is far below the current data rate of our DV media, so we need to optimize the file as we export it, simultaneously dropping the data rate and creating a separate file for use elsewhere, whether on the Web or CD-ROM.

From our discussions about codecs early in the book, we know that there are several issues involved in lowering the data rate of a media file. The initial issue is the codec we choose to utilize for the video. We know that the codec determines not only how much data is included about each frame of video, but other factors, such as the size of the frame and its aspect ratio.

For instance, the DV codec we use for Firewire DV requires that the frame size be 720x480 pixels for NTSC and 720x576 pixels for PAL. It also requires that the frame rate be a standard 29.97 and 25 frames per second respectively. This means that the data rate cannot dip below the 3.6 MB/second and still account for all those pixels.

Other codecs are more flexible. Most software codecs will support different frame sizes and frames rates, and the data rates will adjust lower or higher accordingly. There are a host of other factors for each individual codec as to how it trades off image quality and size for data rate, so there is always a particular codec that is best suited for the job of getting the data rate lower. In addition, it must be remembered that the same codec must live on the machine that exports the file and the machine expected to play the video file back. So selecting the proper codec must involve a little soul-searching as to what the file needs to do and where it needs to go.

For this exercise, we will use an extremely common multimedia codec called Cinepak. This codec is cross-platform, meaning that it is equally functional on both PC and Macintosh systems. It is fairly flexible at offering different frame sizes and frame rates, and it does deliver very small data rates, enabling widespread use in Web and multimedia applications.

STEP 30

After hitting the Option button, you will be presented with a dialog box that lets you optimize each facet of the media file. At the top is the Video section (Figure 7-22).

STEP 31

The default compressor, or codec, is Video (Figure 7-23). Since we want to use Cinepak, click the Settings button. You will be presented with the Compressor dialog box. Click on the Compressor drop-down bar and select Cinepak from the list. When you do, the options for the codec will appear below the bar.

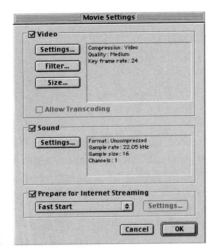

Figure 7-22

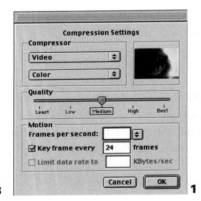

Figure 7-23

STEP 32

First, since we want to shrink the data rate enormously, change the Color Range bar from Millions of Colors to 256 Colors (Figure 7-24). Since the file will not have to include so many different variations of color in each frame, the data rate will shrivel.

STEP 33

Next, either enter 15 in the Frame Rate field or select 15 from the Frame Rate selection bar. Although this means the codec will throw away many frames from the video file, 15 frames per second is still enough to give a pretty good illusion of the moving image, and it will further shrink the size of the exported file radically.

The other two settings involve the way that the codec will operate on the video frames and determine which data is relevant and which can be thrown away. The two methods are

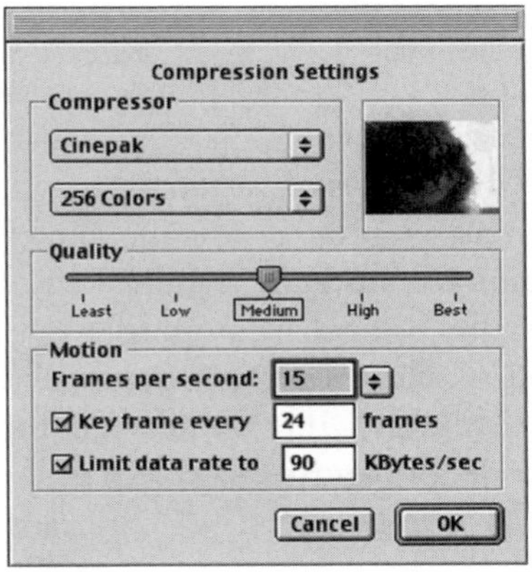

Figure 7-24

referred to as *spatial* and *temporal compression*. Some codecs use one method, some use the other, and still others use both as a means of eliminating excess data from video frames.

Quality or Spatial Compression

The Quality slider represents the amount of spatial compression you are applying to the file. Spatial compression works intuitively, looking at each frame of video and throwing out a certain amount of detail in order to achieve a specific amount of data in the frame. The Quality slider makes no comparisons between individual frames for important or recurrent detail, it just evenly lowers the detail quality for each frame a certain amount based on how low you set the slider.

STEP 34
Set the Quality slider at 75 percent–High (Figure 7-25).

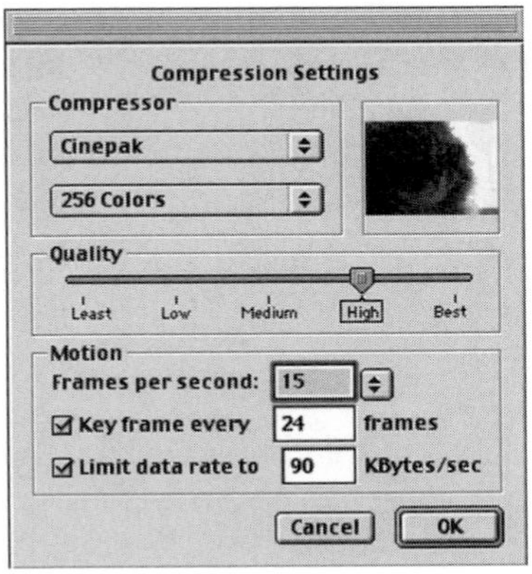

Figure 7-25

Keyframe Every or Temporal Compression

The next setting is called Keyframe Every, and represents the temporal compression technique. This keyframing is not to be confused with the interpolation we worked with in the Motion tab window of a clip, although they do approach their separate tasks from the same universal idea of the amount of change between two frames. Temporal compression sets a keyframe at an interval established by the value you insert in the box. Each keyframe frame is compressed only using your setting for Quality or Spatial compression.

However, all the frames occurring between two keyframes, called *interframes*, are compared to the keyframe. Any differences between the interframes and the keyframe are retained, whereas any similar pixels are discarded. If your video contains areas of the frame that do not change at all, such as static backgrounds around talking heads, much of that data can be thrown out without affecting the image quality of the interframe at all. The result is that the data rate shrinks dramatically as the amount of data in each interframe is lowered.

Of course this compression type is effective only if you do not have a lot of changing imagery in the video footage for the keyframes to keep track of. If you establish keyframes every 90 frames, this means that changes in footage over roughly three seconds will not be tracked. If your keyframes are too far apart, a lot of important movement could be discarded as the interframes get further away from the keyframe. As the distance between keyframes increases, the quality drops.

The option of entering a lower number in the keyframe Every and decreasing the distance between the keyframes and making them closer together results in higher quality interframes. Unfortunately. it also defeats the purpose of the keyframing temporal compression, since the more keyframes you have, the fewer the frames that are processed as interframes and compressed.

The key, of course, is to shoot and select your video with an eye towards the compression issues you'll encounter later. If you are going to compress video for the Web, make sure to get rock-steady shots and don't allow excessive camera or subject movement. Other codecs such as Sorenson offer different mathematical procedures for the compression, but the parameters for codecs are largely limited to spatial and temporal compression.

STEP 35

Leave the Keyframe Every setting at the default value of 24 frames. The final setting in the Compressor box is Limit Data Rate To. This lets you set a ceiling for the data rates, just in case your compression settings are not sufficient to bring the overall data rate down.

STEP 36

Leave the data rate ceiling at 90 KB/second. Click OK and return to the Options box.

Other Factors; Frame size, audio, etc.

Having set the compression, the next setting to adjust is Size. Because the video file is likely to be viewed on a computer screen, it is not necessary to stick religiously to the large 720x480 frame size it currently sports. Cutting the size of the video frames will cut the data rate enormously.

STEP 37

Click the Size button, and you will enter a dialog box labeled Export Size Settings (Figure 7-26). The two checkbox options below are for Use Current Size and Use Custom Size. Click the dot next to Use Custom Size and the Width and Height fields will appear.

You can resize according to whim, which may result in a stretching of the video frame. To choose a new frame size, simply divide the starting dimension in halves or quarters or however small you wish to shrink the frame.

Don't forget that the pixel shape, square or non-square, is determined by the codec in use. The DV codec your footage is originating from used non-square pixels, resulting in a 720x480 or 720x576 frame size. But the codecs you are likely to use with Web and CD-ROM exports will be in square pixels. Therefore the half frame size for this export from DV to Cinepak would be 720x480 to 320x240, that being half the frame size for the square pixel video file the Cinepak compressor generates.

STEP 38

For the Size setting, select 320x240 pixels, then hit OK to return to the Options box. The next section to set is Audio, We do want to include audio with the file, but we definitely need to change the settings from the default uncompressed. Although audio data rates are smaller than video data rates, they can still be far higher than necessary for such uses as Web and CD-ROM, where every byte counts.

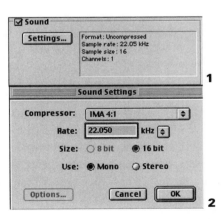

Figure 7-26

Figure 7-27

STEP 39

Click the Settings button in the Sound section of the Options dialog box to enter the Audio Compressor dialog box (Figure 7-27).

STEP 40

For the Compressor bar, select the IMA 4:1. This is only one of a number of excellent audio compressors on the market as well as the ones freely supplied with the QuickTime compression system. Each audio compressor has a specific targeted function, such as optimizing data rate for human voices, midrange music, and many other specific issues. Most audio compressors are expressly for use with Web and multimedia. Because audio data rates are generally so low compared to full-frame, full-motion video editing, we never compress audio there. But for streaming video and other Web and multimedia work, compressing audio is just as integral a process as compressing video.

STEP 41

Setting the sample rate lower to 22.050 KHz will also contribute to a lower data rate. Although 22.05 KHz would be considered unacceptable for inclusion in full-frame, full-motion video, it is more than enough for low-bandwidth options.

STEP 42

With the IMA 4:1 codec, the bit rate is fixed at 16, while you can choose between a Stereo or Mono file. Select Mono and hit OK to return to the Options box.

The last option in the box is for Fast Start. Fast Start refers to file coding that can be inserted in the exported QuickTime file to facilitate Internet streaming functionality. If your QuickTime Movies are to be sent to a server for streaming download, contact the administrator of the server and find out how the Fast Start settings need to be optimized for use. If you have no plans for streaming the video file, you can safely disable the feature.

STEP 43

Hit OK to return to the QuickTime Movie box, check your settings (Figure 7-28), then click OK again to return to the Export dialog box. Now all that remains is to navigate the Save location to a suitable place. This will depend on your Exported file.

If the QuickTime file is of a high data rate and is comparable in size and compression to the DV files you edit with in Final Cut Pro, you'd best save the file to a dedicated media drive just as if it were a Captured clip. If on the other hand the file is a very low data rate clip for the Web, saving it to the Desktop and playing from there will do no harm to your system. Since our exported QuickTime Movie will be heavily compressed with a very low data rate, we will save it to the Desktop.

STEP 44

Navigate the Save location to the Desktop. Name the file, using the .mov suffix in case any of your PC pals want to access it, and then click Save (Figure 7-29).

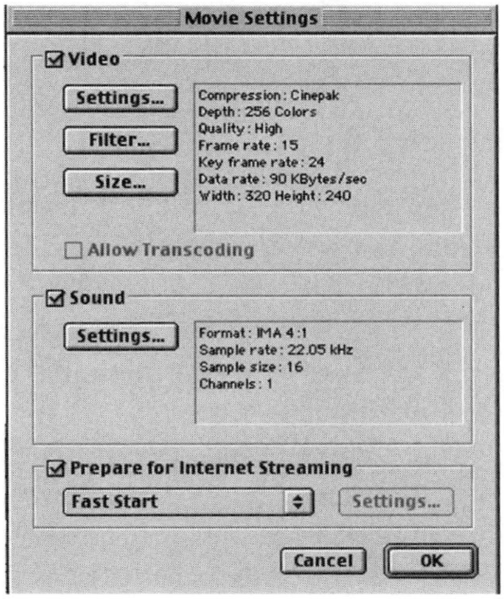

Figure 7-28

Final Cut Pro will begin rendering out your QuickTime file. Depending on the amount of processing and the codec you use, the render can be very short or very long.

STEP 45

Once the render process is complete, you can double-click the exported file on the Desktop to view it in the QuickTime Movie Player application already installed in your Macintosh with the Operating System.

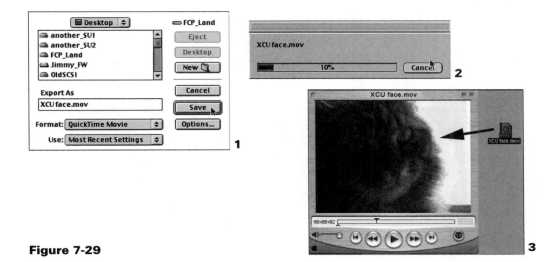

Figure 7-29

Export for loss-less transfer: The Final Cut Pro reference movie option

What is a reference movie and why use it?

Although the previously described Export is much more common, there is another export method that is specially designed for post-production purposes. Post-production is a wide-ranging term that generally refers to anything done to a film or video after it as been shot. The issue for post-production is definitely not reducing the data rate or quality of the editing materials. Quite the opposite; there is a necessity to retain the highest level of quality and fidelity to the film or video while manipulating it.

It often is also necessary to pass sections of a clip or sequence out to be manipulated by another application. There are numerous applications, such as Adobe AfterEffects and Pinnacle System's CommotionPro, which use specialized toolsets to further develop the same matting and keyframing techniques we worked with earlier in the book. Although Final Cut Pro has a great set of compositing tools, professional post-production often requires styling that only specialized applications can deliver.

Other applications like Terran Interactive Cleaner EZ are more streamlined and flexible at producing compressed QuickTime Movies for the Web and multimedia use. Although you could use the previous section's method for exporting compressed QuickTimes, many content producers opt for the different feature sets other applications offer for this activity. These are only a few of the many possible applications that can accept video files for further processing and manipulation.

The material exported to these application needs to maintain the highest fidelity possible. Since the new application will be processing and manipulating the video files, it needs to receive video that is as close to the original as possible. And why shouldn't we be able to fulfill that requirement? We are working with digital media after all, and as we proposed very early in this book, digital media does not lose generations when copied. It should be no problem at all to maintain the highest fidelity to the original when exporting out a copy for either preview or manipulation.

Unfortunately we can't have our cake and eat it too. There is a complication in the premise that digital media cannot lose a generation when being copied, and it has to do with the fact that we are working with codecs, which selectively discard image information. Anytime Final Cut Pro renders media, it applies the codec currently in use in your sequence presets. That means that when you render a section of a sequence, it is compressed using the codec, which in our case would be the Apple DV codec. Compression discards image data, and discarding image data lowers image integrity.

Before the reader begins to feel cheated or alarmed, consider that the second generation of compression with the Apple DV codec yields results of extremely high quality. Any rendering of effects that you complete on the sequence Timeline is liable to blend seamlessly with material that has never been rendered and recompressed, and it really takes something of an expert to track the image degradation. Although the results begin to become apparent in the third, fourth, and fifth renders, the first is usually visually intact. Also remember that each successive render of a section within a sequence actually just re-renders the section, replacing the previous set of render files, so each render there is considered "first-generation" render. So rendering inside a Final Cut Pro sequence, except in extreme cases of images that are very susceptible to compression artifacting, produces visually seamless results.

The problem we are addressing here results from the fact that the export we are proposing in this section is of material that needs to remain free of additional compression. If it is of the utmost importance that we work with the intact original image, we must make sure that the image is not compressed upon export. The media will be exported, manipulated, and rendered in another application and then reimported into Final Cut Pro for inclusion in the project.

If we use the previously described QuickTime export process, our clip will end up having passed through two sets of compression. The first will be compression applied as the QuickTime file is saved using a codec. Then it will likely be compressed a second time as it is rendered in the other application to which we exported the movie. As has been said, any time Final Cut Pro creates media that does not already exist, it must render that media, and rendering with a codec requires compression.

There are two ways to overcome this problem. The first is fairly obvious; if you don't want to involve compression in the export, use a lossless codec! In the first chapter of this book, we described three codecs: Animation, Cinepak, and the Apple DV codec. We have seen that Cinepak and the DV codec are both lossy codecs, in that they lower the data rate of a video file by discarding image data to a greater or lesser degree. However we stated that the Animation codec was a lossless codec. It is not appropriate for Firewire DV editing, because of the enormous data rates it incurs, but its image quality is highest, since it does not discard any data.

So one way around this issue would be to export our file using the Animation codec. Then we would manipulate the exported file in whichever application we want, render the file using the Apple DV codec so that it will return to Final Cut Pro with the native compression of our sequence. We will only have lost one generation, which we have ruled as acceptable. The process is seamless; it gives us the image results we want, plus it makes all of our media available outside of Final Cut Pro, where we can use nearly any application to produce the effects we seek.

Now, this process using the Animation codec works, and it has been a standard method for years in many editing applications. When the image quality going out must be of the highest quality, zero compression has been the rule of thumb. But there are unnecessary inconveniences inherent in this process. First of all, consider that the Animation codec, since it applies no compression to the file, explodes the size of the resulting file. If we export a 5 second clip using the Animation codec, the resulting file may end up nearly 150 megabytes in size. That's pretty enormous, considering that the original clip compressed with the DV codec would be only around 18 megabytes in size.

Once you begin to export even longer clips, you rapidly come to understand why using no compression is problematic. If you are using a Firewire DV editing station, at least part of your original intention was to keep the overall cost of your system down as low as possible. Your 60 GB ATA drives will not hold much material if you are filling them up with media exported as Animation codec files.

Something even more grating on our nerves is the fact that the media we want to export without compression already exists in our Scratch Disk folder. It is already digital, and it is already compressed. Why recompress digital media, when it is already sitting there on the media drive? Creating more copies of the same media on our media drives is a waste of space, particularly if the redundant copies are generated by a codec that explodes the data rate tenfold.

Beyond this, we have already seen that rendering takes time. When you render, Final Cut Pro has to analyze and process each frame. No matter how fast your Macintosh's processor is, that takes time. Decompressing the DV codec source material, then applying the Animation codec, and saving it to disk is a process that will have you looking at the progress bar far longer than you would if you could simply access the relevant parts of the media file already sitting in your Scratch Disk folder.

What if you could simply produce the equivalent of a clip that can function outside of Final Cut Pro? A clip in Final Cut Pro is simply a reference or index back to the original media file back in your Scratch Disk folder. When you work with it in the Viewer or Canvas, you are simply telling Final Cut Pro how you want to work with the media file being indexed by the clip, rather than working with the media itself. This is what makes nonlinear editing possible and what sets digital video editing apart from linear analog tape editing.

Final Cut Pro includes an export feature called Reference Movie, which is a type of exported Final Cut Pro Movie. The reference movie file is treated like a normal QuickTime file by any application that can access such files. But instead of being a stand-alone file that has been processed by a codec, the reference movie acts just like a Final Cut Pro clip and refers back to the media files in Final Cut Pro's Scratch Disk folders. And because the exported reference movie may be a range of clips exported from a sequence, it even accesses any render files that might have been created if you rendered an effect on the sequence Timeline.

Since a stand-alone movie file is not being created, no rendering is required, and the reference file export is nearly instantaneous. And because the file is merely a reference to the media in the Scratch Disk folder, the reference movie file actually takes up very little drive space, in most circumstances, less than a single MB. Thus the reference movie export function allows you to work with material that has not been recompressed with a codec, does not delay you with excessive render times, and takes up nearly zero drive space. This is clearly the best method to use if you want to take your media out to another application.

That said, there are a few tweaks and special rules that apply. Its important to follow the correct process in generating the reference movie to make sure that you are benefiting from all that the feature has to offer. To create the Final Cut Pro reference movie, first create a range of frames that you want to export in a sequence Timeline. Make sure that the range of frames does not require rendering. If it does, simply render it to make it available. If you export material as a reference movie that requires rendering, Final Cut Pro will render the material prior to creating the reference movie and embed the render files in the reference movie itself, resulting in a render process that takes time but does not actually apply to the sequence material you are exporting. In addition, the size of the reference movie itself grows, as the render files are included rather than referenced. Always render your sequence before Exporting a Final Cut Pro Reference Movie.

Create the range of frames for export

This first frame need not be the beginning of the sequence or any particular frame of a clip; it only needs to be the first frame that you think you will want to use.

STEP 46

To create the range of frames you want to export, position the playhead in the sequence Timeline at the first frame of the range that you want to export. Hit the I key to set the In

Figure 7-30

point for the range, which will place an In point indicator above the sequence (Figure 7-30).

STEP 47

Move further down the sequence and select the last frame you want to include in the Export. Hit the O key to establish the Out point of the range.

You will see that there is a lighter gray in the time marker area of the sequence between the In and Out points. This indicates that the area is a range to be used in the coming export action.

The export Final Cut Pro Movie dialog box

STEP 48

Next go to the File drop-down menu, select Export, and in the submenu, choose the top option, Final Cut Pro Movie (Figure 7-31).

When you do so, you will be presented with the Final Cut Pro Movie Export dialog box. You must configure this box correctly to achieve the goal of a true reference movie, since Final Cut Pro will allow you great flexibility in generating an exported movie file.

At the top of the dialog box, you will find the familiar Save location tools. Because the reference movie is not a true media file, it is less imperative that you save it to your media drive, but you generally should do so anyway, for reasons of organization.

STEP 49

After you navigate to the appropriate Save location, name the file, making sure to specify in the file name that it is a reference movie (Figure 7-32).

Its easy to confuse a reference movie with the real thing. Although you wouldn't be throwing any real media away, you could easily screw up your entire workflow by tossing out a seemingly insignificant movie file that seems unrelated to your other work.

Figure 7-31

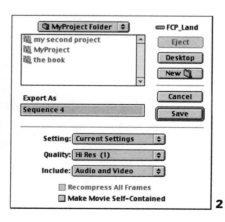

Figure 7-32

You also want to make sure you know to keep the reference files original media file intact. Remember that the reference movie file cannot operate without its indexed media file from the Scratch Disk, so keeping them well identified may help you to avoid throwing away seemingly unused media files that are in fact necessary for the reference movie.

STEP 50

Leave the Setting drop-down bar at Current Settings, since we don't want to change the codec being used, which would result in having to generate a new file. Also leave the Quality drop-down bar set at Hi-Res, since that is the quality we have been using and want to continue using.

The next setting, Include, bears a little discussion. A limitation of the reference movie is that only video files can be indexed; audio files cannot be indexed by a reference movie. If you leave the Include bar set at the default Video and Audio, the resulting reference movie will actually embed the audio tracks, just as it embeds render files that are not prerendered prior to export. Selecting Video Only eliminates this problem, as only the video files will be included in the reference movie.

If you need to include audio in the reference movie, do not hesitate to do so. Remember that audio data rates are very low, and the inclusion of the audio data in the index file will not increase the file size radically. If audio is necessary for use in timing effects while manipulating in another application, go ahead and include it. Just remember that you may want to store the reference movie on your media drive, since it now contains audio files that need to be accessed at the requisite high speeds of a true media file.

STEP 51

Set the Include bar at Video Only. Underneath the Include bar are two checkbox options that warrant special attention. The first to address is the Make Movie Self-Contained box. The default setting for this box is enabled. If this box is enabled, Final Cut Pro will create a stand-alone movie file for the export. But this is inimical to our pursuit here, because we want to create a reference movie that simply indexes the original file. Creating a stand-alone file would be duplicating media that already exists.

The checkbox above Make Movie Self-Contained mirrors this fact, because Recompress All Frames asks if you want to apply a codec to the generated file or to simply copy the files that already exist in the Scratch Disk. If you disable the Make Movie Self-Con-

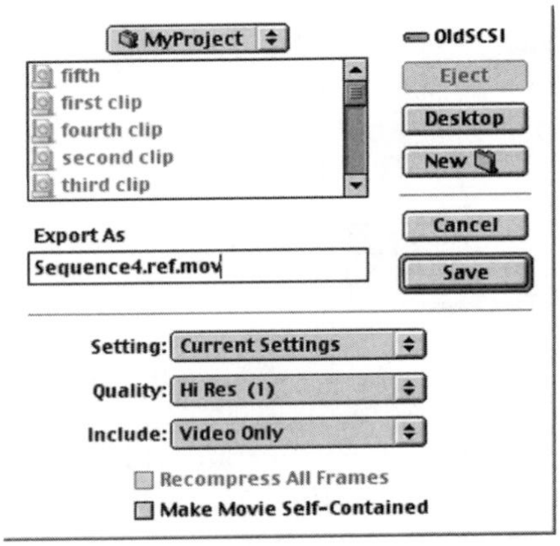

Figure 7-33

tained checkbox, the Recompress All Frames box becomes unavailable, because it is impossible to recompress with a codec when you are not actually processing video frames but merely referencing them.

STEP 52
Deselect Make Movie Self-Contained to create the reference movie. Click OK to save the file (Figure 7-33).

STEP 53
Unless you have chosen to include audio or are making a self-contained stand-alone Final Cut Pro movie, the process should be nearly instantaneous. Hide Final Cut Pro, and locate the saved Final Cut Pro Reference Movie. If you select the reference movie and hit Command-I, you will see that it is indeed quite small in size, probably somewhere in the neighborhood of a few hundred kilobytes (Figure 7-34).

The icon used for Final Cut Pro Movies is the Final Cut Pro slate clapper icon. If you double-click on the Final Cut Pro Movie icon, you will see that it opens the file up as a clip inside of Final Cut Pro instead of opening the QuickTime Movie Player application that the QuickTime movie file we exported earlier opened. This is because the file contains code establishing that it was created by Final Cut Pro. This information is used by the Macintosh OS to decide which application to open when the file is double-clicked. Have no fear; applications that can use QuickTime movie files will recognize the reference movie as such.

Exporting for metadata: Batch Lists and EDLs

As mentioned earlier, there is a third sort of export that has no direct effect on your media. This sort of export merely creates a word processing file that carries important information about the media itself. We call this sort of information *metadata*, because it is, strictly speaking, data about data. It lets us analyze and evaluate our project's resources instead of

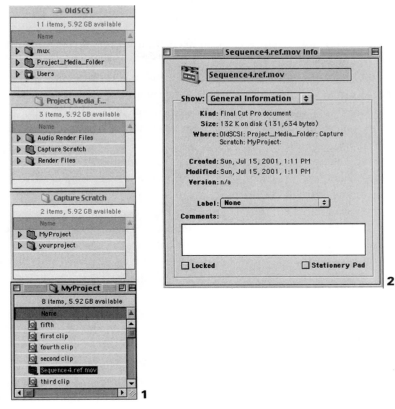

Figure 7-34

directly manipulating them. Aside from backed up project files, Batch Lists and EDLs are the best form of insurance and allow the most resilient system of project organization possible.

To understand what these two metadata types are, we need to think about what a project really is. Although we see the intuitive button-and-menu interface of Final Cut Pro, it is important to remember that a Final Cut Pro project is simply a coded document that describes relationships between media files on your hard drive. It organizes how these media files are played back, in what order, and even allows you to make changes in the way those files appear in playback through the use of effects.

Interface aside, we really do have to think about the Final Cut Pro project as a set of numbers, especially when we have access to the benefits of timecode. If we have captured our media using timecode in the Log and Capture process, then each clip has this timecode data embedded as part of its clip data. As a further consequence of this, each sequence that contains clips with timecode data is simply a visual representation of a list of timecode values that relate to each clip in the sequence. This list gives the order of the clips, the timecode value of the In and Out points of the included clips. and other information about the tracks of the sequence itself. Final Cut Pro can interpret this list as a timeline and display it visually in the interface you see and work with.

One reason it is important for you to keep archived versions of your project files is not out of fear of corrupt files or the desire to return to an earlier stage in the edit, as described earlier in the book. Using timecode values to log and capture your clips enables you to recapture the same clips should they lose contact with the media. Clips can lose their links for various reasons, and if the media is still present on your media drives, Final Cut Pro has decent tools for re-linking them in the Reconnect Media command located in the File drop-down menu.

But what if your media no longer exists on the media drive, if, for example, you accidentally erased the files or if you are re-creating the project long after its completion and removal from the system? Recapturing your media from the original tapes is as easy as selecting the sequence or clip in question and choosing Batch Capture from the File drop-down menu. A clip that has no associated media is referred to as an Offline clip, but your editing application still regards it as a functional set of timecode numbers. Final Cut Pro will take the clips in their Offline status, ask for the appropriate tapes based on the Reel Name information in the clip data and recapture the media it needs to return the clip to Online status.

The Batch List

The batch list is a variation on this concept. A batch list is simply an export of all the information about the clips in your project. Best of all, the information is exported in the format of a standard spreadsheet that can be opened using any common office spreadsheet software. If you don't own Microsoft Excel or any of the other commercially available applications, you can simply use Apple's own Appleworks word processing software that contains a simple spreadsheet function. If you can't find a copy of Appleworks, even SimpleText can be used, although caution must be taken to avoid screwing up the spreadsheet formatting. Spreadsheet columns are really just tab delimits in a regular text file that spreadsheet applications can interpret and display as columns. If you are not using a real spreadsheet database application, be careful not to accidentally change the format.

Take a quick glance at the Browser window.

STEP 54

Change the Browser Item's View mode to As List so that you are viewing the Browser window Project tab items in columns (Figure 7-35).

As you scroll the window to the right, you will see the enormous amount of information that is stored about each clip in the detail columns. Even if your clips are made Offline, this information remains intact. Each type of information is separated into columns that allow it to be analyzed and categorized according to your priorities.

Behind the scenes, this Browser window is a kind of tab-delimited document. The document uses the tab functioning of ordinary word processing files to separate the differ-

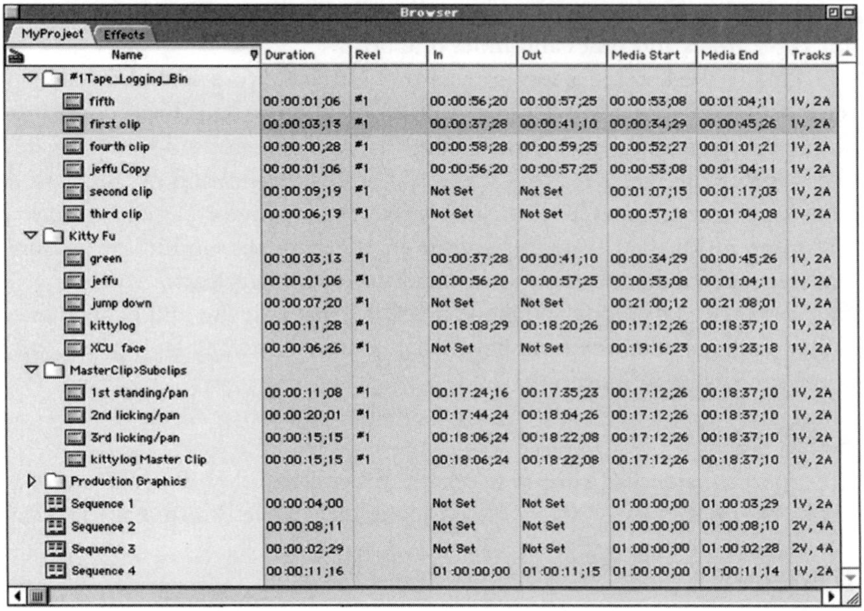

Name	Duration	Reel	In	Out	Media Start	Media End	Tracks
▽ 🗀 #1Tape_Logging_Bin							
fifth	00:00:01;06	#1	00:00:56;20	00:00:57;25	00:00:53;08	00:01:04;11	1V, 2A
first clip	00:00:03;13	#1	00:00:37;28	00:00:41;10	00:00:34;29	00:00:45;26	1V, 2A
fourth clip	00:00:00;28	#1	00:00:58;28	00:00:59;25	00:00:52;27	00:01:01;21	1V, 2A
jeffu Copy	00:00:01;06	#1	00:00:56;20	00:00:57;25	00:00:53;08	00:01:04;11	1V, 2A
second clip	00:00:09;19	#1	Not Set	Not Set	00:01:07;15	00:01:17;03	1V, 2A
third clip	00:00:06;19	#1	Not Set	Not Set	00:00:57;18	00:01:04;08	1V, 2A
▽ 🗀 Kitty Fu							
green	00:00:03;13	#1	00:00:37;28	00:00:41;10	00:00:34;29	00:00:45;26	1V, 2A
jeffu	00:00:01;06	#1	00:00:56;20	00:00:57;25	00:00:53;08	00:01:04;11	1V, 2A
jump down	00:00:07;20	#1	Not Set	Not Set	00:21:00;12	00:21:08;01	1V, 2A
kittylog	00:00:11;28	#1	00:18:08;29	00:18:20;26	00:17:12;26	00:18:37;10	1V, 2A
XCU face	00:00:06;26	#1	Not Set	Not Set	00:19:16;23	00:19:23;18	1V, 2A
▽ 🗀 MasterClip>Subclips							
1st standing/pan	00:00:11;08	#1	00:17:24;16	00:17:35;23	00:17:12;26	00:18:37;10	1V, 2A
2nd licking/pan	00:00:20;01	#1	00:17:44;24	00:18:04;26	00:17:12;26	00:18:37;10	1V, 2A
3rd licking/pan	00:00:15;15	#1	00:18:06;24	00:18:22;08	00:17:12;26	00:18:37;10	1V, 2A
kittylog Master Clip	00:00:15;15	#1	00:18:06;24	00:18:22;08	00:17:12;26	00:18:37;10	1V, 2A
▷ 🗀 Production Graphics							
Sequence 1	00:00:04;00		Not Set	Not Set	01:00:00;00	01:00:03;29	1V, 2A
Sequence 2	00:00:08;11		Not Set	Not Set	01:00:00;00	01:00:08;10	2V, 4A
Sequence 3	00:00:02;29		Not Set	Not Set	01:00:00;00	01:00:02;28	2V, 4A
Sequence 4	00:00:11;16		01:00:00;00	01:00:11;15	01:00:00;00	01:00:11;14	1V, 2A

Figure 7-35

ent informational categories for each clip or sequence in the list. When Final Cut Pro sees this tab delimitation, it displays individual columns. A Final Cut Pro Browser window list can be Exported as a printable Batch List of all your clips and sequences, complete with the entire set of informational columns.

Note: Do not attempt to open a Final Cut Pro Project file in any application other than Final Cut Pro. You might irrevocably damage the Project file. To print the Project file's list, always export a Batch List for use in spreadsheet software.

This can be invaluable for various reasons. A physical hard copy of your clip resources is the most resilient backup you can create. Hard drives can die, and files can corrupt in an instant, taking your hard work with them. But a paper copy can always be accessed easily, even when you are away from your computer. You can log your tapes completely, with no intentions of capturing the clips you log, and simply generate the paper log for your records. This is the best way to keep your tape catalog in order without relying on scribbled or typed lists that can be riddled with mistakes and inconsistencies. When you want that 15-second clip of a water buffalo charging the camera, you can search your tape spreadsheet database or quickly glance through your paper logs rather than shuttling through countless tapes in a frustrating search that puts your tapes through passes that could potentially wear them out.

The batch list can also facilitate the process of communicating information about clips when the logger or director of a production is not the editor and will not be around to inform the editor about the clips intended for use. It's simple enough for tapes to be logged while or just after they are shot. This information is easily logged as Offline clips that are generated as a batch list. The batch list can then be printed out and/or opened directly in

Final Cut Pro by the editor to focus directly on the materials to be used instead of relying on vague notes and miles and miles of tape.

Finally the batch list serves as an easy batch capture tool. A batch capture file, or any spreadsheet file that carries the column structure that Final Cut Pro wants, can be imported, generating the Offline clips in the Browser window that were logged into the file. One of the nicest features is the backwards implication of this. You don't need to log your tapes using Final Cut Pro. Simply start from a preformatted spreadsheet file, and enter the information you want to include in the clip, the minimum amount being the clip name, timecode In and Out points, and a reel name for each. After you've logged an entire tape with this data, you can take that file to Final Cut Pro and open it as clips in a project. Why take Final Cut Pro to a shoot when you can take any laptop, or even a PDA that can run a spreadsheet application?

STEP 55

To export a batch list, simply select the Browser window Project tab, or if you have your clips organized in Bins as you should, double-click the Bin to make it the active window.

STEP 56

Go to the File drop-down menu, select Export, and then choose Batch List (Figure 7-36). The dialog box that appears will let you navigate to a save location. A format drop-down bar allows a choice of Tab Delimited or Formatted Text. For maximum compatibility with spreadsheet applications, select Tab Delimited.

STEP 57

Click Save, and the batch list will appear on the drive where you saved it. Its icon will be a Final Cut Pro slate clapper icon with a little white square indicating that it is a text object.

Opening the file in a spreadsheet application will reveal all your information in columns, complete with the column headers describing what information belongs in each

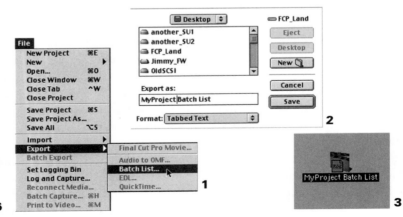

Figure 7-36

column. To change or alter the information in the columns, just work as you normally would with spreadsheet values, then save the spreadsheet again as a Tab Delimited file.

STEP 58

To return the batch list to Final Cut Pro, simply go to File drop-down menu, select Import and, in the submenu, choose Batch List (Figure 7-37). Then navigate the box to the file and click Open. The Offline clips from the batch list will instantly appear in your Project tab.

To make sure that your office spreadsheet application's software generates the ideal batch list from scratch, start with a dummy batch list generated by Final Cut Pro.

STEP 59

Create a new project that as yet contains no clips (Figure 7-38). Then export a batch list from this project. Opening the file in spreadsheet software will reveal the exact same columns and column headers, but with no row content, since there were no clips in the fresh project.

Use this dummy as an initial batch list spreadsheet file whenever you need to log outside of Final Cut Pro.

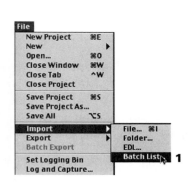

Figure 7-37

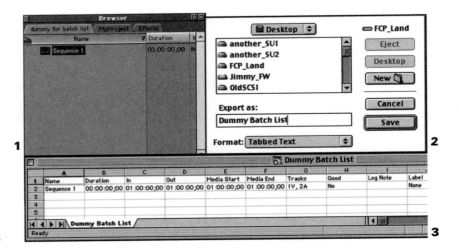

Figure 7-38

Notice that sequences are passed between Final Cut Pro and the batch list, including information about the In and Out points of any selected range. However, since sequences do not correspond directly to any timecode reels, when you import the batch list, those sequences will appear as clips rather than as a sequence. Batch lists are not for archiving sequence or edited information; they specify only the logged clips that compose the intended media to be used there.

The EDL

The batch list, while valuable for clip data, does not address the issue of the sequence metadata. We have seen the ability of Final Cut Pro project files to recapture media that is Offline, giving us a safety net in the event that we need to recapture a project whose media has disappeared. But this is clearly not enough. There are reasons why you might need to have access to a standardized text file version of the sequence, just as you did with the batch list version of your clips.

The first reason is predictable, in view of the comparison with a batch list. There is nothing more resilient than a paper copy of important aspects of your project. Should disaster strike, having a text version of the item can help you reconstruct your work. But another reason for recording such metadata would be the ability to share the sequence with other nonlinear editors besides Final Cut Pro. Although you could open your Final Cut Pro project on any Macintosh that also has Final Cut Pro installed, the day may come when you need to share your sequence with editors using other applications (e.g., Avid or Media 100).

Once again, Final Cut Pro offers a feature that clearly establishes it as a professional editing application. In such a situation, you would need to export an Edit Decision List (EDL). An EDL is a standardized text format that many different professional nonlinear editing applications can access and translate. It is a nearly foolproof Rosetta Stone that uses standard text information and structure to indicate the order and interaction of clips in a timeline.

STEP 60

Select a sequence, and go to the File drop-down menu and select Export. In the submenu that follows, select EDL (Figure 7-39).

Such lists are not used merely to pass edit information between digital nonlinear editing applications. High-end post-production facilities that perform linear tape edits of material using much higher quality copies of the original footage, a process called Online editing, often perform their edits based on EDLs generated by an editor working with Offline quality media that shares the same timecode resources as the post-production house's copy of the media. Edit Decision Lists are the standard method of delivering exact editing information between one machine and another machine or a human editor. Most television and film productions that have passed through a post-production house have followed this process.

Such a text document would need to include far more data than the simple batch list we described earlier. In addition to the individual information that needs to be gathered about each clip in the sequence, the order and interaction between the clips in the sequence must be codified. This information extends beyond the simple ordering that you would expect of a straight cut edit list, but also may need to cover transitions and split (L-cut) edits, which consist of audio cuts and video cuts being staggered such that linked audio and video clips do not necessarily begin or end at the same frame.

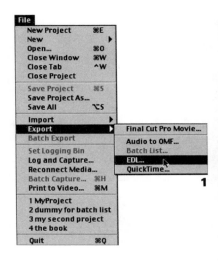

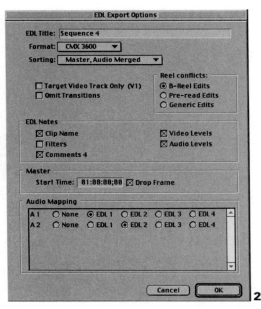

Figure 7-39

This is further complicated by the fact that the text format for such a list must remain relatively basic and simple in its language. Because so many different applications, edit controller machines and humans must be able to translate the text into editing commands, there are limitations on what the EDL can contain. Among other things, this often places a limit on the number of video tracks and audio tracks are allowable within a sequence, as well as which transitions, if any, are allowed.

STEP 61

Select CMX 3600 from the Format drop-down bar. Take a look at the options for this EDL format, and make sure that your sequence does not exceed the limitations of the format (Figure 7-40).

STEP 62

Click OK, and then save the EDL to the Desktop.

STEP 63

Find the SimpleText word processing application in your application folder on the hard drive. Start it, and then open the EDL you have just saved to the Desktop. Take a look at the formatting and the conversion to text that has taken place (Figure 7-41).

There are a number of different formats of EDL, each of which allows for different parameters. The five formats accessible to Final Cut Pro are CMX 340, CMX 3600, Sony 5000, Sony 9100, and the GVG 4 Plus. The names correspond to professional edit con-

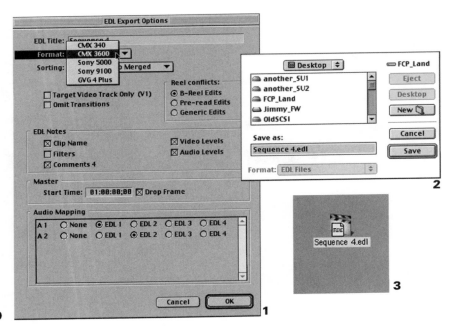

Figure 7-40

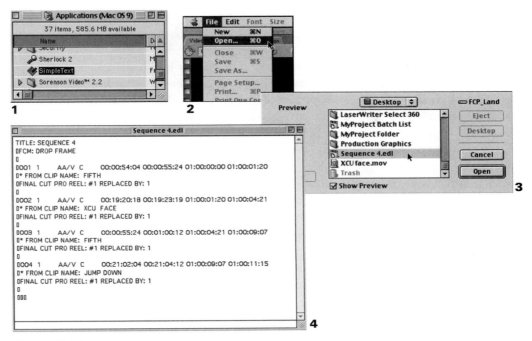

Figure 7-41

trollers that are used by post-production facilities to process the EDLs using linear tape edit systems. Since most professional editing applications also provide coverage for these EDL formats, this ensures that you can pass you Final Cut Pro EDLs to another professional editing application with the reasonable certainty that they will be understood and interpreted into an Offline sequence resembling the one you exported from.

A complete address of the issue of exporting EDLs is beyond the scope of this book. But there are some issues to keep in mind if you should find generating an EDL necessary. The most important is to keep the sequence to be exported as an EDL basic and to avoid using effects, transitions, and multitrack editing techniques that cannot be replicated in the text language of the EDL.

First and foremost, communicate with the potential recipient of your Final Cut Pro–generated EDL, or if you are to be on the receiving end, make sure you know the EDL will work. One way to test this is to pass dummy EDLs between machines to make sure that they can read each others' language. Sometimes, a system may need an extra tab here or space there to get the communication perfected. These are surprises you don't want to walk into at a critical moment in the Offline/Online process, where rental of facilities cost a million dollars a minute.

Limit the sequence to one video track and two audio tracks. Although there are specific uses for a second video track in EDLs, they are mostly limited to superimposing clips, rather than including the sequencing of normal clips. Using the Cutting Station mode from the User Mode tab of the Preferences will help limit the possibility of including non—EDL compatible edits.

Be very careful and organized with your tape Reel Names. Remember that the timecode numbers in the EDL will have no meaning if the proper tape is not associated with it. Also make sure to keep individual clip names under 25 characters in length. Make sure that you calibrate the timecode between your deck and Final Cut Pro, since slight variations in the reading of the tape's timecode can play havoc with the timecode values as they must be used to edit with in the next edit step.

Do not attempt to use nested sequences as a single track for use in an EDL or to use odd transitions unless you are sure they are supported by the EDL you are using. Make sure that all transitions occur between clips in the first video track. Finally, keep a manual count of the actual number of clips in your sequence, and then verify that number within the generated EDL text file.

For further details about the functioning of the various EDL formats and the specific parameters involved in exporting each type, consult the Apple Final Cut Pro manual, as well as any post-production facility or site at which you may need to provide the list.

Epilogue

Once you've worked through the previous chapters, you will have gotten your feet wet in the major areas that characterize digital nonlinear video editing. Even if you move to a system besides Final Cut Pro, you will find that the principles and techniques hold true, even if the toolsets change somewhat. With few exceptions, the processes covered in this book are based on a universal model, and that knowledge can be applied wherever you need it.

As mentioned in the Introduction, this book is definitely not the last word in DV editing. There are many different ways to approach tasks, even from within Final Cut Pro itself. Investigate further and always keep your eyes open for more resources about Final Cut Pro editing techniques. Read the Final Cut Pro Manual, and participate in the various Final Cut Pro communities on the Web. You can always learn from other editors, and the Web provides a great resource for doing so. In addition to many free reviews and instructional articles, most Web sites offer discussion forums these days that can get you an answer to your question in a few seconds flat.

Look around your home community for other film and video artists. You may be surprised to find that there is an active group of people right under your nose with a like-minded drive toward making their own work. It doesn't have to be "art" or "Hollywood" as long as you enjoyed making it and someone enjoyed watching it. Local video screenings offer a chance to get healthy feedback, a process that really improves your editing technique in ways no instructional book ever could. And there's nothing more gratifying than having a roomful of people applaud your efforts.

But the best way to get better at Final Cut Pro is simply to make work, period. With every project you complete, you will become more powerful and gain more control over the interface. Final Cut Pro is a very deep, rich application, and no 400-page book could hope to explore it in all its facets. There is no substitute for practice. Watch television and films, find editing styles or compositing techniques you haven't mastered, and then try them at home. You'd be surprised how simple some complex-looking effects are to create. Soon, you'll be stretching Final Cut Pro to the limits.

Visit the following Web sites for important up-to-the-minute resources about Final Cut Pro and DV editing in general, including additional articles by the author.

- www.2-pop.com: Features news flashes, articles and reviews and a huge collection of specialized discussion forums. The first and greatest of all Final Cut Pro informational Web sites—enough said.

- www.LAFCPUG.org: The Web site of the Los Angeles Final Cut Pro User Group, a group that has had a great deal of influence in the continued development of the application through its contacts with Apple and the rest of the DV industry. This site contains reviews, articles, and a discussion board.
- www.kenstone.net: A really fantastic collection of articles, reviews and a discussion board from this veteran Web teacher. The list of articles grows by the week and addresses nearly every aspect of the Final Cut Pro Firewire DV editor, especially for system maintenance and keeping your Macintosh healthy and happy.
- www.adamwilt.com: The Einstein of DV, Adam Wilt has a site that is a fantastic resource containing many detailed articles about the more technical aspects of DV.
- www.apple.com/finalcutpro/: The Apple Final Cut Pro Web site contains links to relevant software updates for your system as well as fairly current details about Final Cut Pro's support for various DV devices and third-party software.
- www.info.apple.com/usen/finalcutpro/: This site contains Apple's excellent list of troubleshooting tools and quick resources about specific issues in Final Cut Pro. Your question has probably already been answered in detail by an Apple Tech Info Library article, and a link to the article is usually available on this page.

Now, don't forget to have fun while working in Final Cut Pro.

Index